Advances in Novel Phytopharmaceuticals

Advances in Novel Phytopharmaceuticals covers the current aspects of the development of novel herbal formulations and summarizes their type of active components, biological activity, applications, and toxicity associated with novel herbal formulations. Such distinct types of novel herbal formulations are described to have remarkable advantages over conventional formulations of plant actives and extracts. The book also covers the various herbal phytoconstituents used for novel drug delivery applications. These novel herbal formulations can be used more uprightly and with enhanced efficacy by incorporating them into modern dosage forms. This can be accomplished by designing novel drug delivery systems for herbal ingredients. This book provides key information for everyone interested in novel phytodrug delivery systems and drug discovery, including medicinal chemists, cosmetic experts, nutritionists, toxicologists, drug formulators, and healthcare professionals. Students, professors, and researchers working in pharmaceutical sciences and beyond will also find the book useful.

We will also focus on patents filed, technology transfer, market potential, and regulatory aspects of herbal nanomedicines, if any. Research in this area is still in the exploratory stage, and in the market there are no books available that summarize all the contents of the research. As such, we will try to make our full contribution to this area. This book is the continuation of a first volume and readers and the target audience will gain the all the latest knowledge in the subject.

AIMS AND SCOPE

- To understand the basic information and application of phytoconstituents
- To understand the various phytoconstituents-based formulations for disease management
- To understand the herbal drugs used for cosmetic applications
- To understand the phytodrug delivery systems and their applications
- To provide detailed knowledge about the technical aspects of formulation
- To update readers with the latest research progress in this emerging research area
- To provide the details about the patents filed, technology transfer, market potential, and regulatory aspects for the same

Advances in Novel Phytopharmaceuticals

Edited by

Durgesh Nandini Chauhan,
Madhu Gupta,
Vikas Sharma, and
Nagendra Singh Chauhan

CRC Press
Taylor & Francis Group
Boca Raton London New York

CRC Press is an imprint of the
Taylor & Francis Group, an **informa** business

First edition published 2025
by CRC Press
2385 NW Executive Center Drive, Suite 320, Boca Raton FL 33431

and by CRC Press
4 Park Square, Milton Park, Abingdon, Oxon, OX14 4RN

CRC Press is an imprint of Taylor & Francis Group, LLC

ISBN: 978-1-032-29244-1 (hbk)
ISBN: 978-1-032-29245-8 (pbk)
ISBN: 978-1-003-30058-8 (ebk)

DOI: 10.1201/9781003300588

Typeset in Times
by Apex CoVantage, LLC

Contents

Preface ... xi
Editor Biography ... xiii
List of Contributors ... xv

Chapter 1 Application of Drug Delivery Systems for Delivery of Phytoconstituents ... 1

Sara Efigênia Dantas de Mendonça y Araújo, Analara Cordeiro de Macêdo, Brenda Maria Silva Bezerra, Kammila Martins Nicolau Costa, and João Augusto Oshiro Junior

1.1 Introduction ... 1
1.2 Novel Drug Delivery Systems for Phytoconstituents ... 2
1.2.1 Resveratrol ... 2
1.2.2 Hesperidin ... 4
1.2.3 Celastrol ... 5
1.2.4 Daidzein ... 7
1.2.5 Catechin ... 8
1.2.6 Cyanidin ... 9
1.2.7 Naringenin ... 10
1.2.8 Apigenin ... 11
1.2.9 Baicalin ... 13
1.2.10 Rosmarinic Acid ... 14
1.2.11 Emodin ... 15
1.2.12 Stevioside ... 16
1.2.13 Epigallocatechin-3 Gallate ... 17
1.2.14 Fisetin ... 19
1.2.15 Genistein ... 20
1.2.16 Gallic Acid ... 21
1.2.17 Coumaric Acid ... 23
1.2.18 Mangiferin ... 24
1.3 Discussion and Conclusions ... 25
References ... 25

Chapter 2 Novel Drug Delivery System for Resveratrol ... 33

Manhua REN, Jiayi GAI, and Shirui MAO

2.1 Introduction ... 33
2.2 Sources, Physicochemical, and Biological Properties of Resveratrol ... 33
2.2.1 Sources of Resveratrol ... 33
2.2.2 Structure and Physicochemical Properties ... 34
2.2.3 Biological Activities ... 35
2.2.4 Biopharmaceutical Characteristics ... 38
2.3 Application of Different Carrier Systems in Drug Delivery ... 39
2.3.1 Nanocarrier for the Delivery of Resveratrol ... 40
2.3.2 Microcarrier for the Delivery of Resveratrol ... 49
2.3.3 Others ... 51
2.4 Summary and Outlook ... 52
References ... 52

Chapter 3 Novel Drug Delivery System for Gallic Acid 61

Afonso H.S. Júnior, Domingos L.P. Macuvele, and António B. Mapossa

3.1 Introduction 61
3.2 Trends on Drug Delivery System for Gallic Acid 61
3.2.1 Biopolymers 63
3.2.2 Nanoparticles 66
3.2.3 Liposomes 68
3.2.4 Hybrid-Based Materials 69
3.3 Conclusions and Remarks 71
Acknowledgments 71
Conflicts of Interest 71
References 71

Chapter 4 Nanocarrier Systems and Their Application for the Delivery of Different Phytoconstituents 74

Vikas Pandey, Tanweer Haider, Monika Vishwakarma, Sakshi Soni, and Vandana Soni

4.1 Introduction 74
4.2 Various Nanocarrier Systems for the Delivery of Herbal Medicines 74
4.2.1 Nanoemulsions 75
4.2.2 Liposomes 76
4.2.3 Niosomes 78
4.2.4 Nanocrystals 78
4.2.5 Phytosomes 79
4.2.6 Ethosomes 80
4.2.7 Cubosomes 81
4.2.8 Metallic Nanoparticles 82
4.2.9 Dendrimers 83
4.2.10 Solid Lipid Nanoparticles 83
4.2.11 Poly-Lactic-Co-Glycolic Acid Nanoparticles 85
4.2.12 Nanostructured Lipid Carriers 86
4.2.13 Extracellular Vesicles 87
4.2.14 Nano Polymersomes 88
4.2.15 Polymeric Nanoparticles 89
4.3 Future Prospects 90
4.4 Conclusion 90
Abbreviations 90
References 90

Chapter 5 Liposomes and Nanoliposomes as Potential Nano-Enabled Platforms for Enhanced Bioavailability of Phytoconstituents 99

Petra Obioma Nnamani, Ozioma Blessing Onokala, Kruga Owodeha-Ashaka, Ibilola Mary Cardoso-Daodu, Margaret Ilomuanya, Christian Chukwuemeka Mbah, and Anthony Amaechi Attama

5.1 Introduction 99
5.1.1 General Introduction 100
5.1.2 Liposomes and Nanoliposomes 100

5.1.3 Phytoconstituents ... 102
5.2 Liposome- and Nanoliposome-Based Drug Delivery Systems ... 104
5.3 Formulation of Liposomes and Nanoliposomes ... 108
5.3.1 Materials Used in the Preparation of Liposomes and Nanoliposomes ... 108
5.3.2 Synthesis/Preparation of Liposomes and Nanoliposomes ... 109
5.3.3 Characterization of Liposomes and Nanoliposomes ... 111
5.4 Development Challenges of Liposomes and Nanoliposomes as Drug Delivery Vehicles ... 113
5.4.1 Drug Leakage ... 113
5.4.2 Cost and Scale-Up Issues ... 114
5.4.3 Lyophilization of Liposomes ... 114
5.4.4 Quality Control of Liposomes ... 114
5.4.5 Safety and Toxicity Concerns ... 114
5.5 Applications of Liposomes and Nanoliposomes as Drug Delivery Vehicles for Phytoconstituents ... 115
5.5.1 Oral ... 115
5.5.2 Topical ... 116
5.5.3 Parenteral ... 116
5.5.4 Pulmonary ... 117
5.6 Mechanism of Enhancement of the Bioavailability of Phytoconstituents by Liposomes and Nanoliposomes ... 117
5.6.1 Liposomal Formulation of Phytoconstituents in *Salvadora Persica* L. (Salvadoraceae) by the Spraying Technique ... 118
5.6.2 Curcumin Nanoliposome Formulation by the Micelle-to-Vesicle Transition Method ... 118
5.6.3 Encapsulation of Naringenin into Nanoliposomes by Microfluidization and pH-Driven Method ... 119
5.6.4 Formulation of Long-Circulating Daidzein-Loaded Liposomes by Ultrasonication and Lipid Film-Hydration Method ... 119
5.6.5 Formulation of Resveratrol-Loaded Peptide and Sucrose Liposomes ... 119
5.6.6 Formulation of EGCG-Loaded Liposomes with Vitamin E ... 120
5.6.7 Albumin-Coating of Naringenin Liposomes ... 120
5.7 Approval Status of Phyto-Liposomes and Phyto-Nanoliposomes ... 120
5.8 Future Perspectives ... 120
5.9 Conclusions ... 122
Conflict of Interest ... 122
References ... 122

Chapter 6 Liposomes and Exosome-Like Nanoparticles for Curcumin Delivery to the Gut and to Treat Inflammatory Bowel Disease ... 134

Vincenzo De Leo, Lucia Catucci, Anna Maria Maurelli, Valeria Daniello, Massimo Conese, and Sante Di Gioia

6.1 Curcumin and Curcuminoids ... 134
6.2 Curcumin's Biological Effects ... 135
6.3 Inflammatory Bowel Disease ... 138
6.4 Curcumin in IBD ... 141
6.5 Liposome Delivery of Curcumin ... 145

6.6 Exosome Delivery of Curcumin 149
6.7 Conclusion 150
References 151

Chapter 7 Lipid-Based Nanoparticles: SLN and NLC—a Versatile Delivery Vehicle for Phytodrugs 160

S. Mohana Lakshmi and C.K. Ashok Kumar

7.1 Introduction 160
7.1.1 Lipid-Based Nanocarriers 160
7.1.2 Liposomes 160
7.1.3 Solid-Lipid Nanoparticles 161
7.1.4 Nanostructured Lipid Carriers 161
7.1.5 Nanoemulsions 162
7.1.6 Microemulsions 162
7.1.7 Nanocapsules 162
7.1.8 Niosomes 163
7.1.9 Self-Nano- and Microemulsifying Drug Delivery Systems 163
7.2 Nanocarrier-Based Delivery of Phytochemicals 164
7.2.1 NLCs 167
7.2.2 Lipids 168
7.2.3 Surfactants 168
7.2.4 Other Excipients 168
7.2.5 Preparation of NLCs 168
7.3 Conclusion 174
References 174

Chapter 8 The Use of Bioadhesive, Thermoresponsive Nanostructured Systems for the Delivery of Phytopharmaceutical Constituents: Curcumin, Gallic Acid, p-Coumaric Acid and Propolis 179

Sabrina Barbosa de Souza Ferreira, Rafaela Said dos Santos, Camila Felix Vecchi, and Marcos Luciano Bruschi

8.1 Introduction 179
8.2 Definitions and Fundamentals 180
8.2.1 Bioadhesive, Thermoresponsive, Nanostructured Systems 180
8.2.2 Bioactive Agents of Natural Origin 184
8.3 Bioadhesive, Thermoresponsive Systems Containing the Natural Bioactive Agents 187
8.3.1 Strategies 187
8.3.2 Technology 190
8.3.3 Type of Systems 193
8.4 Toxicity and Activity 197
8.4.1 In Vivo Evaluation 197
8.4.2 In Vitro Evaluation 198
8.4.3 Clinical Trials 198
8.5 Patents 200
8.6 Concluding Remarks 201
References 201

Chapter 9 Albumin Nanoparticles as a Versatile Carrier in Phytonanotechnology: Expanded Scope for Phytochemical Delivery........213

Deepika Kumari, Akansha Jha, Purnima Mahato, Shuvadeep Bhowmik, Jatin Kumar Sahu, Madhu Kumari, Bapi Gorain, Pankaj Kumar Singh, Prashant Shukla, and Monika Dwivedi

9.1 Introduction........213
9.2 Phytoconstituents for Healthcare........214
9.3 Approaches for Phytonanotechnology........214
9.4 Albumin as a Versatile Carrier in Phytocompound Delivery........215
9.4.1 Proteins for Drug Delivery........215
9.4.2 Types of Albumins........216
9.4.3 Properties of Albumin Nanoparticles........217
9.4.4 Method of Preparation of Albumin Nanoparticle........217
9.5 Application of Albumin Carrier in Phytochemical Delivery........221
9.5.1 Application of Albumin Nanoparticles in Cancer Treatment........221
9.5.2 Application of Curcumin-Loaded Albumin Nanoparticles for the Treatment of Parkinson's Disease........223
9.5.3 Application of Albumin Nanoparticles in Tissue Regeneration........223
9.5.4 Applications of Albumin Polymeric Nanoparticles........225
9.5.5 Application of Albumin Carrier in Phytochemical Delivery for the Treatment of Liver Fibrosis........226
9.6 Conclusion and Future Aspects of Albumin Nanoparticles in Phytochemical Delivery........226
References........227

Chapter 10 Quality-by-Design Perspectives of Phytosomes for Pharmaceutical Benefits........230

Sumant, Subh Naman, and Ashish Baldi

10.1 Introduction........230
10.2 Components of Phytosomes........230
10.2.1 Phospholipids........230
10.2.2 Phyto-Active Constituents........232
10.3 Preparation of Phytosomes........233
10.4 Method of Preparation of Phytosomes........234
10.4.1 Antisolvent Precipitation Method........234
10.4.2 Rotary Evaporation Technique........234
10.4.3 Solvent Evaporation Technique........235
10.5 Mechanism of Complex Formation in Phytosomes........235
10.6 Characterization of Phytosomes........236
10.7 Implementation of the Concept of QbD in Phytosomal Preparation........236
10.7.1 Step 1 of QbD: Identification of Quality Target Product Profile (QTPP) for Phytosomes........237
10.7.2 Step 2 of QbD: Identification of Critical Quality Attributes (CQA) for Phytosomes........237
10.7.3 Step 3 of QbD: Application of Quality Risk Management (QRM)........238
10.7.4 Step 4: Development of DoE-Based Model and Validation........239
10.7.5 Step 5: Pilot Plant Scale-Up and Commercialization........240
10.7.6 Scale-Up Challenges in Phytosomes........241

10.8 Therapeutic Applications of Phytosomes ... 241
10.9 Conclusion and Future Perspective ... 242
References ... 243

Index ... 247

Preface

Since the ancient era, herbal medicines have been used for the treatment of different health related problems. Also in modern times, herbal medicines are good alternatives to synthetic drugs. Similar to synthetics, they make a better therapeutic part of medicinal system.

Phytopharmaceuticals are plant-based medicines or plant-derived ingredients having medicinal properties. Phytopharmaceuticals are processed from crude herbal drugs that came from unrefined parts of plants, like stems, roots, fruits, and leaves, which are then prepared as herbal extracts (mixture of herbal therapeutic constituents). The final formulations can be a nutraceutical or pharmaceutical product or a health supplement. These plant-based formulations play a significant role in regulating normal body functions by disrupting the life cycle of pathogens, targeting particular receptors, and interrupting pathological pathways. As these formulations are employed for the treatment of disorders like allergies, diabetes, aging, cancer, skin disorders, microbial infections, and many more; the regulation consideration of phytopharmaceuticals becomes essential regarding their dose of therapeutic agent, chemical composition of API/excipients and product standardization. Herbal phytopharmaceuticals comprise at least four bioactives (as assessed qualitatively and quantitatively) from medicinal plant extracts, used for the diagnosis, treatment, mitigation, or prevention of any disease, but somehow it does not include the parenteral administration.

Crude herbal drugs are processed into herbal extracts using milling, grinding, and drying; still, there are various limitations, like poor solubility or permeability, poor target specificity, pH instability, and metabolic inactivity. From the past two decades, considerable development is made in the drug delivery systems to address the various limitations associated with plant medicines. These advanced modern drug delivery systems help us deliver herbal drugs in desired concentrations at specific sites in a controlled or sustained manner with less or no systemic toxicity. These systems are nanosized formulations named as novel drug delivery systems (NDDSs). Some of the formulation that fall under these modern approaches are nanoemulsions, nanoliposomes, ethosomes, transferosomes, nanoparticles, polymeric nanocapsules, nanomicelles, and polymeric micelles.

The ability of novel drug delivery systems to encapsulate, load, and deliver the efficacy of phytochemicals at targeted sites depends on the polymeric material used for the drug delivery systems. This novel drug delivery technology is used to deliver the drug at a specific site for quick and sustained action so that more of the drug is concentrated on the organ, decreasing the toxicity to other parts of body. Polymers with synthetic, semi-synthetic, and natural sources as well as lipids play an important role in the formulation of novel herbal drug delivery systems. These polymers and lipids are biocompatible and biodegradable in nature that may be of plant, animal, and microbial origin like fungi and bacteria.

In the followup of these revolutionary progresses in the drug delivery field, *Advances in Novel Phytopharmaceuticals* is more concrete and precise in discussing the technology advancement related to novel phytopharmaceuticals. The chapters were contributed by the eminent researchers and experts and comprise insights into the most promising moves toward superior delivery system of phytopharmaceuticals.

The first chapter covers the various phytoconstituents and their physicochemical characteristics. In this context, the incorporation of phytoconstituents into macro, micro, and nanosystems is an alternative to temporally and spatially control the phytoconstituent release and improves its pharmacokinetic and pharmacodynamic limitations. Considering this, the chapter unites the phytoconstituents encapsulated into organic and inorganic drug delivery system. We discuss all molecules and present their pharmacological and pharmacokinetic properties. Finally, we exhibit the advances and differences in the release profile and in preclinical activity.

The second chapter is dedicated to the compilation of recent technological advancements about novel drug delivery systems for resveratrol. Resveratrol displays many pharmacological activities and plays an important role in many different diseases. However, its effectiveness is confined to its inherent physical properties, including limited water solubility, lower stability, and high photosensitivity. To more clinically validate resveratrol, many novel drug delivery strategies have been created that could overcome these restrictions.

The third chapter is dedicated to the design considerations for novel drug delivery systems for gallic acid and the different strategies to overcome the limitations associated with gallic acid. Undoubtedly, the development of such drug delivery systems not only for gallic acid but for others compounds with biological activity is of great importance to modern science.

The fourth chapter covers phytoconstituent-loaded nanomedicine and how it improves the clinical ability of phytomedicines. In this context, various available nanoparticulate and nanovesiculate carrier systems are discussed for the treatment of diseases as well as comparisons related to the bioavailability and therapeutic efficacy of phytoconstituent-loaded nanomedicines with native phytoconstituents.

The fifth chapter focuses on how phytobioactive-based liposomes and nanoliposomes increase bioavailability, prolong residence time, and allow for multiple drug loading to enhance therapeutic efficacy and decrease toxicity. This chapter addresses the existing gap and research initiatives to synthesize novel phytonanoliposomes and develop them into dosage forms that may solve the problems associated with conventional delivery systems of phytobioactive compounds.

The sixth chapter presents the role and potential of curcumin and explores its different properties, like anti-inflammatory, anticancer, antioxidant, antimicrobial, and wound healing effects. Liposomes and exosome-like nanoparticles have been conceived to overcome these hurdles and deliver curcumin to intestinal epithelial cells as well as the immune cells in the intestine. This chapter deals with the liposome and exosome-like-nanoparticle delivery systems to treat IBD, highlighting that more in-depth research is warranted to verify if these curcumin-based nanoformulations will have a therapeutic impact.

The seventh covers the new vision on the NLCs and SLNs as potential phytodrug carriers enabling better treatment.

The eighth chapter focuses on the principles, systems, technology, therapeutic approaches, safety, toxicity, and patents of bioadhesive, thermoresponsive nanostructured systems containing curcumin, gallic acid, p-coumaric acid, or propolis. In this context, the most valuable and up-to-date drug delivery systems are discussed with consideration for their benefits.

The ninth chapter presents albumin nanoparticles as a versatile carrier in phytonanotechnology and their expanded scope for phytochemical delivery. Albumin nanocarriers offer a safe delivery mode of phytocompounds with site-specific targeting and augmented bioavailability. This chapter details the numerous uses of albumin nanoparticles in phytochemical delivery and a brief overview of the preparation methods for albumin-based nanocarriers. The potential and prospects of employing albumin-based nanoparticles for phytochemical drug delivery in cancer therapy, Parkinson's disease, and tissue regeneration are also discussed in this chapter.

The last chapter covers the importance of quality-by-design (QbD) perspectives of phytosomes for pharmaceutical benefits. This chapter focuses on the detailed understanding of phytosomes with specific attention toward the implementation of QbD principles involved in successful commercialization of herbal products. In-depth elaboration of devising quality target profiles for phytosomal formulations, critical attributes of components involved, and the impact of process parameters on quality parameters are featured in designing these products at the industrial scale.

The book series *Advances in Novel Phytopharmaceuticals* is focused on the development of new technologies in the form of drug delivery systems for phytopharmaceuticals. Such type of novel herbal system overcomes the poor and/or erratic oral bioavailability of phytoconstituents by enhancing the rate and the extent of solubilization and enhancing the therapeutic effect.

This book provides a consolidated overview on the current status of phytopharmaceutical technology, including the conceptualizations with their future prospects. It is a good compilation for giving knowledge not only to non-experts but also to the eminent personalities in the field of phytopharmaceuticals. Last but not least, we would like to thank all the contributors of this book for their excellent work. It has been a privilege to work with them.

Editor Biography

Durgesh Nandini Chauhan, MPharm, has several years of academic (teaching) experience at institutes in India in pharmaceutical sciences, such as the Columbia College of Pharmacy, Raipur CG, India. She taught subjects as pharmaceutics, pharmacognosy, traditional concepts of medicinal plants, drug delivery, phytochemistry, cosmetic technology, pharmaceutical engineering, pharmaceutical packaging, quality assurance, dosage form designing, and anatomy and physiology. She is a member of the Association of Pharmaceutical Teachers of India, SILAE: Società Italo-Latinoamericana di Etnomedicina (The Scientific Network on Ethnomedicine, Italy), and others. She has attended several workshops, conferences, and symposiums, including the AICTE-sponsored staff development program "Effects of Teaching and Learning Skills in Pharmacy: Tool for Improvement of Young Pharmacy Teachers." She has written more than ten articles published in national and international journals, fifteen book chapters, and editors in ten books. She is also active as a reviewer for several international scientific journals and an active participant at national and international conferences, including Bhartiya Vigyan Sammelan and the International Convention of Society of Pharmacognosy. Recently she joined the Ishita Research Organization, Raipur, India, as a freelance writer and guide for students of pharmacy, Ayurvedic medicine, and science in their research projects. She is interested in data analysis, proposal writing, questionnaire design, designing and developing research questionnaires, and policy. She is actively engaged in accurately monitoring, investigating, and analyzing all statistical information, professionally presenting information and reports, and investigating and documenting key data and information at the request of senior researchers.

Dr. Madhu Gupta has research experience pertaining to drug delivery to nanoformulations for magical molecule delivery, bioligands for targeting of bioactives and drug moiety, biopolymers, cancer nanomedicine, as well as topical delivery that is carried out at the Department of Pharmaceutical Sciences, Dr. H.S. Gour Central University, Sagar. Dr. Gupta has done her BPharm (gold medalist), MPharm., and PhD in pharmaceutics, with experience of more than 15 years in academics, administrative functions, and research in areas in pharmaceutical nanotechnology and targeted drug delivery related to cancer, fungal infection, and psoriasis. Presently she is an associate professor in the Delhi Pharmaceutical Sciences and Research University. Along with doing teaching and research, she is a member secretary of World Class Skill Centre and IQAC in charge with many other managerial works. She is a pioneer scientist in the field of nanotechnology and drug delivery field. She has over 50 research publications to her credit, published in journals of high scientific impact, and has contributed chapters to various renowned books and to several international and national books. She has supervised 15 MPharm students. Dr. Gupta has availed several prestigious fellowships and awards, such as SRF AICTE (NDF) and JRF (AICTE), travel grant awards (DST, ICMR, INSA, DBT, and MPCST), Prof. G.P. Nait Award (2004), and Prof. C.S. Chauhan Award (2004). She has received awards for her research, at BioAsia Innovation Award 2012, Grace India Awards, Youth Education Award 2018, and Young Researcher Award. She is successfully complete one project that is funded by MPCST Bhopal.

Dr. Vikas Sharma is a member of the Divine International Group of Institutions Gwalior (MP), India. Dr. Sharma obtained his BPharm from Jiwaji University Gwalior and MPharm in pharmacognosy and PhD from Dr. H.S. Gour Central University Sagar, India. Sharma has teaching and research experience of more than 12 years. He has made significant contributions in the area of medicinal plant research. He has been working on evidence-based validation of medicinal plants and pharmacological screening of herbal drugs. Dr. Sharma works on marker and biomarker analysis of herbs for quality evaluation. He has also made significant contribution on development of chemical profiling of extracts and formulation through TLC, HPTLC, and LC-MS. He is supervising scientific research at the postgraduate and doctoral levels. He has written several research papers in reputed journals and has contributed several chapters in books on pharmaceutical and Indigenous drugs. He is a member of various professional and academic bodies, such as the Association of Pharmaceutical Teachers of India (APTI) and the Society of Pharmacognosy.

Dr. Nagendra Singh Chauhan obtained his MPharm and PhD from the Department of Pharmaceutical Sciences, Dr. H.S. Gour University, Sagar, MP, India. He has around 20 years of research experience. He is presently working as a senior scientific officer grade I and government analyst at Drugs Testing Laboratory Avam Anusandhan Kendra, Raipur, Chhattisgarh, India. He has professional expertise in natural product isolation and phytopharmacology. He has written more than 75 articles published in national and international journals and 32 book chapters, and he has edited 7 books. He is a recipient of the Dr. P. D. Sethi Memorial National Award for the best HPTLC publication in India in the years 2021 and 2022. He is a member of various professional and academic bodies, such as the Society of Pharmacognosy India, the International Natural Product Sciences Taskforce (INPST), SILAE (Società Italo-Latinoamericana di Etnomedicina [The Scientific Network on Ethnomedicine, Italy]), the Institutional Human Ethical Committee, the Association of Pharmaceutical Teachers of India (APTI), and the Indian Pharmacy Graduates Association.

Contributors

Sara Efigênia Dantas de Mendonça y Araújo
Pharmaceutical Sciences Postgraduate Program
Center for Biological and Health Sciences, State University of Paraíba—UEPB
Campina Grande, Paraíba, Brazil

Anthony Amaechi Attama
Institute for Drug-Herbal Medicine-Excipient Research and Development
University of Nigeria, Nsukka

Ashish Baldi
Pharma Innovation Lab, Department of Pharmaceutical Sciences and Technology
Maharaja Ranjit Singh Punjab Technical University
Bathinda-15001, Punjab, India

Brenda Maria Silva Bezerra
Pharmaceutical Sciences Postgraduate Program
Center for Biological and Health Sciences, State University of Paraíba—UEPB
Campina Grande, Paraíba, Brazil

Shuvadeep Bhowmik
Department of Pharmaceutical Sciences and Technology
Birla Institute of Technology
Mesra, India

Marcos Luciano Bruschi
Postgraduate Program in Pharmaceutical Sciences
Department of Pharmacy
State University of Maringa
Maringa, PR, Brazil

Ibilola Mary Cardoso-Daodu
Faculty of Pharmacy
University of Lagos, Nigeria

Lucia Catucci
Department of Chemistry
University of Bari, A. Moro
Bari, Italy

Massimo Conese
Department of Medical and Surgical Sciences
University of Foggia
Foggia, Italy

Kammila Martins Nicolau Costa
Pharmaceutical Sciences Postgraduate Program
Center for Biological and Health Sciences, State University of Paraíba—UEPB
Campina Grande, Paraíba, Brazil

Valeria Daniello
Department of Medical and Surgical Sciences
University of Foggia
Foggia, Italy

Vincenzo De Leo
Department of Chemistry
University of Bari, A. Moro
Bari, Italy

Analara Cordeiro de Macêdo
Pharmaceutical Sciences Postgraduate Program
Center for Biological and Health Sciences, State University of Paraíba—UEPB
Campina Grande, Paraíba, Brazil

Sante Di Gioia
Department of Medical and Surgical Sciences
University of Foggia
Foggia, Italy

Rafaela Said dos Santos
Postgraduate Program in Pharmaceutical Sciences
Department of Pharmacy
State University of Maringa
Maringa, PR, Brazil

Monika Dwivedi
Department of Pharmaceutical Sciences and Technology,
Birla Institute of Technology
Mesra, India

1 Application of Drug Delivery Systems for Delivery of Phytoconstituents

Sara Efigênia Dantas de Mendonça y Araújo, Analara Cordeiro de Macêdo, Brenda Maria Silva Bezerra, Kammila Martins Nicolau Costa, and João Augusto Oshiro Junior

1.1 INTRODUCTION

Pathologies have always been a concern for humanity, leading to the search for knowledge aimed at prevention and treatment, avoiding morbidity. In the modern era, especially in the 20th century, with advances in the fields of knowledge, such as chemistry, physics, molecular biology, genetics, and engineering, the treatment of diseases has been improved and new drugs have emerged with the ability to act on preventing, curing, and fighting varied diseases. With this, the estimated life span and quality of life of human beings considerable increased. The average life expectancy doubled from 1900 to 2000, going from 32 to 66.3 years old. However, diseases are still a problem, according to the data from the World Health Organization (WHO, 2020, 2023). Between 2019 and 2020, about 30.47 million deaths worldwide were caused by ten major diseases, including cardiovascular disease, stroke, chronic obstructive pulmonary disease, and some types of cancer (Boudoulas et al., 2017).

Chronic obstructive pulmonary disease (COPD), in 2018, was considered as the fourth leading cause of death in the world and has a shared etiology with lung cancer. Lung function is a key point in preventing both COPD and lung cancer, as it is associated with the oxidative stress and inflammatory activities. Thus, phytoconstituents with potent antioxidant or anti-inflammatory activity, such as carotenoids, vitamins A, C, and E, minerals, and flavonoids are important for use in prevention, as they can preserve lung function (Zhai et al., 2018).

In addition, Alzheimer's disease is also a worrisome disease affecting 42.3 million people worldwide in 2020, according to data from the Delphi Consensus study, and it is expected to double by 2040. In one hypothesis, a cause of the dementia is thought to be an alteration of the amyloid pathway and hyperphosphorylation of tau, but its etiology is not yet known, so this factor added to the late onset of symptoms that lead to the impairment in the treatment of patients. Green tea has been shown to contain several phytoconstituents and has several benefits in the treatment of this disease (Cano et al., 2019).

Another disease that stands out is cancer due to the biochemical complexity involved in each case. In this context, although several synthetic biologically active substances are used in the treatment, a wide range is based on or isolated from natural products, such as *Vinca* derivatives and taxanes, which are one of the most effective molecules currently used in chemotherapy. Taxanes act as inhibitors of mitosis by stabilizing microtubules that are developed at the beginning of cell division and are broken when the cell stops dividing; thus, taxanes induce programmed apoptosis (Luz and Adami, 2019). Thus, flora is an important focus in the midst of scientific research, where new molecules that are increasingly selective are sought, with the intention of decreasing toxicity.

DOI: 10.1201/9781003300588-1

value of 53.5 ± 4.1%, suggesting that CS-Gel Nfs has great potential to be used in the treatment and prevention of gastrointestinal cancer (Rostami et al., 2019).

1.2.2 Hesperidin

Hesperidin, (2S)-5-hydroxy-2-(3-hydroxy-4-methoxyphenyl)-7-{[(2S, 3R, 4S, 5S, 6R)-3,4,5-trihydroxy-6-([(2R, 3R, 4R, 5R, 6S)-3,4,5-trihydroxy-6-methyloxan-2-yl] oxymethyl) oxan-2-yl] oxan}-3,4-dihydro-2H-1-benzopyran-4-one (Figure 1.2), is a flavanone glycoside (a subclass of the flavonoids) found mainly in citrus fruits, from which it was first isolated by the French chemist Lebreton. Hesperidin has limited bioavailability due to the rutinoside portion attached to the flavonoids, and its low solubility in water and instability at gastric pH. It has antioxidant, anti-inflammatory, anticarcinogenic and anti-allergic properties, and its antioxidant activity is not only limited to its radical elimination activity but also to its ability to increase antioxidant cell defenses through the ERK/Nrf2 signaling pathway (Parhiz et al., 2015; Tejada et al., 2017).

A new delivery system was developed, where hesperidin was loaded into gold nanoparticles (Hsp-AuNPs), for possible use as an anticancer drug. *In vitro* and *in vivo* toxicity assays showed good results, with no apparent detrimental effects on the liver, spleen, lung, and kidney. The MTT cell viability assay indicated that synthesized Hsp-AuNPs exhibited a higher anticancer activity (higher toxicity) to breast cancer cells MDA-MB-231, with a cytotoxic effect of 88% using a dose of125 μg mL^{-1}, compared to isolated hesperidin with only 45.3%. The cytotoxic effect of Hsp-AuNPs was also less than 10% for normal human breast epithelial cells (HBL-100), which demonstrates the specificity of the treatment. Crystal violet assay against MDA-MB-231 also reinforced these results by demonstrating that Hsp-AuNPs stimulated several morphological changes in the cells, which also formed clusters with fewer extensions and impaired cell-to-cell communication. The tumoricidal analysis of Hsp-AuNPs demonstrated that the treatment increased the functional activity of macrophages, the main contributors to tumor rejection, in mice with Ehrlich ascites tumor cells compared to the control group, leading to deceleration of tumor growth. These data together indicate that the loading of hesperidin into gold nanoparticles has great potential as an alternative treatment for breast cancer (Sulaiman et al., 2020).

To increase bioavailability and intestinal absorption after oral administration of hesperidin, a new transport system was developed by encapsulating hesperidin into alginate microspheres (AP), which were coated with different polymers, these being chitosan (ACP), gelatin (AGP), and PAH/PSS (APPP), to improve the protection of hesperidin and control its release. Characteristic release time (CRT) determined the period within which 50% (CRT50) and 100% (CRT100) of the released cargo was detected in the selected environment for all microparticles. The test determined that ACP exhibited the most efficient release-limiting behavior as it had no release in acidic medium

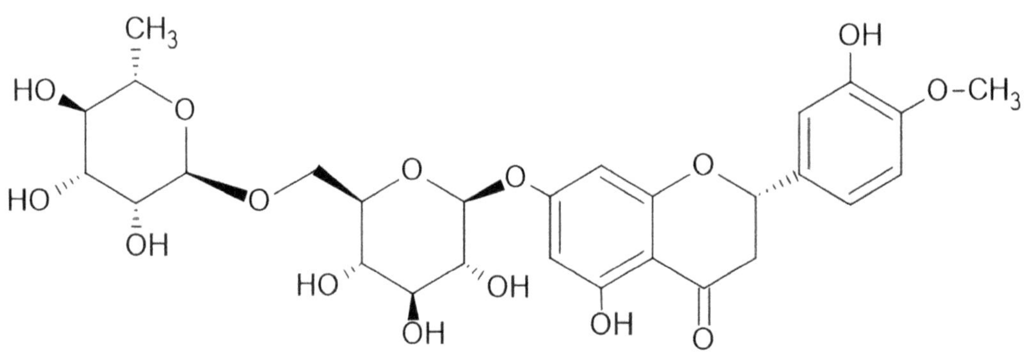

FIGURE 1.2 Chemical structure of hesperidin.

(pH = 2), only 0.2% of the cargo released at neutral pH (pH = 7), and 100% released only at alkaline pH (pH = 10). *In vitro* simulation of the behavior of hesperidin-loaded microspheres during their passage through the stomach (simulated gastric fluid [SGF]) and human intestine (simulated intestinal fluid [SIF]), also exhibited similar results with low cargo release (0, 6–2%) in SGF, and most of the cargo released in SIF, with ACP exhibiting a more controlled cargo release, where most of the hesperidin was released within 120 minutes, while hesperidin alone dissolved entirely within the first 30 minutes of the experiment. Therefore, the results indicate that incorporation of hesperidin into alginate microspheres coated with chitosan is a system that safely delivers hesperidin with controlled release efficiency, improving its oral bioavailability (Tsirigotis-Maniecka et al., 2017).

A transdermal patch has been developed to enhance the therapeutic efficacy of hesperidin in the treatment of venous diseases such as hemorrhoids, varicose veins, and venous stasis, for which adequate therapeutic treatment is not widely available. The patches were tested with different concentrations of the chosen polymers, and formulation 5 (F5), with HPMC E5 and Eudragit concentrations of 450 and 400 mg, respectively, held the highest drug loading, with 98% retention. The surface pH of all patches was in the range close to the pH of the skin, so the transdermal patch should not cause irritation. The *in vitro* release study for F5 demonstrated a controlled release of the drug, with about 93% of the contents released over 6 hours. Therefore, these data indicate that hesperidin transdermal patches may be effective as an alternative treatment for venous disease, avoiding its evolution to the surgical level (Bhalerao et al., 2019).

1.2.3 Celastrol

Celastrol, or (2R, 4aS, 6aR, 6aS, 14aS, 14bR)-10-hydroxy-2,4a, 6a, 6a, 9,14a-hexamethyl-11-oxo-1,3,4,5,6,13,14,14b-octahydropicene-2-carboxylic acid (Figure 1.3), is a triterpenoid that was first extracted in 1936 from *Tripterygium wilfordii*, a plant used in traditional Chinese medicine for the treatment of arthritis and other inflammatory diseases. It has low solubility and bioavailability. It

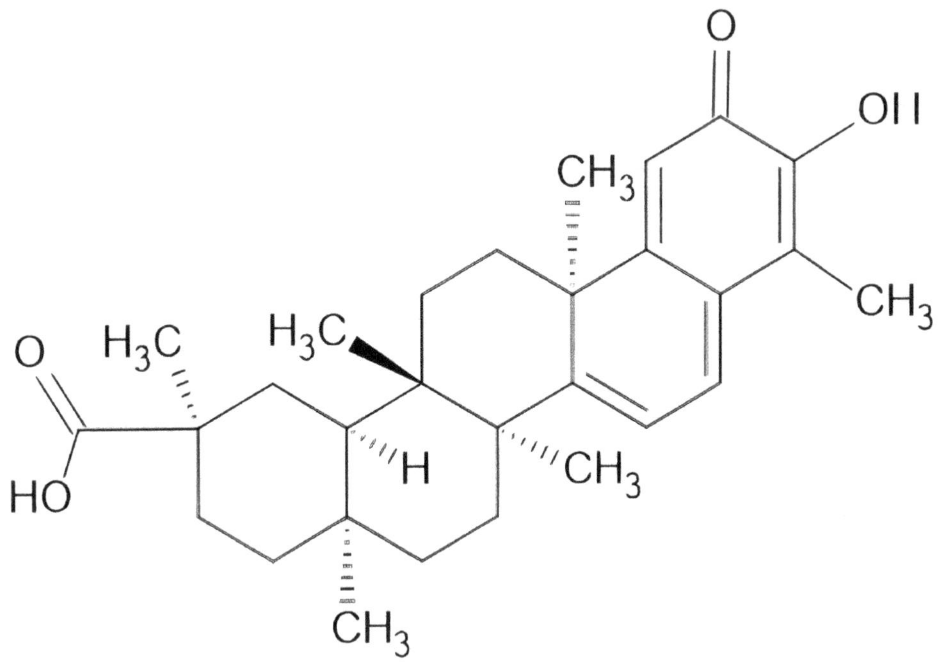

FIGURE 1.3 Chemical structure of celastrol.

Daidzein was loaded in nanostructured lipid carriers, PLGA nanofibers, for transdermal delivery. *In vitro* release was analyzed by comparing two formulations, D-NLCs-nanofibers and D-NLCs-azone-nanofibers (where the permeation enhancer was inserted into the formulation). Based on the data, D-NLCs-azone exhibited a sustained release profile, with release of 25.84% in the first 12 hours, and cumulative daidzein release of 52.43%. D-NLCs-nanofibers, on the other hand, exhibited a rapid release pattern, with a cumulative release of daidzein being 53.68% within 24 hours. The skin permeation profile was estimated *in vitro* for daidzein, D-NLCs, D-NLCs-nanofibers, and D-NLCs-azone-nanofibers, and the results of cumulative amount of drug after 60 hours indicated that D-NLCs-azone-nanofibers exhibited the best profile, with a value 3.78× higher than daidzein. The *in vivo* skin retention study indicated a concentration of D-NLCs-azone-nanofibers of about 180 ng/cm^2 after 24 hours, a higher result than D-NLCs. Based on these findings, D-NLCs-azone-nanofibers are a promising transdermal system (Song et al., 2016).

1.2.5 Catechin

Catechin, (2R, 3S)-2-(3,4-dihydroxyphenyl)-3,4-dihydro-2H-chromene-3,5,7-triol (Figure 1.5), is one of the main flavonoids found in the composition of green tea. It is soluble in water and in ethanol. It plays an important role in protecting against degenerative diseases and in protecting against cardiovascular diseases, as well as exhibiting antibacterial activity (Coşarcă et al., 2019; Cuevas-Valenzuela et al., 2014).

Green tea extract was incorporated into a thermosetting gel system for the treatment of periodontal disease. *In vitro* analyses of the drug showed complete drug release within 36 hours and zones of growth inhibition indicative of sensitivity for strains of *Porphyromonas gingivalis*, *Fusobacterium nucleatum*, *Prevotella intermedia*, and *Actinobacillus actinomycetemcomitans*. *In vivo* analysis showed a significant reduction in plaque index (PI), gingival index (GI), and probing pocket depth (PPD), demonstrating an improvement in periodontal conditions. Thus, the system presents the advantage of local and controlled release of the drug and has proven effective against microorganisms that cause periodontal diseases (Yuvaraja et al., 2016).

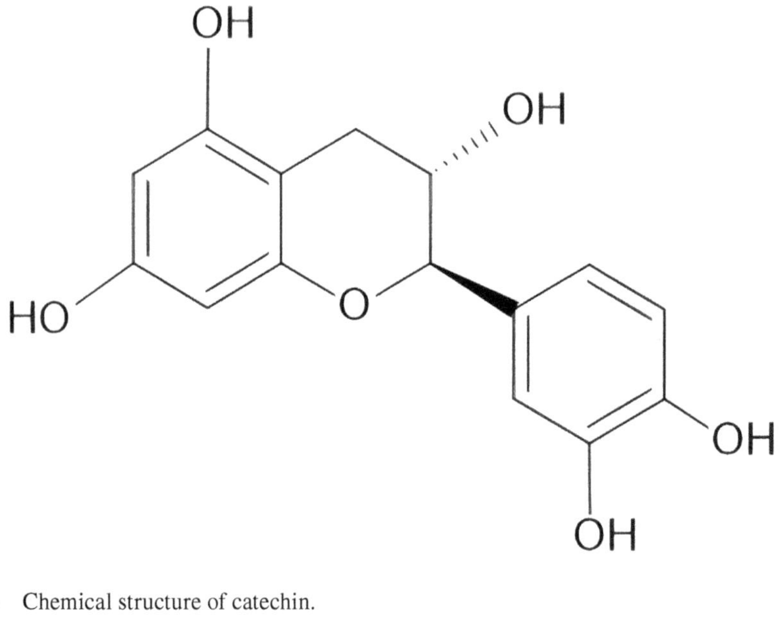

FIGURE 1.5 Chemical structure of catechin.

PLA-PEG nanoparticles were developed as a system for oral delivery of catechin (CATNP) allowing its controlled release and increasing its stability. The *in vitro* profile indicated an initial explosive release of half of the nanoparticle content, followed by a slow and sustained release for CATNP. The system proved stability in both simulated gastric fluid with 73.33% of catechin remaining at the end of 3 hours, and in simulated intestinal fluid with 63.75% at the end of 6 hours. Comparing the radical scavenging ability of CATNP against isolated catechin, a 1.64× increase in DPPH activity was observed. These data suggest that CATNP has potential as a delivery system for catechin (Singh et al., 2017).

To attempt to combat sodium-fluoride-induced oxidative stress, a nanoencapsulated (NP) drug release system was developed with catechin (CH) and sodium metaborate (SMB), a fluoride scavenger. The efficacy of the formulation was tested in combating free radical-mediated oxidative insults produced by sodium fluoride (NaF) in rat models. Treatment with NPCH and NPSMB resulted in a significant decrease in fluoride content compared to the untreated fluoride group, indicating a synergistic action of NP (CH + SMB) on post-fluoride exposure. Furthermore, combined NP therapy (CH + SMB) has proven beneficial to reduce the level of mitochondrial reactive oxygen species (ROS) significantly, as well as protecting the liver, kidney, and brain from fluoride-induced oxidative injury. Thus, the combination therapy in the nanoencapsulated system proved to be able to protect the kidney, liver, and brain from fluoride-induced fluorosis damage (Samanta et al., 2016).

1.2.6 Cyanidin

Cyanidin-3-glucoside (C3G), 2-(3,4-dihydroxyphenyl)chromenylium-3,5,7-triol (Figure 1.6), is one of the most common anthocyanins found naturally in black rice, black beans, purple potatoes, and many fruits and is responsible for their coloration. It is soluble in water, but very unstable, and can be affected by factors, such as pH, temperature, light, oxygen, solvents, enzymes, and metal ions. It acts by reducing damage to the DNA molecule, decreasing the risk of pathologies associated with oxidative stress (Bernardes, 2017; Tan et al., 2019).

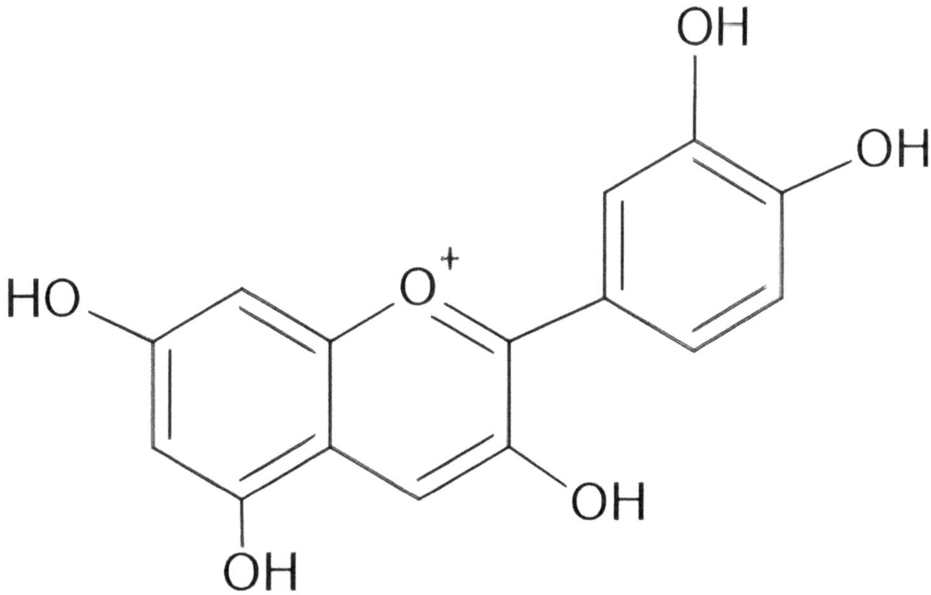

FIGURE 1.6 Chemical structure of cyanidin.

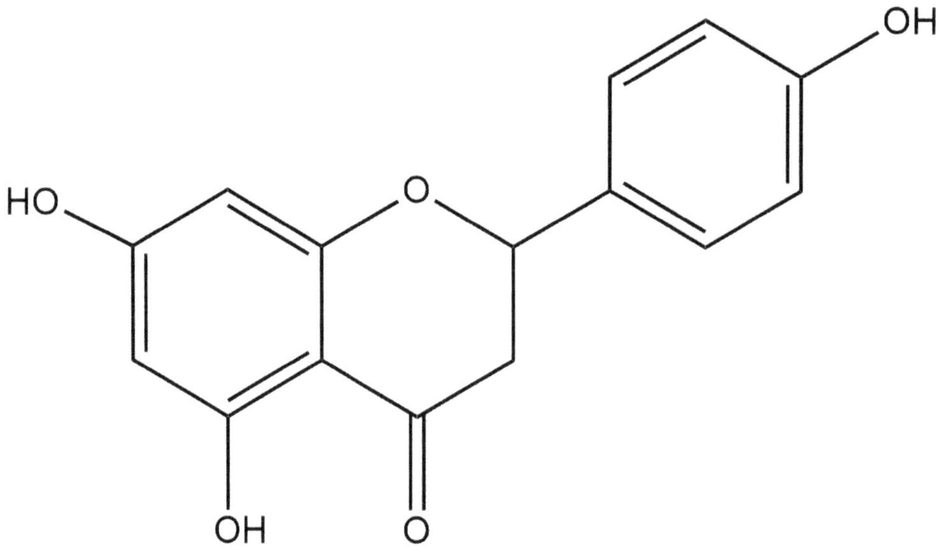

FIGURE 1.8 Chemical structure of apigenin.

crystalline structure resulting from the formation of a salt with arginine, which also showed activity against cervical cancer cells demonstrated by a higher IC_{50} than pure apigenin (Kassem et al., 2021).

Apigenin can also be loaded into microscale release systems. In W/O/W emulsions prepared with this phytoconstituent, particle size ranged from 0.91 µm for the formulations containing orange oil and Tween 80 and 4.45 to 6 µm for those prepared with soybean and orange oil, Tween 80, and PGPR. The release of apigenin was higher in the orange oil-Tween 80 emulsion (53.99%) compared to the other preparations (33.4–38.8%), which corroborates with the particle size. The soybean oil–Tween 80 emulsion proved to be the most stable and *in vivo* pharmacokinetic tests showed a peak plasma concentration equal to 395.47 µg/mL in about 10 hours compared to 43.55 µg/mL of an apigenin suspension. The 1.83% increase in oral bioavailability contributes positively to the pharmacological effects of apigenin and to its therapeutic use (Kim et al., 2016).

Another possibility to improve the solubility of apigenin is its combination with hydroxypropyl-β-cyclodextrin in an O/W microemulsion, which led to an increase in solubility constant *K* values at 30, 37 and 45°C, and this was even higher at the maximum temperature, indicating the endothermic characteristic of the binding of the phytoconstituent to cyclodextrin. In microemulsions S1 and S2 with 85% water and surfactant/oil ratio 12:1 and 6:1, respectively, approximately 0.6% and 0.65% apigenin was released within 50 minutes, while 0.5% was released from formulations S′1 without cyclodextrin in the same period. The antioxidant activity of S1 was also considerably higher than that of S′1, demonstrated by superoxide anion elimination rates of 43.9% and 40.4%, which confirms the ability of the cyclodextrin formulation to improve the release and pharmacological profiles of apigenin (Zhao et al., 2016).

Bovine serum albumin nanoparticles loaded with apigenin measuring about 376 nm increased the solubility of this flavonoid in simulated lung fluid (82–98% dissolved in 5 minutes) and its antioxidant activity, with 38% free radical inhibition compared to 35% of apigenin isolated in the DPPH assay. A similar inhibition result has also been observed for spray-dried nanoparticles with excipients. This reveals the possibility of therapeutic use of this formulation in acute lung injury (Pápay et al., 2017). In turn, a stearate and chitosan-based nanogel loaded with apigenin, revealed improved anticancer activity of the flavonoid, characterized by an increase in the level of apoptosis of K562 chronic myeloid leukemia cells, from 30.34 to 60.21% at 20 µmol/L of isolated apigenin to 38.81 to 66.37% at 10 µmol/L of apigenin-nanogel (Samadian and Hashemi, 2018).

1.2.9 Baicalin

Baicalin, 7-D-glucuronic acid-5,6-dihydroxyflavone (Figure 1.9), is one of the major flavonoids extracted from *Scutellaria baicalensis* roots, which is an herb widely used in traditional Chinese medicine. Baicalin has low bioavailability, is poorly absorbed in the intestine and may interact with co-administered drugs due to the inhibition of CYP450 enzymes. It possesses a variety of pharmacological activities, including antivirus, antioxidative, anticancer, anti-apoptotic and anti-thrombotic ("Baicalin | C21H18O11—PubChem," [n.d.]; Huang et al., 2019c; Zuanget al., 2016).

In order to accelerate wound healing, injectable hydrogel formulations prepared with PBS as solvent and Pluronic® F-127 as matrix, loaded with 0.5 mg/mL to 1.5 mg/mL baicalin, exhibited a rapid initial release of 70% of the phytoconstituent within the first 8 hours, followed by a steady release of the remaining content within 24 hours, ideal for the initial phase of the healing process. The preparations also demonstrated optimal biocompatibility and cell viability above 80%, associated with an antioxidant effect that increased with increasing baicalin concentration. *In vivo* assay showed a reduction of the wound area to 18% in 10 days with the high baicalin concentration hydrogel, compared to an area of 45% when hydrogel without phytoconstituent was used, which confirms the antioxidant potential of this flavonoid in wound treatment (Liu et al., 2020).

On a micro scale, the effect of baicalin on the mineralization of primary human osteoblasts was studied. When encapsulated in calcium polyphosphate microparticles measuring 2.9 µm, there was no impairment of growth and viability of these cells, and an increase of 21.12 nmoles in intracellular Ca^{+2} could be observed after 12 minutes of exposure to 2.1 µg/ml baicalin in the microparticles. In addition, an increase in phospholipase C enzyme activation and osteoblast mineralization was also observed, indicating potential use of baicalin in the treatment of bone disorders, such as arthrosis and osteoporosis (Wang et al., 2017).

PEGylated nanostructured lipid carriers loaded with baicalin measuring 83.9 nm promoted a rapid initial *in vitro* release of baicalin, about 60% in the first 10 hours, followed by constant release of more than 90% in 24 hours. *In vivo*, the nanoformulation was also able to remain 28 hours longer in the plasma circulation and reach the cardiac tissue in larger quantity (36.26 mg/L), where it was

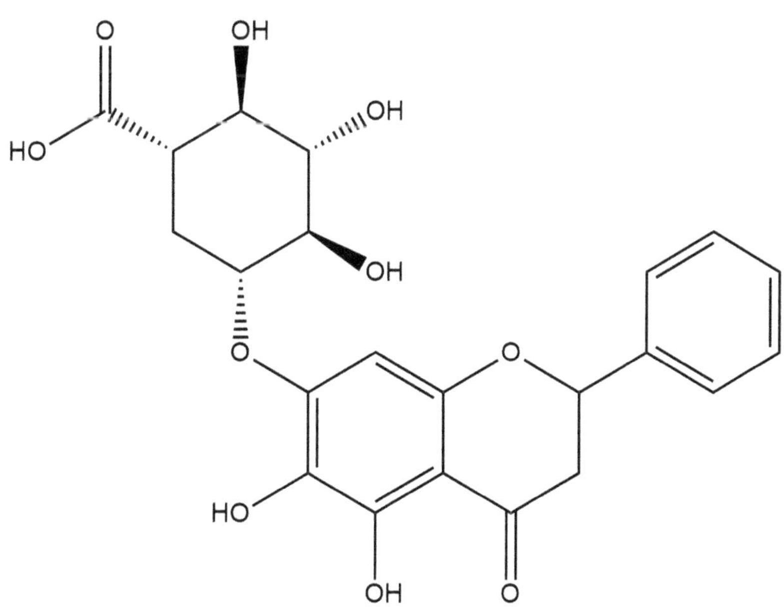

FIGURE 1.9 Chemical structure of baicalin.

used in traditional Chinese medicine. It occurs as orange crystals or powder and its practically insoluble in water but is soluble in ethanol. This compound possesses several pharmacological activities, such as antioxidant, anti-inflammatory, antimicrobial, antidiabetic, anticancer, and anti-metastatic effects, and studies have shown that it is also capable of reversing multidrug resistance (Zheng et al., 2021; Dong et al., 2016; "Emodin | C15H10O5—PubChem" n.d.).

Aiming to increase the solubility of emodin for topical application as an alternative to low oral availability, thermoreversible gels based on poloxamer P407 and P188 were investigated. Most promising formulations containing 10% P407, 30% P188 and 5% PEG400 (co-solvent), and another containing P407/P188 (20/10) increased emodin solubility 100× (0.56 and 0.61 mg/mL, respectively), compared to 10% ethanol and water (0.0065 and 0.0047 mg/mL), and gelled at physiological body temperature (32–36°C). However, *in vitro*, changes in cell morphology and viability were observed after treatment with the gel (Ban et al., 2017).

Polylactic acid microspheres with emodin measuring 9.7 μm promoted a controlled release of anthraquinone, with 23.46% released within 12 hours and 67.06% within 8 days, compared to 55% released within 1 hour from an emodin solution. The size of the microspheres promotes their targeting to the lungs, which was confirmed *in vivo*, with more than 12 μg/g concentrated in this tissue after 1 hour of intravenous administration in mice, indicating the promising capacity to use of this preparation and the exploitation of the effects of emodin for the treatment of lung diseases (Chen et al., 2014).

The suppressive effects of emodin on renal fibrosis were observed in a self-microemulsifying system, which exhibited, *in vivo*, 1.87× higher bioavailability than an emodin suspension, as well as enhanced uptake by human colon carcinoma Caco-2 cells and significantly reduced levels of fibronectin, TGF,-β1 and ICAM-1 in AGES-induced GMCs and NRK-52E cells (Huang et al., 2017). Similarly, to increase the targeting of emodin to HT-29 colon cancer cells, it was loaded into fluorescent or non-fluorescent mesoporous silica nanoparticles and released in high amount within 48 hours of application, triggering apoptotic and autophagy cell death mechanisms (Jänicke et al., 2021).

Emodin loaded in polymeric lipid nanoparticles was released in a controlled manner within 48 hours at normal tissue pH (6–6.9) and in tumor tissue (7.35–7.45) and was internalized by breast cancer MCF-7 cells, in which the apoptosis rate was 35.83%. *In vivo*, the preparations exhibited a tumor inhibition of over 60%, much higher than emodin alone, which inhibited less than 20%, indicating the promising effect of the nanoscale delivery system on the anticancer activity of emodin (Liu et al., 2021).

1.2.12 Stevioside

Stevioside (Figure 1.12) is a diterpene glycoside and the main sweet component of *Stevia rebaudiana* leaves. Despite having a bitter taste in the mouth, it is 250–300× sweeter than sucrose. In addition to being a sweetener, it has antioxidant, anticancer, anti-inflammatory, and antidiabetic effects. It is safe for diabetic and phenylketonuric patients and can be a potent non-cariogenic sucrose substitute (Geuns, 2003; "Stevioside | C38H60O18—PubChem" n.d.).

Pellets loaded with stevioside to regulate blood glucose in diabetes were effective in promoting high release of the phytoconstituent *in vitro*, from 82 to almost 100% released by the end of 1 hour. Following *in vivo* oral administration of formulations containing the stevioside-containing pellets in rats, blood glucose levels reduced from 338.24 to 310.25 mg/dL and 341.28 to 287.24 mg/dL within 24 hours. Moreover, high blood glucose reduction rates were also noticed when the rats were observed for 21 days, which indicates the high antidiabetic potential of the formulation (Dangre et al., 2017).

Another alternative for diabetic patients is a formulation of stevioside in edible jelly containing HPMC K15M and K100M, which promoted higher *in vitro* stevioside release (92.81%) in 6 hours when the polymeric composition was maximal. The *in vivo* antidiabetic effect was demonstrated by

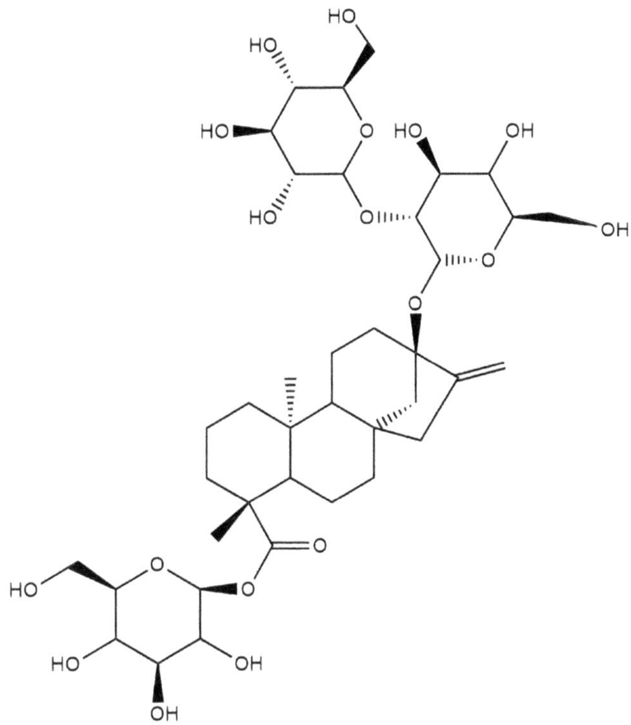

FIGURE 1.12 Chemical structure of stevioside.

the 8% reduction in blood glucose in 5 hours, which, even though it was half of that observed in the metformin-treated group (16%), indicates its potential for management of hyperglycemic episodes (Godbole et al., 2017).

The therapeutic potential of *S. rebaudiana* leaf extract was evaluated after its encapsulation in chitosan nanoparticles. In the *in vivo* assays, rats treated with the preparation showed weight gain, glycated hemoglobin reduction, and glycemia equal to 167.2 mg/dL at the end of 28 days of experimentation, while those receiving the isolated extract had glycemia equal to 195.6 mg/dL. In addition, the antioxidant effect was observed by normalization in the levels of the enzyme catalase, superoxide dismutase, and reduced glutathione, responsible for controlling the production of free radicals, in the liver and kidneys (Perumal et al., 2016).

To increase intestinal absorption and bioavailability of stevioside, polylactic acid (PLA) nanoparticles of Pluronic-F-68 copolymer showed promise since *in vitro* release assay showed initial burst in the first 25 hours, being almost 80% released in 100 hours, followed by controlled release from 200 hours, a profile considered important to achieve high initial concentrations of stevioside in the target tissue and enhance its therapeutic potential, especially the antidiabetic one (Barwal et al., 2013).

1.2.13 EPIGALLOCATECHIN-3 GALLATE

Epigallocatechin-3 gallate (EGCG) is the most abundant catechin, normally extracted from green tea (*Camellia sinensis* var. *assamica*), very popular in China and Japan. This polyphenol is considered responsible for the health benefits provided by this drink, among them antioxidant effects, cancer chemoprevention, and improved cardiovascular health. Furthermore, it has a solubility in water of 23.60 mg/mL, which is a high solubility and a high dissolution rate, leading to a shorter are its low aqueous solubility (10,45 μg/mL), high lipophilicity (logP 3,2), and low oral bioavailability (44.1%) (Kadari et al., 2017; "Fisetin | C15H10O6—PubChem," n.d.).

A dry powder of a sulfobutyl ether-β-cyclodextrin complex (SBE-β-CD) with fisetin, forming microparticles, was developed for the treatment of acute lung lesions and lung cancer. Complexation of fisetin with cyclodextrins proved that besides improving solubility, it increased cytotoxicity against Hela and MC-7 cells. The complex maintained the antioxidant properties of fisetin, as the IC_{50} values in the DPPH assay did not show a relevant difference between complexed and isolated fisetin, where any activity of the cyclodextrins was ruled out. Moreover, by adding leucine to the powder, better fluidity and fraction of fine particles were demonstrated (Mohtar et al., 2016).

Using human serum albumin (HSA) nanoparticles were prepared. The *in vitro* release of fisetin (pH 7.4) showed a higher initial release, followed by a slow, sustained, and controlled release, obtaining 75% of the fisetin released within 72 hours. The DPPH assay proved the antioxidant activity of the conjugate and the hemolytic assay resulted in an irrelevant hemolysis. Thus, it demonstrated that the conjugate was able to improve the biopharmaceutical properties of this phytoconstituent and enhanced its antioxidant potential in anticancer treatment (Ghosh et al., 2016).

Polylactic acid (PLA) nanoparticles demonstrate an ability to improve the delivery efficacy of fisetin in tumor treatment and are non-cytotoxic to different types of cell lines. In an *in vitro* study, isolated fisetin showed a 78% release of the drug in the first hour of incubation, while encapsulated fisetin obtained 35% within 24 hours, having 80% of the drug released in four days; thus, the complex demonstrated a sustained release of the drug, avoiding the accumulation of nanoparticles in the tumor tissue and possible systemic side effects. The antitumor effect was tested *in vivo* using 4T1 xenograft, mouse breast cancer, and when comparing isolated to incorporated fisetin, being administered 40 mg/kg on alternate days for three weeks, it was observed that incorporated fisetin reduced about 50% more tumor volume compared to isolated. Besides not demonstrating great weight reduction among mice for the group that was administered the incorporated fisetin, which proves low potential of systemic toxicity and great potential in anticancer therapy (Feng et al., 2019).

Particulate matter (PM2.5), of size ≤2.5 μm, the main compound of pollution, has the ability to directly cross the lung alveoli, blood, and blood-brain barrier, causing several health harms, such as neuroinflammation occurring in the hypothalamus. Thus, in a study, a nanoemulsion of fisetin was able to decrease inflammation and oxidative stress, relieving signaling activation in PM2.5-induced astrocytes by applying concentrations 5, 10, and 20 μg/ml of fisetin in an *in vitro* test. *In vivo* study, using mice with PM2.5-stimulated metabolic disorder, proved that long-term exposure to PM2.5 is able to increase neuroinflammation, but when treated with fisetin, the disorder was partially restored, highlighting the potential in preventing these syndromes (Xu et al., 2020).

1.2.15 Genistein

Genistein, 4′,5,7-trihydroxy isoflavone (Figure 1.15), is a soy-derived isoflavonoid that has antineoplastic activity. This compound has antitumor activities from its estrogenic, antioxidant, and

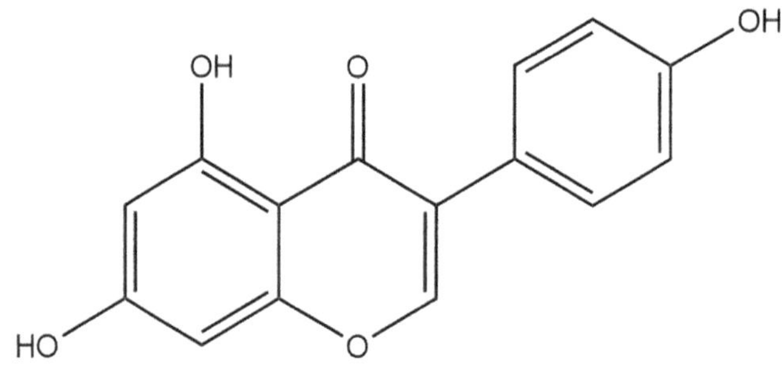

FIGURE 1.15 Chemical structure of genistein.

anti-migration properties and metabolism regulation activities, among others. However, genistein has a very low aqueous solubility and low bioavailability, the latter being caused by first-pass metabolism and enterohepatic recycling, high protein binding content, and low oral absorption (Botet-Carreras et al., 2021; "Genistein | C15H10O5—PubChem," n.d.).

In order to combat human cervical cancer, chitosan nanoparticles loaded with genistein (GCN) and folic-acid-conjugated GCN (FGCN) were developed. The *in vitro* drug release kinetics, performed by dialysis method, where the two systems obtained a 35–40% release profile within 24 hours. FGCN in the *in vitro* cellular uptake study showed a higher cellular uptake in cancer cells compared to GCN, but both had a maximum uptake time of 4 hours, which is explained by targeting the folic acid receptors that are present in these cells. In the *in vitro* cytotoxic activity, the systems were more effective in killing cancer cells than isolated genistein, and among the systems, GCN was the most successful, being attributed to the internalization of nanoparticles that were able to increase the intracellular drug concentration. In addition, in the analysis of apoptosis in 24-hour treated HeLa cells, it was observed that isolated genistein was not able to induce apoptosis in cancer cells, while GCN induced 22% and FGCN 55%, revealing the relevance of the use of chitosan nanoparticles in cancer treatment (Cai et al., 2017).

Another use is in the treatment of human colon cancer, based on the encapsulation of genistein in PEGylated silica nanoparticles (PEG-SiNPs) forming Gen-PEG-SiNPs. Regarding the *in vitro* release, it was observed at pH 7.2, and the range of 5.6 to 7.0, which are the values found for normal and mutated cells, respectively, demonstrated a faster release at more basic pH. Furthermore, when analyzing the aqueous solubility, it was seen that isolated genistein showed 4.098 µg/ml and the complex 9.864 µg/ml, which highlights that the nanoencapsulation was successful in increasing this aspect. In the *in vitro* antioxidant activity assay, it was observed that genistein showed an increased activity in the complex compared to the isolated one, which was justified by the possibility of PEG-SiNPs protecting genistein from oxygen that could be produced during incubation, in addition to the impact of the increased solubility of the complex. Regarding cell viability, it was tested *in vitro* on HT29 cells in 24 hours, showing that genistein inhibited up to 49.17% at 15 µM, but at certain concentrations, it can stimulate proliferation of cancer cells that are estrogen-dependent, while Gen-PEG-SiNPs reduced cell viability by 70%, with IC_{50} of 24.43 µM, which is justified by the presence of PEG that possibly potentiated the antiproliferative property of genistein (Pool et al., 2018).

In neurodegenerative diseases, such as Alzheimer's and Parkinson's, genistein can be used to reduce Aβ deposition, inflammatory damage, and calcium levels, acting in the prevention of these diseases. Thus, to enable an administration capable of crossing the blood-brain barrier, a study used the intranasal route for application of genistein-loaded chitosan NPs. Thus, in the *ex vivo* permeation study in the nasal mucosa, isolated genistein was not able to cross the cell membranes unlike the complex that obtained 60% of the drug found inside the mucosa within 2 hours, demonstrating that the NPs increased the permeability and bioavailability of the phytoconstituent in this route. Furthermore, the complex did not show altered cell viability and apoptotic events in PC12 (Rassu et al., 2019).

For intestinal lymphatic transport genistein has enhanced absorption via solid lipid nanoparticles (SLNs). In an *ex vivo* uptake study of the intestinal mucosa, at a temperature of 25°C, followed by an *in vitro* cellular absorption study, it was seen that SLNs loaded with genistein were visualized as internalized in the intestinal cells, proving the success of the complex, making it possible for the drug to reach the intestinal lymphatic vessels and avoid first-pass metabolism of the phytoconstituent (Obinu et al., 2021).

1.2.16 Gallic Acid

Gallic acid (Figure 1.16) is an endogenous plant polyphenol found in fruits such as grapes and berries, wines, hot-brewed teas, and also oak (*Quercus robur*), chestnut (*Castanea sativa*), and other wooden plant species. This acid is in the form of a slightly yellowish, odorless crystal and has a between the oil and the acid. In the *in vitro* antimicrobial study, the nanogel showed 98.05% growth inhibition of *S. aureus* strains, while the isolated and the loaded essential oil obtained 91.25% and 99.83%, respectively. In the MTT assay, in a series of concentrations from 0–1200 μg/ml of the tested compounds against skin (A-375) and breast (MCF-7) cancer cells, it was noted that the isolated essential oil and nanogel did not have significant results, but the complex reduced cell viability to 50% at all concentrations (Kamal et al., 2021).

Another essential oil of rosemary (*Lippia origanoides*) was loaded to a chitosan-based nanogel and incorporated with p-coumaric acid (CS-pCA) aiming for better antioxidant activity. In the DPPH assay, it was seen that the antioxidant activity of the chitosan and coumaric acid nanogel was improved with the essential oil that resulted in a neutralization percentage of 31.49%, being superior to the nanogel with coumaric acid only and only with the essential oil, which was confirmed by the ABTS assay where the CS-pCA obtained 444.8 trolox mg/nanogel g, which may be associated with a synergism between the compounds of the nanogel (Damasceno et al., 2018).

1.2.18 Mangiferin

Mangiferin (Figure 1.18) is a xanthone belonging to the Anacardiaceae and Gentianaceae families, and can be found in mangoes (*Mangifera indica*) and bee tea. Among its pharmacological activities are antidiabetic, antioxidant, anti-inflammatory, antiviral, and antimicrobial effects and chemotherapy potential. Despite being quite promising, it has disadvantages, such as its low aqueous solubility, which makes its application difficult (Gold-Smith et al., 2016; "Mangiferin | C19H18O11—PubChem," n.d.).

The mangiferin microparticles obtained from the supercritical antisolvent method were shown to possess 4.39× the aqueous solubility and a faster dissolution rate than the isolated phytoconstituent in an *in vitro* study. Regarding the antioxidant activity by DPPH, the microparticles, the isolated mangiferin and the control showed an IC_{50} of 0.040 mg/ml, 0.148 mg/ml, and 0.114 mg/ml, respectively, which shows that the microparticles obtained a higher activity at lower concentration. The results are confirmed by the ABTS radical assay, where the elimination efficiency of the microparticles was 1.33× higher than the isolatedmangiferin. Furthermore, the oral bioavailability test in rats showed that the microparticles increased this parameter by 2.07× compared to the isolated compound (Liu et al., 2020).

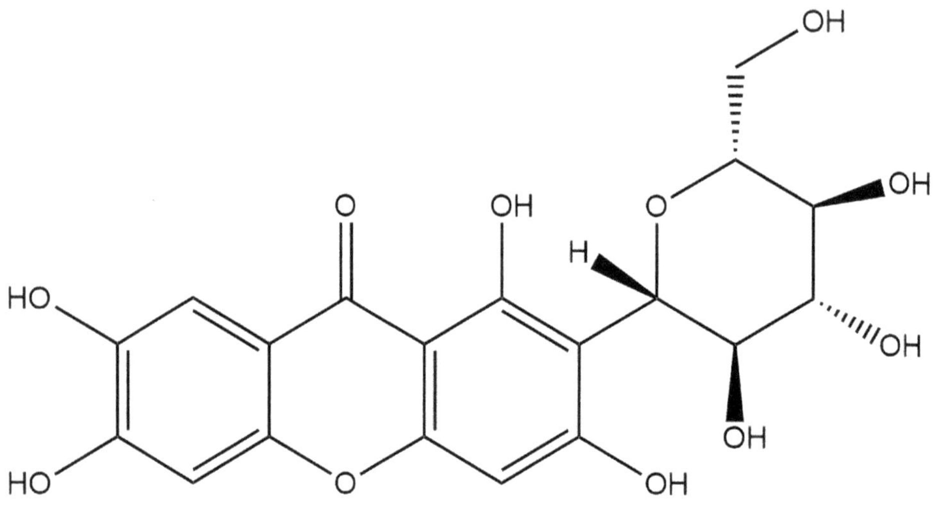

FIGURE 1.18 Chemical structure of mangiferin.

To treat oxidative stress in kidney cells, a study developed the encapsulation of mangiferin into chitosan nanoparticles. In the *in vitro* release study, in an environment simulating the gastrointestinal environment for 8 hours at 37°C, it was observed that isolated mangiferin released a smaller amount than the encapsulated one, which is explained by the possibility that the isolated compound is being degraded in the stomach pH, proving that the nanoparticle protects mangiferin and increases its bioavailability. In the DPPH assay, it was observed that the encapsulated phytoconstituent obtained 68%, being higher than the antioxidant activity of the isolated one, which was 60%. Furthermore, it was observed that the nanoparticles acted synergistically with the antioxidant enzymes that eliminate reactive oxygen species, and thus, the study concluded that the conjugate is a potent antioxidant and is not toxic, being very promising (Samadarsi and Dutta, 2020).

Radioactive gold nanoparticles with mangiferin were developed for use in prostate tumor therapy, in which it was observed that polyphenol associated with gold nanoparticles targets laminin receptors. Thus, mangiferin can direct the complex to prostate tumor cells, which was confirmed by the *in vivo* study in mice where an average of 80.98% of the injected dose of the complex was retained in the tumor in 30 minutes, considered a high initial retention value that is attributed to high selectivity and affinity of the complex phytoconstituent. Assessing the ability of the complex to control tumor size *in vivo*, applying 160 μCi/30 μL of the complex per tumor, it was found that after the 17th day the tumor volume presented a significant decrease and continued to decrease, unlike the control that was treated with saline solution (Al-Yasiri et al., 2017).

1.3 DISCUSSION AND CONCLUSIONS

With the increase of drug resistance and intrinsic resistance linked to microorganisms, the search for alternative medicines has been growing. Natural products have been applied in this environment as an alternative for this and other deficiencies, but a wide range has disadvantages related to low solubility, low stability, and consequently low bioavailability, among other characteristics. These characteristics, which can be circumvented with the aid of materials in a macro-, micro-, and nanotechnological scale, which in addition, conveniently present factors that allow a controlled, prolonged, and continuous release for these molecules, besides potentiating their therapeutic actions.

Secondary metabolites generated by these natural products can present several pharmacological functions. Among the materials discussed in this chapter, the activities that are most present are antitumor activities, being present in 14 of the 18 cited phytoconstituents; antioxidant and anti-inflammatory, present in 13 and 10, respectively; and antimicrobial and antidiabetic, present in 6. In addition, other properties can be found in these compounds, even not being so present in terms of numbers, but they are still of paramount importance, such as kidney and liver protection, metabolism regeneration, and anticonvulsant and antiviral properties, among others.

The next step for phytocompounds in macro, micro and nano scales would be the adaptation of the techniques and norms established for conducting phase 1 of clinical trials, whose results are important in finding alternative treatments of several diseases. This can lead to a significant improvement in phytochemicals' therapeutic effects, a reduction of their adverse effects, and promotion of better treatment adherence by the patient.

REFERENCES

"5,7-Dihydroxy-2-(4-Hydroxyphenyl)Chroman-4-One I C15H12O5—PubChem." (n.d.). Accessed October 18, 2021. https://pubchem.ncbi.nlm.nih.gov/compound/932.

Almeida, L., J. A. Oshiro-Junior, M. Silva, et al. (2019). "Tablet of Ximenia Americana L. Developed from Mucoadhesive Polymers for Future Use in Oral Treatment of Fungal Infections." *Polymers(Basel)* 11 (2): 379. https://doi.org/10.3390/polym11020379.

Alves, A. de C. Soares, R. M. Mainardes, and N. M. Khalil. (2016). "Nanoencapsulation of Gallic Acid and Evaluation of Its Cytotoxicity and Antioxidant Activity." *Materials Science and Engineering: C* 60: 126–134. https://doi.org/10.1016/j.msec.2015.11.014.

Alves, L. P., K. S. Oliveira, J. A. P. Santos, et al. (2020). "A Review on Developments and Prospects of Anti-Inflammatory Administration in Microemulsions." *Journal of Drug Delivery Science and Technology* 60: 102008. https://doi.org/10.1016/j.jddst.2020.102008.

Al-Yasiri, A. Y., et al. (2017). "Mangiferin Functionalized Radioactive Gold Nanoparticles (MGF-198 AuNPs) in Prostate Tumor Therapy: Green Nanotechnology for Production, In Vivo Tumor Retention and Evaluation of Therapeutic Efficacy." *Dalton Transactions* 46 (42): 14561–14571. https://doi.org/10.1039/C7DT00383H.

"Apigenin | C15H10O5—PubChem." (n.d.). Accessed October 15, 2021. https://pubchem.ncbi.nlm.nih.gov/compound/5280443.

Araújo, G. M. F., A. R. A. Barros, J. A. Oshiro-Junior, et al. (2019). "Nanoemulsions Loaded with Amphotericin B: Development, Characterization and Leishmanicidal Activity." *Current Pharmaceutical Design* 25 (14): 1616–1622. https://doi.org/10.2174/1381612825666190705202030.

"Baicalin | C21H18O11—PubChem." (n.d.). Accessed October 15, 2021. https://pubchem.ncbi.nlm.nih.gov/compound/64982#section=IR-Spectra.

Ban, E., M. Park, S. Jeong, T. Kwon, E.H. Kim, K. Jung, and A. Kim. (2017). "Poloxamer-Based Thermoreversible Gel for Topical Delivery of Emodin: Influence of P407 and P188 on Solubility of Emodin and Its Application in Cellular Activity Screening." *Molecules* 22 (2). https://doi.org/10.3390/molecules22020246.

Barros, R. M., M. S. de Oliveira, K. M. N. Costa, et al. (2020). "Physicochemical Characterization of Bioactive Compounds in Nanocarriers." *Current Pharmaceutical Design* 26 (32): 4163–4173. https://doi.org/10.2174/1381612826666200310144533.

Barwal, I., A. Sood, M. Sharma, B. Singh, and S. C. Yadav. (2013). "Development of Stevioside Pluronic-F-68 Copolymer Based PLA-Nanoparticles as an Antidiabetic Nanomedicine." *Colloids and Surfaces B: Biointerfaces* 101: 510–516. https://doi.org/10.1016/j.colsurfb.2012.07.005.

Bernardes, A. L. (2017). Bioacessibilidade in vitro de compostos fenólicos microencapsulados do fruto jussara (Euterpe edulis Martius) e aplicação em sistema-modelo de gelatina. 70 p. Dissertação (Mestrado em Ciência e Tecnologia de Alimentos)—Universidade Federal do Espírito Santo, [S. l.], 2017. Acesso em: 5 November 2021. http://portais4.ufes.br/posgrad/teses/tese_10117_Disserta%E7%E3o%20Final%20de%20Mestrado%20-%20Andressa%20Ladeira%20Bernardes%20%20%20PDF.pdf.

Bhalerao, R. A., M. R. Bhalekar, and M. C. Damle. (2019). "Formulation and Evaluation Transdermal Patch of Hesperidin." *Journal of Drug Delivery and Therapeutics* 9 (4): 311–317. https://doi.org/10.22270/jddt.v9i4.3048.

Botet-Carreras, A., Tamames-Tabar, C., Salles, F., Rojas, S., Imbuluzqueta, E., Lana, H., Blanco-Prieto, M. J., Horcajada, P. (2021). "Improving the Genistein Oral Bioavailability Via Its Formulation Into the Metal–Organic Framework MIL-100 (Fe)." *Journal of Materials Chemistry B* 9 (9): 2233–2239. https://doi.org/10.1039/D0TB02804E.

Boudoulas, K. D., Triposkiadis, F., Stefanadis, C., Boudoulas, H. (2017). "The Endlessness Evolution of Medicine, Continuous Increase in Life Expectancy and Constant Role of the Physician." *Hellenic Journal of Cardiology* 58 (5): 322–330.

Boz, H. (2015). "p-Coumaric Acid in Cereals: Presence, Antioxidant and Antimicrobial Effects." *International Journal of Food Science & Technology* 50 (11): 2323–2328. https://doi.org/10.1111/ijfs.12898.

Budhiraja, A., and G. Dhingra. (2015). "Development and Characterization of a Novel Antiacne Niosomal Gel of Rosmarinic Acid." *Drug Delivery* 22 (6): 723–730. https://doi.org/10.3109/10717544.2014.903010.

Cai, L., R. Yu, X. Hao, et al. (2017). "Folate Receptor-Targeted Bioflavonoid Genistein-Loaded Chitosan Nanoparticles for Enhanced Anticancer Effect in Cervical Cancers." *Nanoscale Research Letters* 12: 509. https://doi.org/10.1186/s11671-017-2253-z.

Cano, A., et al. (2019). "Dual-Drug Loaded Nanoparticles of Epigallocatechin-3-Gallate (EGCG)/Ascorbic Acid Enhance Therapeutic Efficacy of EGCG in a APPswe/PS1dE9 Alzheimer's Disease Mice Model." *Journal of Controlled Release* 301: 62–75. https://doi.org/10.1016/j.jconrel.2019.03.010.

Celep, A. G. S., A. Demirkaya, and E. K. Solak. (2020). "Antioxidant and Anticancer Activities of Gallic Acid Loaded Sodium Alginate Microspheres on Colon Cancer." *Current Applied Physics*. https://doi.org/10.1016/j.cap.2020.06.002.

Chaurasia, S., R. R. Patel, P. Vure, and B. Mishra. (2017). "Oral Naringenin Nanocarriers: Fabrication, Optimization, Pharmacokinetic and Chemotherapeutic Efficacy Assessments." *Nanomedicine* 12 (11): 1243–1260. https://doi.org/10.2217/nnm-2016-0436.

Chen, X., Z. Yang, R. Sun, Z. Mo, G. Jin, F. Wei, J. Hu, W. Guan, and N. Zhong. (2014). "Preparation of Lung-Targeting, Emodin-Loaded Polylactic Acid Microspheres and Their Properties." *International Journal of Molecular Sciences* 15 (4): 6241–6251. https://doi.org/10.3390/ijms15046241.

"cis-p-Coumaric acid | C9H8O3—PubChem." (n.d.). Accessed October 25, 2021. https://pubchem.ncbi.nlm.nih.gov/compound/1549106.

Coşarcă, S., C. Tanase, and D. L. Muntean. (2019). "Therapeutic aspects of catechin and its derivatives–an update." *Acta Biologica Marisiensis* 2 (1): 21–29. https://doi.org/10.2478/abmj-2019-0003.

Costa, K. M. M., M. R. Sato, T. L. A. Barbosa, et al. (2021). "Curcumin-Loaded Micelles Dispersed in Ureasil-Polyether Materials for a Novel Sustained-Release Formulation." *Pharmaceutics* 13 (5): 675. https://doi.org/10.3390/pharmaceutics13050675.

Cuevas-Valenzuela, J., Á. González-Rojas, J. Wisniak, A. Apelblat, and J. R. Pérez-Correa. (2014). "Solubility of (+)-Catechin in Water and Water-Ethanol Mixtures within the Temperature Range 277.6–331.2K: Fundamental Data to Design Polyphenol Extraction Processes." *Fluid Phase Equilibria* 382: 279–285. https://doi.org/10.1016/j.fluid.2014.09.013.

Damasceno, E. T. S., R. R. Almeida, S. Y. B. de Carvalho, G. S. G. de Carvalho, V. Mano, A. C. Pereira, and L. G. de Lima Guimarães. (2018). "Lippiaoriganoides Kunth: Essential Oil Loaded in Nanogel Based on the Chitosan and ρ-Coumaric Aid: Encapsulation Efficiency and Antioxidant Activity." *Industrial Crops and Products* 125: 85–94. https://doi.org/10.1016/j.indcrop.2018.08.074.

Dangre, P. V., V. D. Fating, S. B. Wakodkar, and M. D. Godbole. (2017). "Development of Stevioside Loaded Pellets for Regulating the Blood Glucose Level in Diabetes: In Vivo-In Vitro Characterization." *Journal of Drug Delivery and Therapeutics* 7 (4): 59–65. https://doi.org/10.22270/jddt.v7i4.1461.

Dehghani, M. A., et al. (2020). "Protective effect of gallic acid and gallic acid-loaded Eudragit-RS 100 nanoparticles on cisplatin-induced mitochondrial dysfunction and inflammation in rat kidney." *Biochimica et Biophysica Acta(BBA)-Molecular Basis of Disease* 1866 (12): 165911. https://doi.org/10.1016/j.bbadis.2020.165911.

de Oliveira, M. S., J. A. Oshiro-Junior, et al. (2020). "Polymeric Nanoparticle Associated with Ceftriaxone and Extract of Schinopsis Brasiliensis Engler against Multiresistant Enterobacteria." *Pharmaceutics* 12 (8): 695.

Dong, X., J. Fu, X. Yin, S. Cao, X. Li, L. Lin, and J. Ni. (2016). "Emodin: A Review of Its Pharmacology, Toxicity and Pharmacokinetics." *Phytotherapy Research* 1218 (January): 1207–1218. https://doi.org/10.1002/ptr.5631.

"Emodin | C15H10O5—PubChem." (n.d.). Accessed October 17, 2021. https://pubchem.ncbi.nlm.nih.gov/compound/3220#section=Experimental-Properties.

"Epigallocatechin-3 gallate | C22H18O11—PubChem." (n.d.). Accessed October 25, 2021. https://pubchem.ncbi.nlm.nih.gov/compound/65064.

Fachel, F. N. S., L. R. Michels, J. H. Azambuja, G. S. Lenz, N. E. Gelsleichter, M. Endres, J. N. Scholl, et al. (2020). "Chitosan-Coated Rosmarinic Acid Nanoemulsion Nasal Administration Protects against LPS-Induced Memory Deficit, Neuroinflammation, and Oxidative Stress in Wistar Rats." *Neurochemistry International* 141 (October). https://doi.org/10.1016/j.neuint.2020.104875.

Feng, C., et al. (2019). "Preparation and Optimization of Poly (Lactic Acid) Nanoparticles Loaded with Fisetin to Improve Anti-Cancer Therapy." *International Journal of BiologicalMacromolecules* 125: 700–710. https://doi.org/10.1016/j.ijbiomac.2018.12.003.

Fernandes, L., et al. (2020). "Green Tea Polyphenol Microparticles Based on the Oxidative Coupling of EGCG Inhibit Amyloid Aggregation/Cytotoxicity and Serve as a Platform for Drug Delivery." *ACS Biomaterials Science & Engineering* 6 (8): 4414–4423. https://doi.org/10.1021/acsbiomaterials.0c00188.

"Fisetin | C15H10O6—PubChem." (n.d.). Accessed October 25, 2021. https://pubchem.ncbi.nlm.nih.gov/compound/5281614.

Freag, M. S., W. M. Saleh, and O. Y. Abdallah. (2018). "Self-Assembled Phospholipid-Based Phytosomal Nanocarriers as Promising Platforms for Improving Oral Bioavailability of the Anticancer Celastrol." *International Journal of Pharmaceutics* 535 (1–2): 18–26. https://doi.org/10.1016/j.ijpharm.2017.10.053.

"Gallic Acid | C7H6O5—PubChem." (n.d.). Accessed October 25, 2021. https://pubchem.ncbi.nlm.nih.gov/compound/370.

"Genistein | C15H10O5—PubChem." (n.d.). Accessed October 25, 2021. https://pubchem.ncbi.nlm.nih.gov/compound/5280961.

Geuns, J. M. C. (2003). "Stevioside." *Phytochemistry* 64 (5): 913–921. https://doi.org/10.1016/S0031-9422(03)00426-6.

Ghosh, P., et al. (2016). "Preparation of Albumin Based Nanoparticles for Delivery of Fisetin and Evaluation of Its Cytotoxic Activity." *International Journal of Biological Macromolecules* 86: 408–417. https://doi.org/10.1016/j.ijbiomac.2016.01.082.

Godbole, M. D., D. K. Mahapatra, and P. D. Khode. (2017). "Fabrication and Characterization of Edible Jelly Formulation of Stevioside: A Nutraceutical or OTC Aid for the Diabetic Patients." *Inventi Rapid:Nutraceut* 2 (2): 1–9.

Gold-Smith, F., A. Fernandez, and K. Bishop. (2016). "Mangiferin and Cancer: Mechanisms of Action." *Nutrients* 8 (7): 396. https://doi.org/10.3390/nu8070396.

Gondim, B. L. C., J. A. Oshiro-Junior, F. H. A. Fernandes, et al. (2019). "Plant Extracts Loaded in Nanostructured Drug Delivery Systems for Treating Parasitic and Antimicrobial Diseases." *Current Pharmaceutical Design* 25 (14): 1604–1615. https://doi.org/10.2174/1381612825666190628153755.

Gratieri, T., L. A. G. Pinho, M. A. Oliveira, L. L. Sa-Barreto, R. N. Marreto, I. C. Silva, G. M. Gelfuso, J. de Souza Siqueira Quintans, L. J. Quintans-Junior, and M. Cunha-Filho. (2020). "Hydroxypropyl-β-Cyclodextrin-Complexed Naringenin by Solvent Change Precipitation for Improving Anti-Inflammatory Effect In Vivo." *Carbohydrate Polymers* 231 (December): 115769. https://doi.org/10.1016/j.carbpol.2019.115769.

Hou, W., B. Liu, and H. Xu. (2020). "Celastrol: Progresses in Structure-Modifications, Structure-Activity Relationships, Pharmacology and Toxicology." *European Journal of Medicinal Chemistry* 189: 112081. https://doi.org/10.1016/J.EJMECH.2020.112081.

Hu, C., Q. Wang, C. Ma, and Q. Xia. (2016). "Non-Aqueous Self-Double-Emulsifying Drug Delivery System: A New Approach to Enhance Resveratrol Solubility for Effective Transdermal Delivery." *Colloids and Surfaces A: Physicochemical and Engineering Aspects* 489: 360–369. https://doi.org/10.1016/J.COLSURFA.2015.11.017.

Huang, H., et al. (2019a). "Novel Multi-Phase Nano-Emulsion Preparation for Co-Loading Hydrophilic Arbutin and Hydrophobic Coumaric Acid Using Hydrocolloids." *Food Hydrocolloids* 93: 92–101. https://doi.org/10.1016/j.foodhyd.2019.02.023.

Huang, J., W. Gong, Z. Chen, J. Huang, Q. Chen, H. Huang, and C. Zhao. (2017). "Emodin Self-Emulsifying Platform Ameliorates the Expression of FN, ICAM-1 and TGF-B1 in AGEs-Induced Glomerular Mesangial Cells by Promoting Absorption." *European Journal of Pharmaceutical Sciences* 99: 128–136. https://doi.org/10.1016/j.ejps.2016.12.012.

Huang, S., Q. Xue, J. Xu, S. Ruan, and T. Cai. (2019b). "Simultaneously Improving the Physicochemical Properties, Dissolution Performance, and Bioavailability of Apigenin and Daidzein by Co-Crystallization with Theophylline." *Journal of Pharmaceutical Sciences* 108 (9): 2982–2993. https://doi.org/10.1016/j.xphs.2019.04.017.

Huang, T., Y. Liu, and C. Zhang. (2019c). "Pharmacokinetics and Bioavailability Enhancement of Baicalin: A Review." *European Journal of Drug Metabolism and Pharmacokinetics* 44 (2): 159–168. https://doi.org/10.1007/s13318-018-0509-3.

Jain, K. K. Drug Delivery Systems, Methods in Molecular Biology, vol. 2059, © Springer Science+Business Media, LLC, part of Springer Nature 2020. cap. 1 p. 37–39. https://doi.org/10.1007/978-1-4939-9798-5_2.

Jänicke, P., C. Lennicke, A. Meister, B. Seliger, L. A. Wessjohann, and G. N. Kaluđerović. (2021). "Fluorescent Spherical Mesoporous Silica Nanoparticles Loaded with Emodin: Synthesis, Cellular Uptake and Anticancer Activity." *Materials Science and Engineering C* 119: 111619. https://doi.org/10.1016/j.msec.2020.111619.

Ji, P., T. Yu, Y. Liu, J. Jiang, J. Xu, Y. Zhao, Y. Hao, Y. Qiu, W. Zhao, and C. Wu. (2016). "Naringenin-Loaded Solid Lipid Nanoparticles: Preparation, Controlled Delivery, Cellular Uptake, and Pulmonary Pharmacokinetics." *Drug Design,Development and Therapy* 10: 911–925. https://doi.org/10.2147/DDDT.S97738.

Kadari, A., et al. (2017). "Enhanced Oral Bioavailability and Anticancer Efficacy of Fisetin by Encapsulating as Inclusion Complex with HPβCD in Polymeric Nanoparticles." *DrugDelivery* 24 (1): 224–232. https://doi.org/10.1080/10717544.2016.1245366.

Kamal, I., et al. (2021). "Chemotherapeutic and Chemopreventive Potentials of ρ-Coumaric Acid–Squid Chitosan Nanogel Loaded with Syzygiumaromaticum Essential Oil." *International Journal of Biological Macromolecules* 188: 523–533. https://doi.org/10.1016/j.ijbiomac.2021.08.038.

Kassem, F. A., A. E. Abdelaziz, and G. M. El Maghraby. (2021). "Ethanol-Assisted Kneading of Apigenin with Arginine for Enhanced Dissolution Rate of Apigenin: Development of Rapidly Disintegrating Tablets." *Pharmaceutical Development and Technology* 26 (6): 693–700. https://doi.org/10.1080/10837450.2021.1922441.

Kim, B. K., A. R. Cho, and D. J. Park. (2016). "Enhancing Oral Bioavailability Using Preparations of Apigenin-Loaded W/O/W Emulsions: In Vitro and In Vivo Evaluations." *Food Chemistry* 206: 85–91. https://doi.org/10.1016/j.foodchem.2016.03.052.

Laddha, A. P., S. Murugesan, and Y. A. Kulkarni. (2020). "In-Vivo and In-Silico Toxicity Studies of Daidzein: An Isoflavone from Soy." *Drug and Chemical Toxicology*: 1–9. https://doi.org/10.1080/01480545.2020.1833906.

Leite, F. G., J. A. Oshiro-Junior, L. A. Chiavacci, and B. G. Chiari-Andréo. (2019). "Assessment of an Anti-Ageing Structured Cosmetic Formulation Containing Goji Berry." *Brazilian Journal of Pharmaceutical Sciences* 55. https://doi.org/10.1590/s2175-97902019000217412.

Li, R., Y. Li, J. Zhang, Q. Liu, T. Wu, J. Zhou, and J. He. (2020). "Targeted Delivery of Celastrol to Renal Interstitial Myofibroblasts using Fibronectin-Binding Liposomes Attenuates Renal Fibrosis and Reduces Systemic Toxicity." *Journal of Controlled Release*. https://doi.org/10.1016/j.jconrel.2020.01.017.

Liu, G., Z. Bao, and J. Wu. (2020). "Injectable Baicalin/F127 Hydrogel with Antioxidant Activity for Enhanced Wound Healing." *Chinese Chemical Letters* 31 (7): 1817–1821. https://doi.org/10.1016/j.cclet.2020.03.005.

Liu, H., Y. Zhuang, P. Wang, T. Zou, M. Lan, L. Li, F. Liu, T. Cai, and Y. Cai. (2021). "Polymeric Lipid Hybrid Nanoparticles as a Delivery System Enhance the Antitumor Effect of Emodin In Vitro and In Vivo." *Journal of Pharmaceutical Sciences* 110 (8): 2986–2996. https://doi.org/10.1016/j.xphs.2021.04.006.

Liu, M., et al. (2020). "Solubility, Antioxidation, and Oral Bioavailability Improvement of Mangiferin Microparticles Prepared Using the Supercritical Antisolvent Method." *Pharmaceutics* 12 (2): 90. https://doi.org/10.3390/pharmaceutics12020090.

Lu, Y., Y. Liu, J. Zhou, D. Li, and W. Gao. (2020). "Biosynthesis, Total Synthesis, Structural Modifications, Bioactivity, and Mechanism of Action of the Quinone-Methide Triterpenoid Celastrol." *Medicinal Research Reviews*. https://doi.org/10.1002/med.21751.

Luz, A. V. da and E. R. Adami. (2019). USO DOS TAXANOS NO TRATAMENTO DO CÂNCER DE MAMA. Maringá: UNICESUMAR—Centro Universitário de Maringá. Acesso em: 21 de outubro de 2019. https://rdu.unicesumar.edu.br/bitstream/123456789/4145/1/Andressa%20Val%20da%20Luz.pdf.

"Mangiferin | C19H18O11—PubChem." (n.d.). Accessed October 25, 2021. https://pubchem.ncbi.nlm.nih.gov/compound/5281647.

Meng, S., L. Sun, L. Wang, Z. Lin, Z. Liu, L. Xi, and Y. Zheng. (2019). "Loading of Water-Insoluble Celastrol into Niosome Hydrogels for Improved Topical Permeation and Anti-Psoriasis Activity." *Colloids and Surfaces B: Biointerfaces* 110352. https://doi.org/10.1016/j.colsurfb.2019.11035.

Mohtar, N., et al. (2017). "Design and Development of Dry Powder Sulfobutylether-β-Cyclodextrin Complex for Pulmonary Delivery of Fisetin." *European Journal of Pharmaceutics and Biopharmaceutics* 113: 1–10. https://doi.org/10.1016/j.ejpb.2016.11.036.

Nadeem, M., M. Imran, T. A. Gondal, A. Imran, M. Shahbaz, R. M. Amir, and M. W. Sajid. (2019). "Applied Sciences Therapeutic Potential of Rosmarinic Acid: A Comprehensive Review." *Applied Science* 9 (3139).

Nagle, D. G., D. Ferreira, and Y. D. Zhou. (2006). "Epigallocatechin-3-Gallate (EGCG): Chemical and Biomedical Perspectives." *Phytochemistry* 67 (17): 1849–1855. https://doi.org/10.1016/j.phytochem.2006.06.020.

Neves, A. R., M. Lucio, J. L. C. Lima, and S. Reis. (2012). "Resveratrol in Medicinal Chemistry: A Critical Review of Its Pharmacokinetics, Drug-Delivery, and Membrane Interactions." *Current Medicinal Chemistry* 19 (11): 1663–1681. https://doi.org/10.2174/092986712799945085.

Obinu, A., et al. (2021). "Transmucosal Solid Lipid Nanoparticles to Improve Genistein Absorption Via Intestinal Lymphatic Transport." *Pharmaceutics* 13 (2): 267. https://doi.org/10.3390/pharmaceutics13020267.

Oshiro-Junior, J. A., C. Rodero, G. Hanck-Silva, et al. (2020). "Stimuli-Responsive Drug Delivery Nanocarriers in the Treatment of Breast Cancer." *Current Medical Chemistry* 27 (15): 2494–2513. https://doi.org/10.2174/0929867325666181009120610.

Pápay, Z. E., A. Kósa, B. Böddi, Z. Merchant, I. Y. Saleem, M. G. Zariwala, I. Klebovich, S. Somavarapu, and I. Antal (2017). "Study on the Pulmonary Delivery System of Apigenin-Loaded Albumin Nanocarriers with Antioxidant Activity." *Journal of Aerosol Medicine and Pulmonary Drug Delivery* 30 (4): 274–288. https://doi.org/10.1089/jamp.2016.1316.

Parhiz, H., A. Roohbakhsh, F. Soltani, R. Rezaee, and M. Iranshahi, M. (2015). "Antioxidant and Anti-Inflammatory Properties of the Citrus Flavonoids Hesperidin and Hesperetin: An Updated Review of Their Molecular Mechanisms and Experimental Models." *Phytotherapy Research* 29 (3): 323–331. https://doi.org/10.1002/ptr.5256.

Park, S. H., H. S. Shin, and S. N. Park. (2018). "A Novel PH-Responsive Hydrogel Based on Carboxymethyl Cellulose/2-Hydroxyethyl Acrylate for Transdermal Delivery of Naringenin." *Carbohydrate Polymers* 200: 341–352. https://doi.org/10.1016/j.carbpol.2018.08.011.

Pei, K., et al. (2016). "p-Coumaric Acid and Its Conjugates: Dietary Sources, Pharmacokinetic Properties and Biological Activities." *Journal of the Science of Food and Agriculture* 96 (9): 2952–2962. https://doi.org/10.1002/jsfa.7578.

Perumal, V., T. Manickam, K. S. Bang, P. Velmurugan, and B. T. Oh. (2016). "Antidiabetic Potential of Bioactive Molecules Coated Chitosan Nanoparticles in Experimental Rats." *International Journal of Biological Macromolecules* 92: 63–69. https://doi.org/10.1016/j.ijbiomac.2016.07.006.

Pool, H., et al. (2018). "Development of Genistein-PEGylated Silica Hybrid Nanomaterials with Enhanced Antioxidant and Antiproliferative Properties on HT29 Human Colon Cancer Cells." *American Journal of Translational Research* 10 (8): 2306.

Rassu, G., et al. (2019). "Intranasal Delivery of Genistein-Loaded Nanoparticles as a Potential Preventive System against Neurodegenerative Disorders." *Pharmaceutics* 11 (1): 8. https://doi.org/10.3390/pharmaceutics11010008.

"Rosmarinic Acid | C18H16O8—PubChem." (n.d.). Accessed October 19, 2021. https://pubchem.ncbi.nlm.nih.gov/compound/5281792#section=Density.

Rostami, M., M. Ghorbani, M. Aman Mohammadi, M. Delavar, M. Tabibiazar, and S. Ramezani. (2019). "Development of Resveratrol Loaded Chitosan-Gellan Nanofiber as a Novel Gastrointestinal Delivery System." *International Journal of Biological Macromolecules* 135: 698–705. https://doi.org/10.1016/J.IJBIOMAC.2019.05.187.

Salehi, B. P. V. T. Fokou, M. Sharifi-Rad, P. Zucca, R. Pezzani, N. Martins, and J. Sharifi-Rad. (2019). "The Therapeutic Potential of Naringenin: A Review of Clinical Trials." *Pharmaceuticals* 12 (1): 1–18. https://doi.org/10.3390/ph12010011.

Salehi, B., A. Venditti, M. Sharifi-Rad, D. Kręgiel, J. Sharifi-Rad, A. Durazzo, M. Lucarini, et al. (2019). "The Therapeutic Potential of Apigenin." *International Journal of Molecular Sciences* 20 (6). https://doi.org/10.3390/ijms20061305.

Samadarsi, R., and D. Dutta. (2020). "Anti-Oxidative Effect of Mangiferin-Chitosan Nanoparticles on Oxidative Stress-Induced Renal Cells." *International Journal of Biological Macromolecules* 151: 36–46. https://doi.org/10.1016/j.ijbiomac.2020.02.112.

Samadian, N., and M. Hashemi. (2018). "Effects of Apigenin and Apigenin-Loaded Nanogel on Induction of Apoptosis in Human Chronic Myeloid Leukemia Cells." *Galen Medical Journal* 7 (0): 1008. https://doi.org/10.22086/gmj.v0i0.1008.

Samanta, A., S. Chanda, B. Bandyopadhyay, and N. Das. (2016). "Establishment of Drug Delivery System Nanocapsulated with an Antioxidant (+)-Catechin Hydrate and Sodium Meta Borate Chelator against Sodium Fluoride Induced Oxidative Stress in Rats." *Journal of Trace Elements in Medicine and Biology* 33: 54–67. https://doi.org/10.1016/J.JTEMB.2015.09.003.

Sanna, V., et al. (2017). "Targeted Nanoparticles Encapsulating (−)-Epigallocatechin-3-Gallate for Prostate Cancer Prevention and Therapy. *Scientific Reports* 7: 41573. https://doi.org/10.1038/srep41573.

Shishir, M. R. I., H. Suo, X. Liu, Q. Kang, J. Xiao, M. Wang, F. Chen, and K. W. Cheng. (2021). "Development and Evaluation of a Novel Nanofibersolosome for Enhancing the Stability, In Vitro Bioaccessibility, and Colonic Delivery of Cyanidin-3-O-Glucoside." *Food Research International* 149: 110712. https://doi.org/10.1016/J.FOODRES.2021.110712.

Silva, J. M., M. S. C. Nobre, S. L. Albino, et al. (2020). "Secondary Metabolites with Antioxidant Activities for the Putative Treatment of Amyotrophic Lateral Sclerosis (ALS): Experimental Evidences." *Oxidative Medicine and Cellular Longevity* 2020: 5642029. https://doi.org/10.1155/2020/5642029.

Silva, S. B. da, D. Ferreira, M. Pintado, and B. Sarmento. (2016). "Chitosan-Based Nanoparticles for Rosmarinic Acid Ocular Delivery-In Vitro Tests." *International Journal of Biological Macromolecules* 84: 112–120. https://doi.org/10.1016/j.ijbiomac.2015.11.070.

Singh, N. A., A. K. A. Mandal, and Z. A. Khan. (2017). "Fabrication of PLA-PEG Nanoparticles as Delivery Systems for Improved Stability and Controlled Release of Catechin." *Journal of Nanomaterials*. https://doi.org/10.1155/2017/6907149.

Song, J., X. Fan, and Q. Shen. (2016). "Daidzein-Loaded Nanostructured Lipid Carriers-PLGA Nanofibers for Transdermal Delivery." *International Journal of Pharmaceutics* 501 (1–2): 245–252. https://doi.org/10.1016/J.IJPHARM.2016.02.003.

Stelmakiene, A., K. Ramanauskiene, and V. Briedis. (2015). "Release of Rosmarinic Acid from Semisolid Formulations and Its Penetration through Human Skin Ex Vivo." *Acta Pharmaceutica* 65 (2): 199–205. https://doi.org/10.1515/acph-2015-0012.

"Stevioside | C38H60O18—PubChem." (n.d.). Accessed October 20, 2021. https://pubchem.ncbi.nlm.nih.gov/compound/442089.

Sulaiman, G. M., H. M. Waheeb, M. S. Jabir, et al. (2020). "Hesperidin Loaded on Gold Nanoparticles as a Drug Delivery System for a Successful Biocompatible, Anti-Cancer, Anti-Inflammatory and Phagocytosis Inducer Model." *Scientific Report* 10: 9362. https://doi.org/10.1038/s41598-020-66419-6.

Tan, J., Y. Li., D. X. Hou and S. Wu. (2019). "Os efeitos e mecanismos da cianidina-3-glicosídeo e seus metabólitos fenólicos na manutenção da integridade intestinal." *Antioxidants* 8 (10): 479. https://doi.org/10.3390/antiox8100479.

Tejada, S., S. Pinya, M. Martorell, X. Capó, J. A. Tur, A. Pons, and A. Sureda. (2017). "Potential Anti-Inflammatory Effects of Hesperidin from the Genus Citrus." *Current Medicinal Chemistry* 24. https://doi.org/10.2174/0929867324666170718104412.

Trotta, V., B. Pavan, L. Ferraro, S. Beggiato, D. Traini, L. G. Des Reis, S. Scalia, and A. Dalpiaz. (2018). "Brain Targeting of Resveratrol by Nasal Administration of Chitosan-Coated Lipid Microparticles." *European Journal of Pharmaceutics and Biopharmaceutics* 127: 250–259. https://doi.org/10.1016/J.EJPB.2018.02.010.

Tsirigotis-Maniecka, M., Ł. Lamch, I. Chojnacka, R. Gancarz, and K. A. Wilk. (2017). "Microencapsulation of Hesperidin in Polyelectrolyte Complex Microbeads: Physico-Chemical Evaluation and Release Behavior." *Journal of Food Engineering* 214: 104–116. https://doi.org/10.1016/J.JFOODENG.2017.06.015.

Vatankhah, E., S. Hamedi, and O. Ramezani. (2020). "Surfactant-Assisted Incorporation of Rosmarinic Acid into Electrosprayed Poly(Lactic-Co-Glycolic Acid) Microparticles with Potential for Cosmetic and Pharmaceutical Applications." *Polymer Testing* 81: 106180. https://doi.org/10.1016/j.polymertesting.2019.106180.

Verma, S., A. Singh, and A. Mishra. (2013). "Gallic Acid: Molecular Rival of Cancer." *Environmental Toxicology and Pharmacology* 35 (3): 473–485. https://doi.org/10.1016/j.etap.2013.02.011.

Wang, H., X. Li, H. Yang, J. Wang, Q. Li, R. Qu, and X. Wu. (2020). "Nanocomplexes Based Polyvinylpyrrolidone K-17PF for Ocular Drug Delivery of Naringenin." *International Journal of Pharmaceutics* 578 (February): 119133. https://doi.org/10.1016/j.ijpharm.2020.119133.

Wang, H., Y. Xiao, H. Wang, Z. Sang, X. Han, S. Ren, R. Du, X. Shi, and Y. Xie. (2019). "Development of Daidzein Nanosuspensions: Preparation, Characterization, In Vitro Evaluation, and Pharmacokinetic Analysis." *International Journal of Pharmaceutics* 566: 67–76. https://doi.org/10.1016/J.IJPHARM.2019.05.051.

Wang, Q., W. Liu, J. Wang, et al. (2019). "Preparation and Pharmacokinetic Study of Daidzein Long-Circulating Liposomes." *Nanoscale Research Letters* 14: 321. https://doi.org/10.1186/s11671-019-3164-y.

Wang, X. H., Y. W. Guo, E. Tolba, M. Kokkinopoulou, M. Wiens, H. C. Schröder, and W. E. G. Müller. (2017). "Two-Armed Activation of Bone Mineral Deposition by the Flavones Baicalin and Baicalein, Encapsulated in Polyphosphate Microparticles." *American Journal of Chinese Medicine* 45 (3): 533–555. https://doi.org/10.1142/S0192415X1750032X.

Wei, Y., J. Liang, X. Zheng, C. Pi, H. Liu, H. Yang, Y. Zou, Y. Ye, and L. Zhao. (2017). "Lung-Targeting Drug Delivery System of Baicalin-Loaded Nanoliposomes: Development, Biodistribution in Rabbits, and Pharmacodynamics in Nude Mice Bearing Orthotopic Human Lung Cancer." *International Journal of Nanomedicine* 12: 251–261. https://doi.org/10.2147/IJN.S119895.

World Health Organization (WHO). Hypertension. *WHO*: 16 Mar. 2023. Available at: https://www.who.int/news-room/fact-sheets/detail/hypertension. Accessed on: 7 Nov. 2021.

World Health Organization (WHO). The Top 10 Causes of Death. *WHO*: 09 dec. 2020. Available at: https://www.who.int/news-room/fact-sheets/detail/the top-10-causes-of-death. Accesseds on: 07 nov. 2021.

Wong, C. Y., H. Al-Salami, and C. R. Dass. (2018). "Microparticles, Microcapsules and Microspheres: A Review of Recent Developments and Prospects for Oral Delivery of Insulin." *International Journal of Pharmaceutics* 537 (1–2): 223–244. https://doi.org/10.1016/j.ijpharm.2017.12.036.

Wu, C. F., J. Y. Yang, F. Wang, and X. X. Wang. (2013). "Resveratrol: Botanical Origin, Pharmacological Activity and Applications." *Chinese Journal of Natural Medicines* 11 (1): 1–15. https://doi.org/10.1016/S1875-5364(13)60001-1.

Wu, Y. R., et al. (2017). "In Vitro Study on Anti-Inflammatory Effects of Epigallocatechin-3-Gallate-Loaded Nano-and Microscale Particles." *International Journal of Nanomedicine* 12: 7007. https://doi.org/102147/IJN.S146296.

Xu, M. X., et al. (2020). "Fisetin Nanoparticles Protect against PM2. 5 Exposure-Induced Neuroinflammation by Down-Regulation of Astrocytes Activation Related NF-κB Signaling Pathway." *Journal of Functional Foods* 65: 103716. https://doi.org/10.1016/j.jff.2019.103716.

Xu, Q., A. Zhou, H. Wu, and Y. Bi. (2019). "Development and In Vivo Evaluation of Baicalin-Loaded W/O Nanoemulsion for Lymphatic Absorption." *Pharmaceutical Development and Technology* 24 (9): 1155–1163. https://doi.org/10.1080/10837450.2019.1646757.

Yuvaraja, M., N. R. Reddy, P. M. Kumar, K. S. Ravi, and N. Alqahtani. (2016). "Thermoreversible Gel for Intrapocket Delivery of Green Tea Catechin as a Local Drug Delivery System: An Original Research." *Journal of Advanced Pharmaceutical Technology & Research* 7 (4): 139. https://doi.org/10.4103/2231-4040.191422.

Zhai, T., et al. (2018). "Potential Micronutrients and Phytochemicals against the Pathogenesis of Chronic Obstructive Pulmonary Disease and Lung Cancer." *Nutrients* 10 (7): 813. https://doi.org/10.3390/nu10070813.

Zhang, J., X. Liang, X. Li, Z. Guan, Z. Liao, Y. Luo, and Y. Luo. (2015). "Ocular Delivery of Cyanidin-3-Glycoside in Liposomes and Its Prevention of Selenite-Induced Oxidative Stress." *Drug Development and Industrial Pharmacy* 42 (4): 546–553. https://doi.org/10.3109/03639045.2015.1088867.

Zhang, S., J. Wang, and J. Pan. (2016). "Baicalin-Loaded PEGylated Lipid Nanoparticles: Characterization, Pharmacokinetics, and Protective Effects on Acute Myocardial Ischemia in Rats." *Drug Delivery* 23 (9): 3696–3703. https://doi.org/10.1080/10717544.2016.1223218.

Zhao, X., Z. Wang, and X. Li. (2016). "Preparation, In-Vitro Release and Antioxidant Potential of Formulation of Apigenin with Hydroxypropyl-β-Cyclodextrin Modified Microemulsion." *Journal of Inclusion Phenomena and Macrocyclic Chemistry* 86 (1–2): 93–102. https://doi.org/10.1007/s10847-016-0644-x.

Zheng, Q., S. Li, X, Li, and R. Liu. (2021). "Advances in the Study of Emodin: An Update on Pharmacological Properties and Mechanistic Basis." *Chinese Medicine* 16 (1): 1–24. https://doi.org/10.1186/s13020-021-00509-z.

2 Novel Drug Delivery System for Resveratrol

*Manhua REN, Jiayi GAI, and Shirui MAO**

*Corresponding author: School of Pharmacy, Shenyang Pharmaceutical University, Shenyang 110016, China; Email: maoshirui@syphu.edu.cn

2.1 INTRODUCTION

Resveratrol is a polyphenolic molecule that occurs naturally in a range of plants, including soybeans, peanuts, berries, grapes, and pomegranates (Arora and Jaglan 2018), and it was initially obtained by Michio Takaoka in 1939 from the roots of *Veratrum grandiflorum* (Takaoka 1939). The presence of resveratrol was found in *Polygonum cuspidatum* in 1963. Then, in 1992, Renaud and de Lorgeril proposed the "French paradox" based on epidemiological data from the French population, demonstrating a low prevalence of coronary heart disease despite massive saturated fat intake (Renaud and Lorgeril 1992). According to these findings, the protective effect of the French population with a fairly low risk of cardiovascular disease despite a high saturated fat intake is thought to be the relatively high red wine consumption, which is characterized by its richness in polyphenolic compounds, such as resveratrol, catechins, and quercetin. Since then, resveratrol has gained extensive attention worldwide, and a variety of beneficial biological activities have been revealed, including anti-inflammatory (El-Ghazaly et al. 2020; Lomholt et al. 2018), antioxidant (Kong et al. 2019), anticancer (Honari et al. 2019; Kisková and Kassayová 2019; Xiao et al. 2018), cardiovascular protective (Arinno et al. 2021) and neuroprotective effects (Gomes et al. 2018).

Although the positive properties of resveratrol have led to its widespread use in the food and pharmaceutical industries, there are still some limitations to its clinical application, including low oxidative stability, high photosensitivity, and poor water solubility. Fortunately, various new drug delivery systems, including polymer nanoparticles, lipid nanoparticles, emulsions, and nanocrystals, have been introduced. They provide new strategies for the effective delivery of resveratrol and promote its use in clinical therapies.

2.2 SOURCES, PHYSICOCHEMICAL, AND BIOLOGICAL PROPERTIES OF RESVERATROL

2.2.1 Sources of Resveratrol

Natural resveratrol comes from grape skins, peanut skins, soybeans, berries, pomegranates, and other plants, as shown in Figure 2.2. Resveratrol is synthesized when exposed to some unfavorable conditions such as injury, ultraviolet irradiation, or fungal or other pathogenic infection (Armijo et al. 2016; Hasan and Bae 2017). It is a plant antibiotic with antibacterial and antiviral defense functions. Liswidowati et al. (1991) believed that before fungal infection, basically no stilbene could be produced in plant cells. Even if there is, the amount is small and unstable. When plants are exposed to biotic and abiotic stresses, many complex responses occur, including stress response signal transduction pathways and adaptive mechanisms involving primary and secondary metabolism. In plants, resveratrol is produced through the phenylpropanoid pathway

DOI: 10.1201/9781003300588-2

from phenylalanine or tyrosine (Hasan and Bae 2017). Under the action of phenylalanine tyrosine ammonia lyase or ammonia lyase, both tyrosine and phenylalanine lead to cinnamic acid (CA) production, which subsequently produces coumarin-CoA catalyzed by 4-coumarin-CoA ligase. Then, under the effect of polyketide synthase III (stilbene synthase) in plants, resveratrol is formed by connecting coumarin acyl-CoA with three malonyl-CoA units (Jeandet et al. 2019; Riccio et al. 2020).

The synthesis of resveratrol in plants is affected by both biological and abiotic factors. On one hand, resveratrol synthesis is reduced in plant cells when the level of key enzymes decreases (Fornara et al. 2008). On the other hand, the synthesis of resveratrol is affected by many external factors. UV irradiation at 260–270 nm is most favorable for its production (Douillet-Breuil et al. 1999). In addition, resveratrol accumulation increased under external stimuli, such as microbial infection, fertilizer, ozone, metal salts, and fungicide induction (Hasan and Bae 2017).

Currently, techniques for deriving resveratrol from plants include enzyme-assisted ultrasonic extraction, ultrasonic-assisted extraction, supercritical enzymatic extraction, organic solvents extraction, multi-stage countercurrent extraction, fluid/CO_2 extraction, pressurized liquid extraction, sequential molecularly imprinted solid-phase extraction, and ionic liquid-based salt-induced liquid-liquid extraction (Xiang-Tao Huang et al. 2019).

However, due to the extremely low content of resveratrol in plants and the inconvenience of extraction, separation, and purification, chemical synthesis and biotechnology synthesis have become the main ways to obtain resveratrol. Chemical synthesis methods include the Perkin reaction (Perkin 1868), Heck reaction (Heck and Nolley 1972), Wittig-Horner reaction (Hilt and Hengst 2007; Ianni and Waldvogel 2006), and McMurry reaction (Ephritikhine 1999). Biosynthesis of resveratrol by bacteria includes yeast, *Corynebacterium glutamicum*, and *Escherichia coli* (Thapa et al. 2019).

2.2.2 Structure and Physicochemical Properties

Resveratrol (3,4′,5-trihydroxystilbene), abbreviated as RES, is a phytoalexin generated naturally in plants, featuring a melting point of 253–255°C and a molecular weight of 228.25 g/mol (Aggarwal et al. 2004). This molecule is constructed by linking two aromatic rings with three hydroxyl groups via a methylene bridge which has two stereoisomers, cis and trans configurations, as shown in Figure 2.1. In spectrophotometric analysis, trans-resveratrol has a maximum absorbance of approximately 304 nm, while cis-resveratrol has a maximum absorbance of 286 nm (Camont et al. 2009). Compared with cis-resveratrol, trans-resveratrol has higher biological activity (Silva et al. 2013). But the structural stability of trans-resveratrol is not satisfactory. Influenced by factors such as light wavelength and time, temperature, pH, and molecular state, trans-resveratrol is prone to conformation transition or degradation to other substances (Franciosoa et al. 2014). Because of the structural characteristics of resveratrol, it has unique physical and chemical properties, such as limited

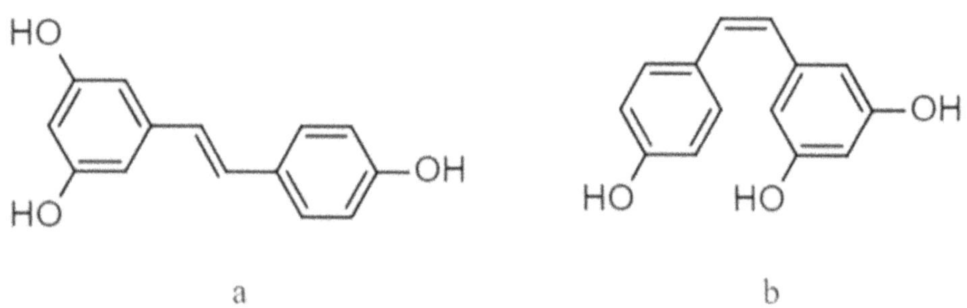

FIGURE 2.1 Structure of trans-resveratrol (a) and cis-resveratrol (b).

oxidative stability, high photosensitivity, and poor solubility in water. Meanwhile, in the process of studying the structure-activity relationship of resveratrol and its derivatives, it was found that the biological activities of its resveratrol, such as anti-inflammatory, antioxidant, and antitumor abilities, are influenced by molecular structural groups, including the amount and location of hydroxyl groups, benzene ring, and double-bond modifications (Szekeres et al. 2010; Tian and Liu 2020), which are important for guiding the development of resveratrol derivatives and improving clinical efficacy.

2.2.3 Biological Activities

Natural products, which are a significant source of biological activity for drug development, exhibit a broad spectrum of physical activities in preventing, protecting, and treating various diseases. Resveratrol, a natural source of polyphenol, exhibits a variety of biological effects, as illustrated in Figure 2.2, including antioxidant, anti-inflammatory, antibacterial, antiviral, anticancer, cardiovascular protection, and neuroprotection properties, and it has been extensively used in food, cosmetics, and pharmaceutical products.

2.2.3.1 Antioxidation Effect

Reactive oxygen species (ROS), such as singlet oxygen, hydroxyl radicals, superoxide anion radicals, and other radicals, are thought to contribute to diabetes, cancer, aging, cardiovascular, and neurological diseases (Biswas et al. 2017; Zia et al. 2022). Excess ROS could cause great harm by attacking polyunsaturated fatty acids on cell membranes, resulting in lipid peroxidation and cellular damage (Gu et al. 2021). Based on the hydroxyl group and conjugated double-bond system in its structure, resveratrol has a strong antioxidant activity. Resveratrol can act as a primary antioxidant, scavenging free radicals by reacting with them and as a secondary antioxidant preventing oxidative stress, for example, by increasing the activity of antioxidant enzymes (Alonso et al. 2017).

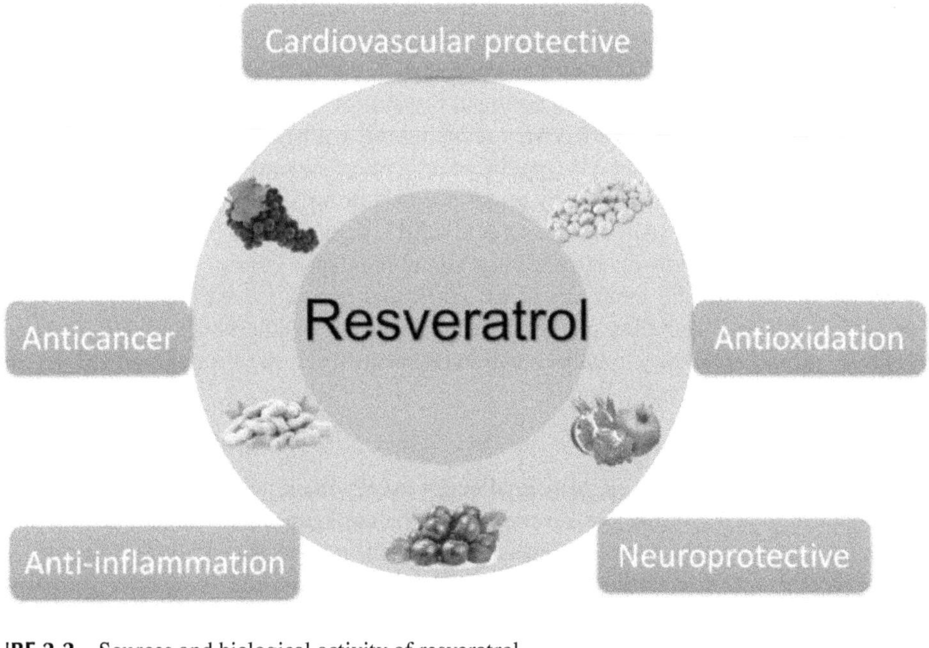

FIGURE 2.2 Sources and biological activity of resveratrol.

Resveratrol exhibits antioxidant activity through various signaling pathways, including the Nrf2 signaling pathway, the NF-κB signaling pathway and the SIRT1 signaling pathway. This can be summarized as follows: (1) As a transcription factor, nuclear factor-erythroid 2-related factor 2 (Nrf2) is stimulated to regulate the expression of antioxidant genes and protect cells from oxidative stress damage. Fortunately, resveratrol is able to activate Nrf2 via cellular signaling pathways, including AMP-activated protein kinase (AMPK) and phospho inositide 3-kinase/protein kinase B (PI3K/AKT). (2) Resveratrol inhibits the activating of nuclear factor kappa B (NF-κB) by tumor necrosis factor (TNF) and H_2O_2 in a dose- and time-dependent manner, which results in enhanced generation for pro-inflammatory cytokines. (3) Silent information regulator 1 (SIRT1) responds to ROS-induced damage primarily through regulating fork head box (FOXO) and peroxisome proliferator-activated receptor gamma coactivator 1-alpha (PGC-1a) regulators transcription factors, which together create transcriptional complexes that promote the production of antioxidant enzymes and scavenge ROS. Resveratrol exhibits antioxidant effects through the stimulation of AMPK to activate SIRT1 (Gu et al. 2021; Truong et al. 2018).

Due to its good antioxidant activity, resveratrol can not only treat various diseases caused by ROS but also used in food packaging (Busolo and Lagaron 2015) and cosmetics (Alonso et al. 2013).

2.2.3.2 Anti-inflammatory Effect

Inflammation is a naturally occurring defensive action in the body against injury or infection, and it contributes to the maintenance of tissue homeostasis under stress conditions. However, excessive and uncontrollable inflammation may lead to various critical chronic diseases, such as metabolic diseases, neurodegenerative diseases, cardiovascular diseases, cancer, and respiratory disorders (Deng and Liu 2021). As a natural medicine, resveratrol has potential anti-inflammatory effects and has been shown to play an important role in the anti-inflammatory treatment of many diseases (de Sá Coutinho et al. 2018).

Studies have demonstrated that resveratrol can regulate inflammatory reactions through various signaling pathways, which include the arachidonic acid (AA) pathway, activator protein-1 (AP-1) pathway, nuclear factor kappa B (NF-κB), and mitogen-activated protein kinase (MAPK) pathway (Meng et al. 2021). The inhibition of the AA pathway has an important effect among the anti-inflammatory pathways which are promoted by polyphenols. Nonsteroidal anti-inflammatory drugs (NSAIDs) are basic anti-inflammatory drugs that work by inhibiting cyclooxygenases 1 (COX-1) and cyclooxygenases 2 (COX-2) during arachidonic acid synthesis. COX-2 contributes to inflammation, but COX-1 is a homodimer that plays a role in homeostatic mechanisms, and inhibiting it can cause gastrointestinal problems. In contrast, resveratrol specifically inhibits COX-2 activity and transcriptional processes, thus effectively reducing the synthesis of prostaglandin E2 (PGE2) (de Sá Coutinho et al. 2018). In the AP-1 pathway, resveratrol suppresses COX-2 by inhibiting the AP-1-mediated gene expression induced by tumor necrosis factor (TNF) or 12-myristate 13-acetate (PMA) (Meng et al. 2021). In addition, Zhang et al. (2017) demonstrated that resveratrol decreased the phosphorylated production of extracellular signal-regulated kinase (ERK) and p38 in the MAPK signaling pathway and inhibitors of κB (IκB) and p65 in the NF-κB signaling pathway. Since MAPK and NF-κB signaling pathways can regulate the synthesis of tumor necrosis factor alpha (TNF-α) pro-inflammatory cytokines and interleukin 1β (IL-1β), they can inhibit the inflammatory response in mastitis.

2.2.3.3 Anticancer Effect

Cancer treatment is currently facing many challenges. As the main method of clinical application, chemotherapy has shown a number of adverse effects, including renal toxicity, hepatotoxicity, cardiotoxicity, and easy recurrence and metastasis. Therefore, it is necessary to introduce new optimized anticancer drugs. Resveratrol, a natural polyfunctional molecule, is reported to potentially act as a beneficial synergistic and chemoprotective ingredient in cancer chemotherapy to lower the associated negative effects and enhance treatment efficacy (Xiao et al. 2018). Meanwhile, it also improves the sensitiveness of drug-resistant tumor cells to cytotoxicity and offers protection to healthy cells against the side effects of conventional drugs (Ren et al. 2021).

Considerable evidence supports that resveratrol exhibits broad preventive and therapeutic effects on different kinds of tumors, such as colorectal (Honari et al. 2019), breast (Tang et al. 2017), and lung cancers (Yu-hua Lin et al. 2021). Resveratrol plays an anticancer role through multiple pathways. Studies have indicated that resveratrol can induce autophagy and subsequently promotes cell apoptosis by regulating the SIRT1/AMPK pathway (Zhao et al. 2017). And through targeting integrins, p53, lactate dehydrogenase (LDH), and focal adhesion kinase (FAK), resveratrol causes apoptosis (Brockmueller et al. 2021). Besides, resveratrol inhibits tumor site angiogenesis by reducing the level of vascular endothelial growth factor (VEGF) in cancer cells, thereby inhibiting the growth and proliferation of cancer cells (Tino et al. 2016). And resveratrol decreases the uptake of glucose and glycolysis by acting on glucose transporter 1 (Glut 1), 6-phosphofructo-1-kinase (PFK1), hypoxia-inducible factor 1-alpha (HIF-1α), reactive oxygen species (ROS), pyruvate dehydrogenase (PDH), and the Ca^{2+} calmodulin kinase kinase B/AMP-activated kinase (CamKKB/AMPK) pathway, the subsequent energy limitation leads to both proliferation and metastasis suppression and apoptosis induction (Brockmueller et al. 2021). Resveratrol has also been shown to promote the levels of anticancer cytokines and inhibit the infiltration and polarization of immunosuppressive cells, thus enhancing the immune system to fight cancer. It also improves cancer cells' susceptibility to death signals provided by anticancer immune cells (Chen and Musa 2021). In general, resveratrol appears to have tremendous potential in fight against cancer.

2.2.3.4 Cardiovascular Protective Effect

Since the "French paradox" reflects the positive influence upon the cardiovascular system caused by resveratrol, many studies have focused on its mechanism of action in relation to cardiovascular disease. And these researches suggest that resveratrol's cardiovascular protective effects are linked to a variety of mechanisms, such as the increase of nitric oxide (NO) synthesis in endothelial cells, the reduction of oxidative stress and inflammation, and the prevention of atherosclerosis and cardiac hypertrophy.

Inflammation plays an important role in cardiovascular disease. It induces the generation of multiple inflammatory factors, which can cause vascular apoptosis and alter vascular wall permeability, thus causing a range of cardiovascular diseases. Resveratrol can inhibit some pro-inflammatory cytokine synthesis by inhibiting the NF-κB signaling pathway (Wiciński et al. 2018; Zhang et al. 2012).

NO is a vasodilator that suppresses platelet aggregation and prevents thrombosis, as well as multiple anti-atherosclerotic properties (Gal et al. 2021). Large amounts of ROS produced by oxidative stress can inactivate NO oxidation. Resveratrol enhanced NO production via inducing endothelial NO synthase (eNOS) expression in endothelial cells, increasing the enzymatic activity of eNOS, and inhibiting the decoupling of eNOS (Gal et al. 2021; Xia et al. 2017). Endothelial cells produce key vasoactive chemicals, such as NO, prostacyclin, and vasoconstrictive endothelin-1 (ET-1), and they contribute to important impacts on the cardiovascular system. Resveratrol can raise NO levels and improve endothelial cell and vascular function by reducing cholesterol and triglyceride levels and upregulating eNOS expression (Gal et al. 2021).

Additionally, resveratrol can prevent atherosclerotic production by inhibiting prostaglandin E2 (PGE2) production by regulating the activity of COX-2. And it can prevent myocardial mitochondrial damage and left ventricular (LV) dysfunction through the activation of SIRT1 and the upregulation of the expression of antioxidant genes including catalase (CAT), glutathione (GSH), and superoxide dismutase (SOD) (Arinno et al. 2021). Resveratrol can also regulate myocardial hypertrophy and fibrosis by activating AMPK and inhibiting protein kinase B/mammalian target of rapamycin (Akt/mTOR) pathway (Chen et al. 2019).

Based on its positive impacts on the cardiovascular system through various routes, resveratrol has important therapeutic value for heart failure, myocardial infarction, atherosclerosis, and hypertension (Li et al. 2016; Raj et al. 2021).

2.2.3.5 Neuroprotective Effect

The neuroprotective effect of resveratrol has been gradually valued by researchers in recent years. Resveratrol can overcome the blood-brain barrier, allowing it to enter brain tissue and exert preventive and therapeutic effects on several neurodegenerative diseases, such as Parkinson's disease (PD) (Lin et al. 2018), Alzheimer's disease (AD) (Gomes et al. 2018), depression (Karam and Radwan 2020), and epileptic (Chuang et al. 2019).

Resveratrol could protect neurons from oxidative stress. It can scavenge free radicals and metals (e.g., copper) and enhance the activity of endogenous antioxidant enzymes and heme oxygenase 1 (HO1), which is resistant to neuronal damage caused by oxidative anti-stress. Additionally, resveratrol inhibits the production of quinone reductase 2 (QR2), a cytosolic enzyme that promotes the creation of potentially harmful activated quinone and reactive oxygen species (ROS) (Bastianetto et al. 2015).

In Alzheimer's disease, the production and accumulation of Aβ peptides is the main cause of pathogenesis. *In vitro* and *in vivo* studies have demonstrated that ROS can increase the production of β-amyloid (Aβ). Resveratrol can inhibit Aβ production and aggregation by activating the SIRT1, thereby improving mitochondrial dysfunction, inhibiting neuronal apoptosis and AD development (Han et al. 2020). At the same time, resveratrol-mediated SIRT1 overexpression could significantly reduce Aβ and NF-κB signaling-mediated activation of microglia. Overexpression of microglia can lead to the production of inflammatory factors as well as the occurrence of neuroinflammation (Zhang et al. 2015). Resveratrol has also been demonstrated to enhance autophagy, promote cell survival, and reduce neuronal apoptosis after nervous system damage (Bar-Yosef et al. 2019; Lin et al. 2018).

The multiple protective mechanisms of resveratrol make it a potential drug for neurological disorders.

2.2.4 Biopharmaceutical Characteristics

Resveratrol is a BCS II compound because of its hydrophobic nature and low water solubility but high membrane permeability (Peng et al. 2018). According to the study of Walle et al. (2004), approximately 70% of orally administered resveratrol (25 mg) is quickly (<30 min) absorbed and metabolized in humans, with peak plasma resveratrol and metabolite levels of approximately 2 μM and a half-life of 9.2 hours. Two peak plasma concentrations occur at around 1 hour and 6 hours after oral administration, suggesting hepatic-intestinal circulation. During blood transport, lipophilic resveratrol readily binds to serum proteins, including hemoglobin, albumin, and lipoproteins, to promote distribution in vivo. It was demonstrated that resveratrol accumulation is the highest in the liver (Neves et al. 2012).

Resveratrol bioavailability is limited on one hand by the low rate of dissolution in biological fluid. The rapid metabolism of resveratrol in the human body, on the other hand, results in low free drug concentrations in the blood. Resveratrol occurs extensive phase I (reduction, oxidation, and hydrolysis) and phase II (sulfate, glucuronide, and methyl conjugations) metabolism quickly after consumption (Neves et al. 2012). And the gut and liver are the main organs of resveratrol metabolism. The UDP-glucuronosyltransferase (UGT), which is widespread in the human liver microsomes (HLMs), and the human sulfotransferase (SULT) differently catalyze the glucuronidation and sulfation of resveratrol in the human body. Resveratrol-3-O-glucuronide and resveratrol-40-O-glucuronide are the major glucuronides of resveratrol. And resveratrol-3-O-sulfate, resveratrol-4′-O-sulfate, and resveratrol-3,4′-O-disulfate are identified as the primary sulfates of resveratrol (Wang and Sang 2018).

It was shown that the resveratrol sulfate metabolites may act as a repository for resveratrol and the resveratrol monosulfate has the potential to regenerate resveratrol and deliver it to target cells (Patel et al. 2013). Apart from glucuronidated and sulfated metabolites, gut microbes also play an important role in the metabolic process of resveratrol. And dihydrogen resveratrol has been determined as the first microbial-derived metabolite (Wang and Sang 2018; Wenzel and Somoza 2005). Resveratrol and its sulfated metabolites are widely expressed in the intestine and feces, while glucuronidated metabolites are prevalent in the small intestine (Bowei Zhang et al. 2021). And studies have shown that resveratrol is mainly excreted in urine and feces (Wenzel and Somoza 2005).

Importantly, resveratrol has stereoselective inhibitory effects on cytochrome P450 enzymes, including CYP1A2, CYP2C19, CYP3A, CYP2B6, and CYP2E1, and interactions should be noted when co-administered with other drugs (Hyrsova et al. 2019; Ji et al. 2021).

2.3 APPLICATION OF DIFFERENT CARRIER SYSTEMS IN DRUG DELIVERY

Many solutions are proposed to address the problem of poor resveratrol bioavailability, such as using precursor drugs, using non-oral routes, and using a variety of delivery systems (de Vries et al. 2018; Peng et al. 2018). Among them, the development of novel delivery systems is an important aspect of improving resveratrol bioavailability that can stabilize and protect trans-resveratrol from degradation, improve its water solubility, establish sustained release, and target trans-resveratrol to key areas through nanocarrier and microcarrier-based delivery systems. The types and structures of different nanocarriers and microcarriers can be seen in Figure 2.3.

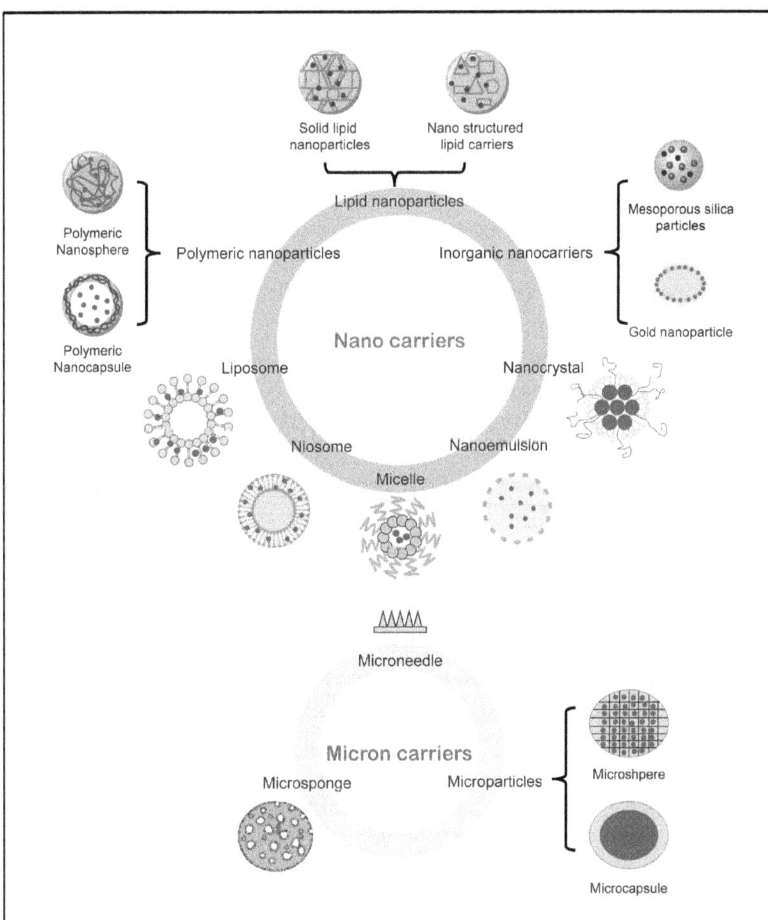

FIGURE 2.3 Types and structures of different nanocarriers and microcarriers.

2.3.1 Nanocarrier for the Delivery of Resveratrol

In recent years, nanotechnology has become a hot topic in pharmaceutical field. Numerous studies have demonstrated that nano delivery methods can significantly improve the stability and transmembrane transport of resveratrol, resulting in increased bioavailability and improved clinical efficacy. There are two types of nano drug delivery systems: nanocarrier medications and nanocrystal drugs. Both of them have their benefits and drawbacks. Drugs are embedded or encapsulated in lipid or polymer-based carriers to form nanocarrier drugs, including lipid nanoparticles, polymeric nanoparticles, liposomes, niosomes, micelles, nanoemulsion, and inorganic carrier drugs. These nanocarrier drugs have the advantage of preserving biological activity and extending or controlling drug release. In comparison, nanocrystals are carrier-free drugs with the advantages of high drug loading and avoiding toxicity and hemolysis from excipients. The application of the nanocarrier delivery system for resveratrol is shown in Table 2.1

2.3.1.1 Lipid Nanoparticles

Lipid nanoparticles (LNPs) mainly include solid lipid nanoparticles (SLNs) and nanostructured lipid carriers (NLCs), which play an important role in resveratrol delivery.

As the first generation (Pandita et al. 2014) of lipid nanoparticles, solid lipid nanoparticles (SLNs) are formed as a solid colloidal drug delivery system by stabilizing solid lipids in an aqueous dispersion with an emulsifier. Because of their submicron size, biocompatibility, controlled drug release, ease of manufacturing and scaling up, SLNs have identified as an important approach to enhance the oral bioavailability of lipophilic drugs (Pandita et al. 2014). However, SLN has some limitations where lipids tend to form a perfect crystalline lattice during storage, leading to drug leakage and low drug loading (Das et al. 2012). And NLCs, as the second generation of lipid nanoparticles, can overcome the drawbacks of SLNs. NLCs are a composite of solid and liquid lipids, including isopropyl myristate, ethyl oleate, glyceryl dioleate, and glyceryl tricaprylate, with less ordered lipids that can disrupt the original perfect lattice structure, thus reducing drug leakage and improving the stability of nanoparticles (Elmowafy and Al-Sanea 2021; García-Pinel et al. 2019).

Prepared by high-pressure homogenization (HPH), emulsification-solvent evaporation, emulsification-solvent diffusion, and microemulsification, SLNs and NLCs are widely employed as carriers for various drug molecules, including lipid-soluble and water-soluble drugs, oligonucleotides, peptides, genes, and vaccines (Khosa et al. 2018).

As an excellent resveratrol delivery strategy, lipid nanoparticles have been found to optimize solubility, stability, bioactivity, and bioavailability, as well as prolong and regulate drug release. Ahmad et al. (2016) successfully encapsulated resveratrol into solid lipid nanoparticles. The X-ray diffraction and differential scanning calorimetry (DSC) spectra indicated that the crystallinity of resveratrol loaded SLNs was lowered compared to that of free resveratrol, while the higher the crystallinity, the lower the solubility. Meanwhile, resveratrol-loaded lipid nanoparticles had stronger free radical scavenging ability, with about 60% of superoxide radicals scavenged within 2 hours and $88.32 \pm 0.731\%$ scavenged within 72 hours, while the maximum scavenging effect of free resveratrol was $50.03 \pm 2.023\%$ within 72 hours. In addition, the encapsulation of SLN increased the stability of resveratrol, allowing resveratrol to avoid rapid metabolism and longer circulation time compared to pure resveratrol. SLNs also increased the stability of resveratrol, reduces drug degradation and provides a long shelf life.

Various investigations have been conducted to explain the processes of increasing the oral bioavailability of LNPs. Due to their tiny particle size and exposure to the lipid component, LNPs improve their adhesion and residence time in the gastrointestinal system. In the presence of lipase, triglycerides in LNPs can be transformed into surface-active mono- and diacylglycerols, which induce bile secretion, resulting in enhanced micelle formation and bioavailability (Khosa et al. 2018; Lin et al. 2017). Lipid nanoparticles also could enhance drug absorption by the lymphatic system via microfold cells (M cells) in the intestinal membrane and bypass the first pass metabolism (Lin et al. 2017).

TABLE 2.1
The Application of Nanocarrier Delivery System for Resveratrol

Type of Nanocarriers	Concrete Carriers	Particle Size (nm)	Preparation Method	Functions	References
Lipid nanoparticles	NLCs	132 ± 12 nm	Melt emulsification-probe sonication method	Enhanced biodistribution of resveratrol in the brain	(Rajput and Butani 2019)
	SLNs	112.5 ± 10.3 nm	Emulsification and low-temperature solidification	Enhanced mitochondrial function and anti-fatigue	(Sun et al. 2019)
	C-RES-FER-FA-SLNs	174 ± 5 nm	Solvent evaporation method and hot homogenization method	Enhanced toxicity to cancer cells	(Senthil Kumar et al. 2020)
	DPPC-LNs	123.7 nm	Thin-film hydration-ultrasonic dispersion method	Enhanced resveratrol concentration and residence time inside the lungs	(Zerong Li et al. 2020)
	SLNs	271.13 nm	Emulsification-diffusion method	Inhibited Dox-induced cardiotoxicity	(Zhang et al. 2019)
Polymeric nanoparticles	RES-PUE-PLGA NPs	238–274 nm	Oil-in-water emulsion method	Enhanced the chemotherapeutic efficacy in spinal cord injury	(Chen et al. 2020)
	RES-CAB NPs	248.5 nm	Solvent evaporation method	Accelerated wound healing	(Amanat et al. 2020)
	PNPs	330 nm	Nebulization, followed by ionotropic	Effectively targeted the colon	(Prezotti et al. 2020)
	RES-NPs	102.7 nm	Gelation in $AlCl_3$ solution	Effectively inhibited VEGF expression without cytotoxicity, enhancing the therapeutic potential of resveratrol	(Bhatt et al. 2020)
	MCTD-NPs	95 nm	Nanoprecipitation method	Targeting mitochondria in ischemic cardiomyocytes	(Cheng et al. 2019)
Liposome and niosome	Ultradeformable liposomes	120–130 nm	Thin-film hydration technique	Showing potential for vitiligo treatment	(Doppalapudi et al. 2017)
	Liposomes	77.50 nm	Thin-film evaporation method	Improved the antioxidant effects	(MeiguiHuang et al. 2019)
	Pectin-coated liposomes	102–122 nm and 140 nm	Film dispersion method-ultrasonic method	Reduced resveratrol leak rate	(Shao et al. 2018)
	Niosomes	214.0–331.9 nm	Thin-film hydration (TFH), ether injection method (EIM)	Extended anti-inflammatory therapeutic effect	(Negi et al. 2017)
	Niosomes	96.1–0.9 nm	Film hydration method	Protected resveratrol from photoisomerization and maintain its biological activity	(Machado et al. 2019)

(Continued)

TABLE 2.1 *(Continued)*
The Application of Nanocarrier Delivery System for Resveratrol

Type of Nanocarriers	Concrete Carriers	Particle Size (nm)	Preparation Method	Functions	References
Micelle	PAM-Chol formed self-assembled micelles	120.4 ± 17.9 nm	Oil-in-water emulsion/solvent evaporation method	Treating acute lung injury	(Gyeungyun Kim et al. 2018)
	Micelles	48.4 nm	Film dispersion method	Help corneal wound healing	(Mengshuang Li et al. 2020)
	RES/MYP-loaded CAS MCs	125 ± 5.34 nm	Film hydration method	Against MCF-7 breast cancer cells	(El-Far et al. 2018)
	RES-TPGS micelles (RES:TPGS = 1.7)	11.9 ± 0.1	Solvent diffusion evaporation method	Resveratrol and TPGS combined as nutritional supplements for chronic hepatitis	(Zuccari et al. 2021)
	Neuronal mitochondria-targeted micelles	42.6 ± 0.53 nm	Solvent diffusion method	Relieve oxidative stress for delayed progression of Alzheimer's disease	(Yang et al. 2020)
Inorganic nanocarriers	Silica-PAMAM Nanoparticles	212–574 nm	Sol-gel method	Inhibited inducible nitric oxide synthase (iNOS)	(Kececiler-Emir et al. 2021)
	Mesoporous silica nanoparticle	60 nm	Rotary evaporation method	Resistant against melanoma	(Marinheiro et al. 2021)
	Mesoporous silica nanoparticles	120 nm	Sol-gel method	Enhanced therapeutic efficacy in gastric carcinoma	(Hu et al. 2019)
	GNPs	100 nm	Seed growth method	Antitumor	(Lee et al. 2019)
	AuNPs	170 ± 30 nm		Prevented the development, promotion, and progression of breast cancer	(Venditti et al. 2020)
Nanocrystal	RES nanocrystals adsorbed to the PCL	800 nm	Microfluidization method	Improved solubility	(Karakucuk and Tort 2020)
	RES nanocrystal self-assembled microspheres	2.95–5.99 µm	Nano spray dryer method	Increased NAD+ bioavailability and reduced injury caused by cardiac ischemia/reperfusion for mice	(Nie et al. 2021)
	Nanocrystals	110.28 ± 12.55 nm	Sonication method	Improved oral bioavailability	(Singh et al. 2017)
	Nanosuspension	46.3 nm	Antisolvent precipitation technique	Improved solubility and bioavailability	(Kuk et al. 2019)
	Nanosuspension	96.9 nm	Antisolvent precipitation method	Improved solubility and bioavailability	(Dong et al. 2018)

C-RES-FER-FA-SLNs: chitosan-coated-trans-resveratrol (RES) and ferulic acid (FER) loaded solid lipid nanoparticles (SLNs) that conjugated with folic acid (FA); DPPC: dipalmitoylphosphatidylcholine; PUE: puerarin; PLGA: poly(lactic-co-glycolic acid); CAB: cellulose acetate butyrate; MCTD: multistage continuous targeted strategy; PNP: polymeric nanoparticles; PAM-Chol: cholesterol-conjugated polyamidoamine; MYP: Monascus yellow pigment; CAS MCs: casein micelles; TPGS: D-α-tocopheryl-polyethylene-glycol-succinate; PAMAM: [olyamidoamine; GNPs/AuNPs: gold nanoparticles.

Meanwhile, LNPs are considered as a good transdermal delivery system due to their high affinity for the stratum corneum and small particle size. LNPs are advantageous in providing protective drug activity, increased local drug accumulation, reduced skin irritation, and controlled drug release (Qi et al. 2012). Miao et al. (2021) prepared a nanostructured lipid carrier gel loaded with resveratrol (RES-NLC-gel) for skin UV protection. The fluorescence signal generated by coumarin 6 labeled RES-NLC-gel and RES-gel was utilized to evaluate the penetration and accumulation of RES-NLC-gel in the skin. The results showed that after topical treatment, coumarin-6-labeled hydrogel and coumarin-6-labeled NLC gel were mainly distributed in the epidermis. In addition, the fluorescence intensity of coumarin-6-labeled NLC gel in the epidermis was higher than that of coumarin-6-labeled hydrogel, indicating that the RES-NLC-gel was effectively accumulated in the epidermis. This was attributed to the biocompatibility and high affinity of NLC to the stratum corneum.

2.3.1.2 Polymeric Nanoparticles

Nanoparticles are classified as nanocapsules and nanospheres, which vary in composition and structural organization. Nanocapsules are composed of a polymeric shell around an oily core in which the medicine is dissolved and/or adsorbed to the polymeric shell. On the other hand, nanospheres, which do not include oil, are created by a polymeric matrix in which the medicine may be maintained or adsorbed physically and uniformly. Notably, polymeric nanoparticles are drug delivery systems with diameters less than 100 nm, often between 100 and 500 nm (Castro et al. 2020). The production of polymeric nanoparticles involves the use of a variety of natural and synthetic polymers. Natural polymers include gelatin, carrageenan, chitosan, and albumin. And the main synthetic polymers are lactic-co-glycolic acid (PLGA), polylactic acid (PLA), and polyglycolic acid (PGA). Among them, many polymers are biocompatible, especially natural polymers (Ahmadi et al. 2019). These polymers contribute to the water solubility, biocompatibility, and biodegradability of nanoparticles on the one hand and to their protection from blood and gastrointestinal effects on pH and temperature on the other. (Castro et al. 2020).

As a delivery vehicle for resveratrol, polymeric nanoparticles provide a variety of benefits, including enhanced solubility and bioavailability, as well as prolonged and controlled drug release. Liu et al. (2020) composed resveratrol-loaded zein/pectin nanoparticles and assessed the bioavailability of encapsulated and unencapsulated resveratrol utilizing in vitro (Caco-2 cell uptake and transmembrane transport) and in vivo techniques (oral administration to rats). It was demonstrated that encapsulated resveratrol had significantly more cellular absorption and transmembrane transport than unencapsulated resveratrol. In addition, animal feeding experiments demonstrated that encapsulating resveratrol enhanced its bioavailability and prolonged its presence in plasma. In another study, Iglesias et al. (2019) reported that the resveratrol/polymer ratio can influence the delayed release performance of polymer nanoparticles, with higher polymer ratios resulting in slower release of resveratrol.

In addition, the application of resveratrol-loaded polymer nanoparticles for therapeutic strategies can achieve both passive targeting effects through enhanced permeation and retention (EPR) effect and active targeting by intra- and extracellular stimulation, including pH, temperature, light, and magnetic field (Castro et al. 2020).

Dai et al. (2017) prepared magnetic RES-loaded lignin nanoparticles (AL/RES/Fe3O4 NPS) by self-assembly of lignin with bioactive molecules resveratrol and ferric tetroxide magnetic nanoparticles. When mice were treated with AL/RES/Fe_3O_4 NPs exposed to an external magnetic field, the intra-tumor drug content was 6.7 and 6.2 times higher than that of free resveratrol treatment after 24 and 48 hours, respectively. After 48 hours, the resveratrol content in the tumor remained at a high level, showing high bioavailability, low toxicity, and a good tumor targeting effect, which was attributed to the EPR and magnetic targeting effects.

Interestingly, resveratrol, a natural polyphenol, can create hyperbranched polymeric nanocarriers via click reactions with acyl alkynyl monomers. Vinyl ether bonds are formed during the process, which can be dissolved under acidic tumor microenvironment (TME) conditions (extracellular pH 6.8, endosomes and lysosomal pH 5.0). This grants pH sensitivity to the polymer nanoparticles, which can be applied to tumor-targeted treatment (Ning Zhang et al. 2021).

Due to the numerous advantages of polymeric nanoparticles, they are excellent carriers for the delivery of resveratrol. However, polymer NPs have some limitations, such as high cost, time consumption, blood instability, with very limited drug delivery capacity. Therefore, they are very difficult to expand into mass production (Moshawih et al. 2019).

2.3.1.3 Liposomes and Niosomes

Liposomes are vesicular structures composed of phospholipids and cholesterol with particle sizes ranging from 50 to 1000 nm that can be created via reverse phase evaporation technique, microfluidic method, electroformation method, thin-film hydration, and supercritical fluids (SCFs) method (Maja et al. 2020). Cholesterol, in general, can be incorporated in the bilayer and contribute to the regulation of membrane fluidity. Different cholesterol concentrations and liposome manufacturing methods result in the formation of small unilamellar vesicles (SUV) (20–100 nm), large unilamellar vesicles (LUV) (100–1000 nm), and multilamellar vesicles (MLV) (500–5000) (Juszkiewicz et al. 2020).

Liposomes are potential drug carriers where lipid drugs can be encapsulated in the outer lipid bilayer while hydrophilic medications remain in the aqueous core to avoid deactivation and degradation (Juszkiewicz et al. 2020). Loading resveratrol into liposomes has been found to protect and enhance its bioactivity and bioavailability, prolong and control drug release, and improve therapeutic efficacy. Caddeo et al. (2021) prepared RES-loaded liposomes with a particle size of approximately 80 nm and a high encapsulation rate (83%) using the commercial liposome platform Pronanosome LIPO-N. Antioxidant colorimetric assays (ferric reducing antioxidant power (FRAP) and 2,2-diphenyl-1-picrylhydrazyl (DPPH) showed that the high antioxidant activity of resveratrol was not affected by its incorporation into the vesicular system. Additionally, resveratrol liposomes showed cytocompatibility with fibroblasts and protected skin cells against oxidative stress more efficiently than free resveratrol by decreasing endogenous and chemically produced reactive oxygen species. Thus, the formulation, which is based on a commercial liposome platform, is a strategy capable of increasing the biological activity, efficacy, and stability of resveratrol in vitro. In another study, Zhao et al. (2020) loaded the RES into peptides and sucrose liposomes (PSL) for breast cancer treatment. RES-loaded PSL was found to have more antitumor activity than free resveratrol, with sustained drug release and reduced resveratrol toxicity in mice.

By attaching PEG to the liposome surface due to increased spatial obstruction, liposome uptake by the reticuloendothelial system can be decreased, hence prolonging circulation time (Juszkiewicz et al. 2020). Caddeo et al. (2018) prepared PEG-modified liposomes by ultrasound of phospholipid water dispersion with two surfactants, poly(ethylene glycol)-poly(propylene glycol)-poly(ethylene glycol) triblock copolymer (PEG-PPG-PEG) and tocopherol PEG1000 succinate (TPGS). PEG assembly on the liposomal bilayer is achieved by involving two PEG surfactants in the process of vesicle formation, making the hydrophobic portion (PPG-PPG-PPG block and the alkyl chain of TPGS) a component of the bilayer, with the hydrophilic portion (PEG chain) projected on one side or both sides of the bilayer. The PEG-modified liposomes exhibited excellent biocompatibility and stability, exhibiting no drug leakage or resveratrol degradation after two months of storage and exhibiting a slow and stable release rate.

Additionally, it has been demonstrated that Fe_3O_4-modified resveratrol liposomes can not only release medications constantly and slowly but can also target them actively using the effect of an external magnetic field (Wang et al. 2018). Due to the fact that resveratrol can only be loaded between the lipophilic phospholipid bilayer, an excess of the compound may impair the liposomal structural integrity, resulting in decreased drug transport ability. Simultaneously, limited shelf life and cold storage conditions are current issues (Moshawih et al. 2019).

Niosomes, which are structurally similar to liposomes, are bilayer structures generated by non-ionic surfactants self-assembling in an aqueous media. Non-ionic surfactants, which feature a hydrophilic head and a hydrophobic tail, have lower toxicity, higher biocompatibility, and a more stable structure than cationic, anionic, and amphoteric surfactants, making them an excellent vehicle for both hydrophilic and lipophilic medicines. Niosomes are mainly prepared by the ether injection method (EIM), reverse phase evaporation (REV), thin-film hydration method (TFH), microfluidization method, supercritical carbon dioxide fluid ($scCO_2$), and transmembrane pH gradient (Ag Seleci et al. 2016). Similar to liposomes, niosomes are composed of three types of vesicles: small unilamellar vesicles (SUV) (10–100 nm), large unilamellar vesicles (LUV) (100–3000 nm), and multilamellar vesicles (MLV) (Thabet et al. 2022). In comparison to liposomes, niosomes have superior chemical stability, are less prone to deterioration, and are relatively easy to preserve. However, the physical stability issues such as aggregation and drug leakage during niosome storage and the lack of acceptable sterilization methods require more investigation (Ag Seleci et al. 2016).

As a lipid-soluble drug, resveratrol is mainly present in the middle of the bilayer film of niosomes. Due to its high biocompatibility, resistance to degradation, increased bioavailability, and controlled drug release, the resveratrol-loaded niosome is a suitable carrier for resveratrol transport. Machado et al. (2021) encapsulated trans-resveratrol in a niosome constructed from Tween 80 and Span 80. They compared the antioxidant and photoprotective properties of the niosome containing resveratrol to those of the free resveratrol using radical tests (2,2′-Azinobis(3-ethylbenzthiazoline-6-sulfonic acid) [ABTS] and 2,2-diphenyl-1-picrylhydrazyl [DPPH]) and UV irradiation, respectively. Additionally, they examined the effect of dodecanol (Dod), a membrane stabilizer similar to cholesterol, on drug release. The results indicate that Tween80:Span80 niosomes and Tween80:Span80:Dod niosomes preserve the overall antioxidant capability of resveratrol. Due to the ability of Dod to enhance drug release, Tween80:Span80:Dod niosomes have a higher antioxidant capacity than Tween80:Span80 niosomes. Meanwhile, both Tween80:Span80 and Tween80:Span80:Dod niosomes showed significant photoprotection, with a lower reduction in trans-resveratrol than free resveratrol.

2.3.1.4 Micelles

Polymeric micelles are unique structures with different forms (spherical micelles, cylindrical micelles, inverted micelles, etc.) formed by amphiphilic materials containing hydrophilic and lipophilic segments (Kaur et al. 2021). The self-assembly of biophilic diblock copolymers, triblock copolymers, or graft copolymers is usually responsible for forming polymeric micelles. When the concentration of these copolymers exceeds the critical micelle concentration (CMC), the hydrophobic blocks are insoluble in water and aggregate inward, while hydrogen bonding with water causes the hydrophilic blocks to extend outward, thus forming micelles. Amphiphilic polymers have lower CMC than low-molecular-weight surfactants with more stable properties. Characteristic parameters of block copolymers, including molecular weight, degree of polymerization, block length, block dispersion, hydrophilic-lipophilic balance (HLB) values, and type of block interaction, have essential effects on micelle traits, such as CMC, size, and shape (Aziz et al. 2016; Kaur et al. 2021).

Typically, polymeric micelles have a hydrophilic shell and a hydrophobic core. As a drug carrier, micelles can wrap lipophilic drugs like resveratrol in the hydrophobic core, improving water solubility, protecting drug bioactivity, increasing drug stability, and controlling drug release. The small size of polymeric micelles allows them to avoid clearance by the mononuclear phagocytic system in the liver and bypass the filtration of inter-endothelial cells in the spleen, thus prolonging blood circulation and increasing their penetration into the tumor vessels via EPR effect. Drugs can be physically captured to achieve drug loading with techniques such as dialysis, mechanical dispersion, co-solvent evaporation, and oil-in-water emulsion, as well as loaded into micelles by chemical conjugation (Aziz et al. 2016; Lu and Park 2013).

Zuccari et al. (2021) prepared RES-D-α-tocopheryl-polyethylene-glycol-succinate (TPGS) micelles. TPGS is constructed by esterifying polyethylene glycol 1000 with D-tocopherol via a succinate bridge with hydrophilic and hydrophobic fragments. On the one hand, TPGS is a water-soluble form of vitamin E and can serve as a vitamin E supplement; on the other hand, TPGS has the ability to self-assemble and form micelles beyond the CMC, allowing for the delivery of drugs in nanoform. RES-TPGS micelles did not affect the antioxidant activity of resveratrol as determined by the DPPH assay. Resveratrol is able to reduce liver inflammation and can work synergistically with TPGS to treat chronic liver disease. The physical protection of hydrophilic shells to resveratrol in the micelle not only enhances water solubility and bioavailability but also prevents premature drug breakdown in the gastrointestinal tract (GI) and improves drug circulation time.

The hydrophobic core is usually composed of biodegradable polymers, including poly (ε-caprolactone) (PCL), poly (lactic acid) (PLA) and poly (l-lysine), whereas the hydrophilic shell is usually composed of hydrophilic polymers, such as polyethylene glycol (PEG) (Lu and Park 2013). Natural polymers, including peptides, polysaccharides, and proteins, have multiple advantages as carriers, including high drug load, good biocompatibility and biodegradability, extended circulation time, nontoxicity, and non-immunogenicity, and the degradation products are nontoxic and easily absorbable (Atanase 2021).

Like most delivery carriers, micelles can also achieve targeted drug release by surface modification. Yang et al. (2020) prepared the neuronal mitochondrial-targeted micelles (CT-NM) composed of poly(glycol)-b-poly (lactide) (PEG-PLA) on the surface modified with brain neuron-specific binding peptide C3 and mitochondrial-targeted triphenylphosphonium (TPP). CT-NM can be delivered explicitly to mitochondria in central neurons to treat Alzheimer's disease (AD). The results show that CT-NM has strong blood-brain barrier (BBB) penetration and high uptake of neurons, which can be localized to the brain mitochondria, effectively reducing oxidative stress, preventing neuronal damage, and restoring memory function. CT-NM also exhibits high bioavailability and long blood circulation time, making it an ideal drug delivery system for resveratrol targeted delivery.

After the polymer micelle injection into the human body, the micelle structure is susceptible to destruction and drug release is under the influence of blood dilution, pH, ionic strength, and plasma protein. While cross-linking can fix the micelle structure and is a vital strategy to increase the stability of the drug-loaded polymer micelle (Meihua Lin et al. 2021). Mocanu et al. (2017) prepared shell cross-linked micelles composed of dextran with octadecyl hydrophobic end groups to load resveratrol. The release profile of resveratrol showed that the release from cross-linked dextran micelles was slower than that from free resveratrol, suggesting that they could be a good vehicle for the controlled release of resveratrol.

Apart from the in vivo stability issues with polymeric micelles, more research is needed into the toxicity and degradation kinetics of polymeric materials and the in vivo fate of multifunctional modified polymeric micelles (Kaur et al. 2021).

2.3.1.5 Nanoemulsion

Nanoemulsions are nanoparticles ranging in size from 0.1 to 500 nm that combine the oil and water phases via the action of an emulsifier (Arora and Jaglan 2018). They are classified into three types: oil-in-water (O/W), water-in-oil (W/O), and multiple emulsions (W/O/W and O/W/O). O/W is a nuclear shell structure composed of an oil phase core and a polar shell that disperses oil in water. In contrast, when water droplets are dispersed in oil, W/O emulsions are formed. Complicated systems can be created by adding an emulsion within another emulsion, known as a double emulsion, resulting in water-in-oil-in-water (W/O/W) or oil-in-water-in-oil (O/W/O) emulsions. As a hydrophobic drug, resveratrol is often dissolved in the oil phase and wrapped in a polar layer to create an O/W emulsion (Wilson et al. 2021). As a vehicle for resveratrol transport, nanoemulsions offer a few advantages, including increased drug stability, absorption rate, and bioavailability. Zhou et al. (2021) developed soybean oil nanoemulsions containing three polyphenols, curcumin, resveratrol, and quercetin and examined the parameters affecting the solubility and stability of nanoemulsions. The three nanoemulsions have a high encapsulation efficiency (75–87%). Stability testing under simulated gastrointestinal conditions showed that resveratrol nanoemulsions had the maximum stability, with 100% remaining stable, but quercetin and curcumin had 44% and 92% remaining under gastrointestinal conditions, respectively. The authors conclude that nanoemulsions are incredibly stable for medicines with a high partition coefficient (log D). In addition, resveratrol nanoemulsions have the highest bioaccessibility among the three, demonstrating the effectiveness of the nanoemulsions delivery approach. The encapsulation of resveratrol in multiple emulsions has also been reported (Wang et al. 2017).

In the case of oral administration, small droplets of the nanoemulsions give it a large specific surface area and a high gastrointestinal digestion efficiency. Free fatty acids formed during lipid digestion combine with bile salts to produce micelles that encapsulate lipophilic substances, facilitating their absorption by intestinal epithelial cells (Qianyuan Liu et al. 2019; Zhou et al. 2021).

While in transdermal drug delivery, nanoemulsions exhibit high permeability and high bioavailability. Kotta et al. (2021) optimized resveratrol nanoemulsions by response surface methodology and applied it to the nasal mucosa to investigate the brain targeting. The ultrasonic method was applied to manufacture the resveratrol nanoemulsions formed by two non-ionic surfactants, Cremophor EL and Pluronic P107, and coconut oil. The optimized resveratrol nanoemulsions showed a uniform spherical shape with a particle size of 110.37 ± 2.16 nm, a PDI of 0.194 ± 0.003, and a zeta potential of −21.13 ± 1.628 mV. They also made resveratrol suspensions and examined the two preparations' differences in drug release and mucosal permeability in vitro. The in vitro release of resveratrol nanoemulsions was higher, at 92.66 ± 3.45%, compared to 54.01 ± 1.56% for suspension. Nanoemulsions have a nasal permeability that is about double that of suspension. At the same time, the nanoemulsions had a higher blood concentration and bioavailability than the suspension and could transfer more resveratrol to the brain. Nanoemulsions outperform suspensions in terms of characteristics. The presence of surfactants and the small size of nanoemulsions are thought to be the reasons.

Self-emulsifying drug delivery system (SEDDS) or self-microemulsifying drug delivery system (SMEDDS), as the Type III (Pouton 2000) and Type IV (Constantinides 1995; Singh et al. 2019) lipid drug delivery systems, were proposed in the 1990s. These two drug delivery systems are not exclusive to emulsions. On the one hand, there is no aqueous phase in the formulation, which comprises an oil phase, emulsifier, co-emulsifier, and drug. On the other hand, they spontaneously emulsified to produce an oil-in-water (O/W) emulsion with the aid of gastrointestinal peristalsis and digestive fluid (Desai et al. 2012).

Considering self-emulsifying and self-microemulsifying drug delivery systems provide several benefits, including improved drug solubility and bioavailability, they have been widely employed in the administration of resveratrol. Vasconcelos et al. (2019) found that SEDDS formulations, Lauroglycol® 90/Labrasol®/Capryol® PGMC (12.5/75.0/12.5) (Lau/Lab/Cap) and Tween® 80/Transcutol®/Imwitor® 742 (33.3/33.3/33.3) (T80/Trans/Imw), were more permeable than free drugs in a Caco-2 cell monolayer permeability test for resveratrol. In addition, resveratrol was absorbed more rapidly in SEDDS, which resulted in shorter Tmax values (30 minutes vs. 2 hours) and twice the Cmax of the free drug. And the small size of the SEDDS provides a large surface area of interaction with the intestinal mucosa, enhances the absorption, and reduces the metabolism and efflux of resveratrol, which reduces the metabolic effects and improves its pharmacokinetic behavior.

2.3.1.6 Inorganic Nanoparticles

Inorganic nanoparticles are formed by inorganic materials including gold, silica, silver, and quantum dots in a size range of 1 to 1,000 nm. Because of their unique physicochemical properties, demonstration of superparamagnetic and quantum confinement effects, good storage stability, and biocompatibility, inorganic nanomaterials are used for a variety of applications, such as drug, protein, and nucleic acid delivery (Luther et al. 2020), cell and tissue imaging and labeling (Dokyoon Kim et al. 2018), and disease diagnosis and treatment (Xu et al. 2018). As promising drug delivery platforms, mesoporous silica (MSNs) and gold nanoparticles (AuNPs) are widely utilized for resveratrol delivery.

Due to its enormous surface area, superior biocompatibility, and simplicity of surface modification, mesoporous silica (MSNs) has become an attractive nano container for drug delivery systems. The large-area pore structure can increase the solubility of less soluble pharmaceuticals while also increasing the nanocarriers' loading capacity and release time. Furthermore, the pore support of MSN protects the medication against degradation and premature metabolism before release (Ahmadi et al. 2019; Chen et al. 2016).

These MSN advantages are well demonstrated in resveratrol delivery. Marinheiro et al. (2021) prepared mesoporous silica nanoparticles loaded with resveratrol, spherical porous structures with an average particle size of 69.1 ± 5.2 nm. The loading rate was >93%, and the encapsulation promoted the amorphization of resveratrol and improved its solubility and in vitro release compared to the unencapsulated resveratrol. Additionally, the pH sensitivity of mesoporous silica nanoparticles (Perris et al. 2021) was also revealed, allowing for the specific delivery of resveratrol to tumor locations. They examined the release of resveratrol mesoporous silica nanoparticles and free resveratrol at pH 7.4 (physiological pH) and pH 5.2 (an acidic environment in some tumor tissues). The results showed that the release of resveratrol tocopherol PEG1000 succinate tocopherol PEG1000 succinate in mesoporous silica nanoparticles was higher than that of bulk resveratrol, with a pH-dependent release rate of pH 5.2, significantly faster than that of pH 7.4.

The morphology, mesostructures, structural properties (e.g., pore size, pore volume and surface area), and structural parameters (e.g., channel structure, cavity topology) of mesoporous silica all affect drug loading and release. For example, hollow mesoporous silica has a large inner pore and many outer mesopores, low density, a large specific surface area, and high drug-loading capacity compared to others (Chen et al. 2016). However, larger pores are not better. Large holes facilitate the development of resveratrol crystals, which would lead to a decrease in solubility and bioavailability (Marinheiro et al. 2021).

Silanol-rich surfaces can easily be functionalized and linked to a variety of ligands, reducing the abrupt release of resveratrol, or enabling targeted resveratrol delivery. Shen et al. (2018) developed a resveratrol delivery system based on mesoporous silica nanoparticles (MSNPs) coated in polylactic acid (PLA) and bound to the low-density lipoprotein receptor-ligand peptide (LDLR). On the one hand, PLA coating can prevent or decrease resveratrol burst release; on the other hand, ROS can trigger resveratrol release by speeding up PLA degradation, resulting in a ROS-responsive release system. Furthermore, the LDLR ligand peptide bound to the corresponding receptors in cerebrovascular endothelial cells, significantly enhancing their transcytosis across the blood-brain barrier (BBB).

Along with mesoporous silica nanoparticles, numerous metal nanoparticles containing resveratrol have been identified, most notably gold nanoparticles. Due to their biocompatibility, non-cytotoxicity, antibacterial qualities, ease of manufacture, and adjustable particle size, they are frequently utilized in imaging, radiosensitization, photothermal therapy, antimicrobial treatment, and drug delivery (Lopes et al. 2019). AuNPs can be linked to drug molecules through physical absorption, covalent bonding, or ionic bonding. Additionally, AuNPs can also be modified by PEG, peptide, amino acid conjugation, or oligonucleotides to achieve long-cycle or targeted release. And the drug is released in response to internal stimuli (glutathione, pH, and enzymes) and external stimuli (e.g., light) (Amina and Guo 2020). RES-loaded gold nanoparticles have been found to enhance resveratrol uptake and targeting in cells. Kamal et al. (2018) prepared Technetium-99m-labeled resveratrol-loaded gold nanoparticles (RES-AuNP) and characterized and evaluated their cancer-targeting effects in HT29 colon cancer cells and animal cancer models. The results showed that the uptake of ^{99m}Tc-RES-AuNP by cancer cells was significantly higher than that of ^{99m}Tc-AuNP and ^{99m}Tc-resveratrol, and ^{99m}Tc-RES-AuNP was better able to target colon adenocarcinoma compared with ^{99m}Tc-resveratrol. Meanwhile, the small particle size and anionic properties of gold nanoparticles conferred low reticuloendothelial system uptake and low cytotoxicity.

Due to the inertness of gold nanoparticles, they are less harmful than other lipid carriers and metal nanoparticles (Kamal et al. 2018). However, they face resource scarcity and expensive costs.

2.3.1.7 Nanocrystals

Nanocrystals are composed of nanomedicine and stabilizers with particle sizes ranging from 1 to 1,000 nm (Junghanns and Müller 2008), which could be dispersed in aqueous or non-aqueous media to form nanosuspensions (Desai et al. 2012). Compared to other nanocarrier drugs, nanocrystals contribute no other excipients to the particle size, have a higher drug loading and could avoid the toxicity problem caused by the excipients. According to the Noyes-Whitney equation (Equation 1), nanoscale drugs have a high surface-area-to-volume ratio that can significantly increase drug solubility and effectively overcome the poor absorption of BCS II drugs such as resveratrol.

$$\frac{dc}{dt} = \frac{D \times A \times (Cs - Cx)}{h}. \tag{1}$$

dc/dt represents the dissolution velocity (change in concentration over time), *D* represents the diffusion coefficient, *A* represents the surface area, *h* represents the diffusional distance, *Cs* represents the saturation concentration, and *Cx* represents the bulk concentration.

However, small particle sizes complicate nanosuspension stability; the smaller the particle size, the larger the surface energy, resulting in particle aggregation, Ostwald ripening, and sedimentation (Li et al. 2021). Therefore, the presence of stabilizers, including surfactants or hydrophilic polymers, is very important. Surfactants include lecithin, Tween80, Pluronic-68, and Pluronic-127, and polymers include cellulose derivatives, d-α-tocopherol polyethylene glycol 400 succinate (TPGS), polyvinylpyrrolidone (PVP), and polyethylene glycol (PEG) (Lu and Park 2013; Singh et al. 2017). These stabilizers adsorb to the surface of the particle, lower surface tension, generate repulsive forces, and induce spatial or electrostatic stability, thus ensuring long-term stability. It has been shown that drugs with high hydrophobicity can form highly stable nanosuspensions. This is because the hydrophobic functional group of the stabilizer has a strong attraction to the hydrophobic drug, resulting in the hydrophobic drug being more easily wrapped by the stabilizer and better stabilized than the hydrophilic drug (Li et al. 2021). This is undoubtedly very favorable for hydrophobic resveratrol. Additionally, previous studies of our group have demonstrated that the type and concentration of stabilizers in resveratrol nanosuspension can influence particle morphology, mobility, pulmonary deposition, and *in vitro* drug release. Thus, the *in vivo* behavior of resveratrol could be adjusted by selecting different types or concentrations of stabilizers (Qiaoyu Liu et al. 2019).

From the perspective of pharmacokinetics, nanocrystals can dissolve quickly to saturation solubility, improving absorption concentration gradients and permeability, and decreasing the effects of food in oral administration. And the adhesion of nanocrystals to the gastrointestinal mucosa can improve bioavailability (Macedo et al. 2021). Due to increased drug loading, better solubility, and superior pharmacokinetic features, nanocrystals play a crucial role in enhancing the bioavailability and therapeutic efficacy of resveratrol, thereby becoming a promising drug delivery system for resveratrol. Singh et al. (2017) successfully prepared trans-resveratrol nanocrystals (t-RES-NCs) using the probe sonication method. The optimized t-RES-NCs had a regular and rod-like morphology, a particle size of 110.28 ± 12.55 nm, a high ζ-potential (-32.96 ± 3.85 mV), and a high drug content (90%). In vivo pharmacokinetics experiments revealed a considerable increase in the $AUC_{0\text{-}t}$ (3.5-fold) and C_{max} (2.2-fold) of t-RES-NCs compared to regular t-RES, showing that NCs have a higher oral bioavailability. The results of in vitro cellular efficacy against MDA-MB-231 cells indicated that generated t-RES-NCs were more efficacious and potent ($p < 0.05$) than plain t-RES, showing that nanocrystals increased antitumor activity of resveratrol.

Despite the numerous benefits of nanosuspensions, maintaining an even particle size distribution, good surface characteristics, and physical stability remain significant issues.

2.3.2 Microcarrier for the Delivery of Resveratrol

Based on the information collected so far, microparticles, microneedles, and microsponges have been reported as micron-level delivery systems for resveratrol, as summarized in Table 2.2.

TABLE 2.2
The Application of the Microcarrier Delivery System for Resveratrol

Type of Microcarriers	Particle Size	Preparation Method	Function	References
Alginate—$CaCl_2$ microspheres	244.03–217.3 μm	Ionotropic gelation process	Delayed release of resveratrol	(Cho et al. 2014a)
Chitosan—TPP microspheres sodium	160–206 μm	Ionotropic gelation process	Controlled release of resveratrol	(Cho et al. 2014**b**)
Microneedle		Micromolding technique	Improved transdermal delivery of resveratrol	(Aung et al. 2021)
Microsponges	400–900 μm	Quasi-emulsion technique	Enhanced efficacy in colon disease	(Gandhi et al. 2020)
Eudragit® microparticles	>420 μm		Improved solubility and stability of resveratrol	(Pignatello et al. 2019)
Poly-(ε-caprolactone) Microcapsules	1.96–29.76 μm	Ultrasonic Atomization	Anti-inflammatory	(Luzardo-Álvarez et al. 2019)
Microencapsulation	8.2 μm	Spray-drying technology	Microcapsules exhibit good encapsulation of resveratrol	(Cardoso et al. 2019)
Eudragit-coated pectin-alginate microparticles	1,450 μm	Ionotropic gelation method	Anti-gastric	(Gartziandia et al. 2018)

TPP: tripolyphosphate

2.3.2.1 Microparticle

Microparticulate drug delivery system (MDDS) includes microspheres and microcapsules with particle sizes ranging from 1 to 1000 μm. Microspheres are spherical solid particles formed by drug dispersion or adsorption in polymeric materials, and drug adsorption on the surface of microspheres can easily cause explosive drug release. In contrast, microcapsules are small vesicle-like structures created by employing the medicine as the capsule core and encasing it in a polymeric shell (Bale et al. 2016).

Microparticles as drug delivery carriers have many advantages (Bale et al. 2016): (1) improve the solubility of drugs; (2) mask the foul odor of drugs, reduce toxicity, and increase compliance; (3) improve the stability of drugs from the stomach and other external environments; (4) improve bioavailability and reduce side effects; (5) delay and control drug release; and (6) achieve targeted drug transport.

Due to the many positive aspects of MDDS, it has been commonly applied in the delivery of resveratrol. Cho et al. (2014b) prepared chitosan-TPP microspheres loaded with resveratrol by the ionic cross-linking method using chitosan and sodium tripolyphosphate (TPP) of different molecular weights and various concentrations. In vitro release studies showed a slow-release trend for all samples at all polymer concentrations, and a delayed release was detected even after 30 minutes, indicating that the polymer microspheres could successfully control the drug release rate. Additionally, in vitro release profiles also revealed that the burst release of medicine decreased as the TPP content increased.

In another study, Eudragit-coated pectin-alginate microparticles had gastric resistance and were able to improve the stability of resveratrol significantly. This is owing to the resistance of pectin to stomach and intestine enzymes, consequently preventing resveratrol from being metabolized until it reaches the distal gut. In addition, alginate has the property of prolonging the residence time of particles at the administration site and achieving controlled release due to its biological/adhesive properties. The Eudragit® FS-30D coating contributes to controlled drug release by avoiding the stomach and dissolving in the intestinal fraction at pH 7 (Gartziandia et al. 2018).

2.3.2.2 Microneedle

The combination of microneedles (MNs) and resveratrol can build a unique and effective transdermal drug delivery system (TDDS) that exerts local or systemic effects, thus overcoming the rapid oral metabolism of resveratrol.

Microneedle devices are made by placing micron-sized needles (usually 150–1,500 m long, 50–250 m wide, and 1–25 m at the tip) on a small area that punctures the stratum corneum of the skin to create microscopic channels perpendicular to the epidermis. The penetration depth is approximately 10–20 μm and does not penetrate the dermis, resulting in no pain (Azmana et al. 2020). Apart from that, the damage caused by nanometer-sized microneedles is minimal. Microneedle manufacture involves the use of a diverse range of materials, including metals, ceramics, glass, silicon, and natural and synthetic biodegradable polymers. Their low cost, ease of production, elasticity, excellent mechanical properties, biocompatibility, and biodegradability make polymeric materials an ideal material for microneedle delivery.

The primary mechanism of drug delivery by microneedle is by piercing the skin and increasing the penetration and bioavailability of the drug. Microneedles have shown excellent performance in the transdermal delivery of resveratrol. Aung et al. (2021) prepared dissolved microneedles (DMNs) loaded with resveratrol using Eudragit E100, Eudragit RS100, and polyvinylpyrrolidone-31K90 (PVP-K90) polymer as materials by micro forming technique. When applying the maximum pressure (10.769 N) commonly applied when people patch, no cracks appeared at the bottom of the patch and microneedles, showing good mechanical properties. In comparison to the gel formulation, DMNs containing resveratrol accumulated more in the skin, which was connected with micropores caused by microneedles. As a result, it is a great delivery strategy for increasing the efficiency of medications with low solubility and bioavailability.

2.3.2.3 Microsponge

Microsponge is a porous polymer system consisting of microspheres and tiny sponge-like particles with large porous surfaces, with particle sizes ranging from 5–300 μm and average pore size of 0.25 μm, which can be further integrated into different drug dosage forms, including tablets, emulsions, gels, and capsules. The tiny pores will inhibit drug penetration, increase drug stability, and prolong drug retention and release (Junqueira and Bruschi 2018; Singhvi et al. 2019), which is critical for extending the action duration and efficacy of resveratrol.

Gandhi et al. (2020) prepared RES-loaded microsponges using the quasi-emulsion technique. The optimized RES-loaded microsponge was compressed into a tablet using pectin as a matrix by Box-Behnken design. The RES-loaded microsponges showed a spherical porous structure that allowed controlled drug release with only 10% of the drug being released at gastric PH during the first 2 hours. Due to the swelling properties of pectin, it has more pronounced slow-release properties compared to microsponges. Pharmacodynamic studies in a rat model of ulcerative colitis showed better efficacy of drug-loaded microsponges and matrix tablets than pure resveratrol. This study demonstrates the advantages of microsponges in delivering resveratrol.

2.3.3 Others

In addition to nano- and microcarrier-based drug delivery systems for resveratrol, other methods for efficient delivery of the drug have been discovered. For example, in traditional Chinese medicine, the combination of the liver meridian guide drug vinegar-baked Radix Bupleuri (VBRB) with resveratrol can form a site-direct delivery mode to enhance the liver targeting of resveratrol, which is a simple, inexpensive, and easy-to-implement method compared to targeted delivery systems (Zhao et al. 2009).

2.4 SUMMARY AND OUTLOOK

Resveratrol has become an outstanding nutritional supplement with a variety of pharmacological properties, including cancer prevention, cardiovascular protection, antioxidant activity, anti-inflammatory activity, and neuroprotection. Nonetheless, it has low solubility, low oxidative stability, and high photosensitivity, which limits its use in clinical trials. Numerous nano- and microcarriers have been designed to overcome their inherent pharmacokinetic restrictions and demonstrate improved performance. Nanoparticle-sized medication formulations are advantageous for enhancing the solubility of insoluble pharmaceuticals and protecting them from renal filtration and fast metabolism. Additionally, nano drug delivery methods can enable passive targeting via EPR and active targeting via ligand binding. Depending on the formulation and structure of the drug, microcarriers can also enhance the solubility and bioavailability of resveratrol at the micron level and control drug release.

However, most research on resveratrol formulations has been conducted using in vitro models, such as cell lines. Additional preclinical and clinical research is necessary to establish the therapeutic effectiveness of resveratrol in combination with nano- and microcarrier-based formulations. Additionally, other obstacles persist for various nano- and microcarrier delivery systems, including excipient toxicity, low drug-loading capacity, and difficulty scaling up manufacturing processes. All of the mentioned problems need to be tackled before it can get into clinical stages.

REFERENCES

Aggarwal, Bharat B., Bhardwaj, Anjana, Aggarwal, Rishi S. et al. "Role of Resveratrol in Prevention and Therapy of Cancer: Preclinical and Clinical Studies." *Anticancer Research* 24 (2004): 2783–840.

Ag Seleci, Didem, Seleci, Muharrem, Walter, Johanna-Gabriela, Stahl, Frank, and Scheper, Thomas. "Niosomes as Nanoparticular Drug Carriers: Fundamentals and Recent Applications." *Journal of Nanomaterials* 2016 (2016): 1–13.

Ahmad, Iqbal, Anwar, Mohammed, Akhter, Sohail et al. "Supercritical Fluid Technology-Based Trans-Resveratrol SLN for Long Circulation and Improved Radioprotection." *Journal of Pharmaceutical Innovation* 11 (2016): 308–322.

Ahmadi, Zahra, Mohammadinejad, Reza, and Ashrafizadeh, Milad. "Drug Delivery Systems for Resveratrol, a Non-Flavonoid Polyphenol: Emerging Evidence in Last Decades." *Journal of Drug Delivery Science and Technology* 51 (2019): 591–604.

Alonso, Cristina, Martí, M., Barba, C. et al. "Skin Permeation and Antioxidant Efficacy of Topically Applied Resveratrol." *Archives for Dermatological Research* 309 (2017): 423–431.

Alonso, Cristina, Martí, Meritxell, Martínez, Vanessa et al. "Antioxidant Cosmeto-Textiles: Skin Assessment." *European Journal of Pharmaceutics and Biopharmaceutics* 84 (2013): 192–199.

Amanat, Shabnam, Taymouri, Somayeh, Varshosaz, Jaleh, Minaiyan, Mohsen, and Talebi, Ardeshir. "Carboxymethyl Cellulose-Based Wafer Enriched with Resveratrol-Loaded Nanoparticles for Enhanced Wound Healing." *Drug Delivery and Translational Research* 10 (2020): 1241–1254.

Amina, Sundus Jabeen, and Guo, Bin. "A Review on the Synthesis and Functionalization of Gold Nanoparticles as a Drug Delivery Vehicle." *International Journal of Nanomedicine* 15 (2020): 9823.

Arinno, Apiwan, Apaijai, Nattayaporn, Chattipakorn, Siriporn C., and Chattipakorn, Nipon. "The Roles of Resveratrol on Cardiac Mitochondrial Function in Cardiac Diseases." *European Journal of Nutrition* 60 (2021): 29–44.

Armijo, Grace, Schlechter, Rudolf, Agurto, Mario et al. "Grapevine Pathogenic Microorganisms: Understanding Infection Strategies and Host Response Scenarios." *Frontiers in Plant Science* 7 (2016): 382.

Arora, Divya, and Jaglan, Sundeep. "Therapeutic Applications of Resveratrol Nanoformulations." *Environmental Chemistry Letters* 16 (2018): 35–41.

Atanase, Leonard Ionut. "Micellar Drug Delivery Systems Based on Natural Biopolymers." *Polymers* 13 (2021): 477.

Aung, Nway Nway, Ngawhirunpat, Tanasait, Rojanarata, Theerasak et al. "Enhancement of Transdermal Delivery of Resveratrol Using Eudragit and Polyvinyl Pyrrolidone-Based Dissolving Microneedle Patches." *Journal of Drug Delivery Science and Technology* 61 (2021): 102284.

Aziz, Zarith, Ahmad, Akil, Mohd-Setapar, Siti et al. "Recent Advances in Drug Delivery of Polymeric Nano-Micelles." *Current Drug Metabolism* 18 (2016): 16–29.

Azmana, Motia, Mahmood, Syed, Hilles, Ayah Rebhi et al. "Transdermal Drug Delivery System through Polymeric Microneedle: A Recent Update." *Journal of Drug Delivery Science and Technology* 60 (2020): 101877.

Bale, Swarna, Khurana, Amit, Reddy, A. Shiva Shankar, Singh, Mandip, and Godugu, Chandraiah. "Overview on Therapeutic Applications of Microparticulate Drug Delivery Systems." *Critical Reviews™ in Therapeutic Drug Carrier Systems* 33 (2016): 309–361.

Bar-Yosef, Tamara, Damri, Odeya, and Agam, Galila "Dual Role of Autophagy in Diseases of the Central Nervous System." *Frontiers in Cellular Neuroscience* 13 (2019): 196.

Bastianetto, Stéphane, Ménard, Caroline, and Quirion, Rémi. "Neuroprotective Action of Resveratrol." *Biochimica et Biophysica Acta (BBA)—Molecular Basis of Disease* 1852 (2015): 1195–1201.

Bhatt, Priyanka, Fnu, Gulimirerouzi, Bhatia, Deepak, Shahid, Amna, and Sutariya, Vijay kumar. "Nanodelivery of Resveratrol-Loaded PLGA Nanoparticles for Age-Related Macular Degeneration." *AAPS PharmSciTech* 21 (2020): 291.

Biswas, Silpak, Das, Rintu, and Ray Banerjee, Ena. "Role of Free Radicals in Human Inflammatory Diseases." *AIMS Biophysics* 4 (2017): 596–614.

Brockmueller, Aranka, Sameri, Saba, Liskova, Alena et al. "Resveratrol's Anti-Cancer Effects through the Modulation of Tumor Glucose Metabolism." *Cancers* 13 (2021): 188.

Busolo, M. A., and Lagaron, J. M. "Antioxidant Polyethylene Films Based on a Resveratrol Containing Clay of Interest in Food Packaging Applications." *Food Packaging &Shelf Life* 6 (2015): 30–41.

Caddeo, Carla, Lucchesi, Daniela, Fernàndez-Busquets, Xavier et al. "Efficacy of a Resveratrol Nanoformulation Based on a Commercially Available Liposomal Platform." *International Journal of Pharmaceutics* 608 (2021): 121086.

Caddeo, Carla, Pucci, Laura, Gabriele, Morena et al. "Stability, Biocompatibility and Antioxidant Activity of PEG-Modified Liposomes Containing Resveratrol." *International Journal of Pharmaceutics* 538 (2018): 40–47.

Camont, Laurent, Cottart, Charles-Henry, Rhayem, Yara et al. "Simple Spectrophotometric Assessment of the Trans-/Cis-Resveratrol Ratio in Aqueous Solutions." *Analytica Chimica Acta* 634 (2009): 121–128.

Cardoso, Teresa, Gonçalves, Antónia, Estevinho, Berta N., and Rocha, Fernando. "Potential Food Application of Resveratrol Microparticles: Characterization and Controlled Release Studies." *Powder Technology* 355 (2019): 593–601.

Castro, Karine Cappuccio de, Costa, Josiel Martins, and Campos, Maria Gabriela Nogueira. "Drug-Loaded Polymeric Nanoparticles: A Review." *International Journal of Polymeric Materials and Polymeric Biomaterials* 71 (2020): 1–13.

Chen, Chen, Zou, Lei-Xin, Lin, Qiu-Yue et al. "Resveratrol as a New Inhibitor of Immunoproteasome Prevents PTEN Degradation and Attenuates Cardiac Hypertrophy after Pressure Overload." *Redox Biology* 20 (2019): 390–401.

Chen, Libo, and Musa, Ahmed Eleojo. "Boosting Immune System against Cancer by Resveratrol." *Phytotherapy Research* 35 (2021): 5514–5526.

Chen, Shizhu, Hao, Xiaohong, Liang, Xingjie et al. "Inorganic Nanomaterials as Carriers for Drug Delivery." *Journal of Biomedical Nanotechnology* 12 (2016): 1–27.

Chen, Wei, Zhao, Zongqin, Zhao, Shuchao, Zhang, Lizhong, and Song, Qimin "Resveratrol and Puerarin Loaded Polymeric Nanoparticles to Enhance the Chemotherapeutic Efficacy in Spinal Cord Injury." *Biomed Microdevices* 22 (2020): 69.

Cheng, Ying, Liu, Dao-zhou, Zhang, Chang-xiong et al. "Mitochondria-Targeted Antioxidant Delivery for Precise Treatment of Myocardial Ischemia–Reperfusion Injury through a Multistage Continuous Targeted Strategy." *Nanomedicine: Nanotechnology,Biology and Medicine* 16 (2019): 236–249.

Cho, Ah Ra, Chun, Yong Gi, Kim, Bum Keun, and Park, Dong June. "Preparation of Alginate–Cacl2 Microspheres as Resveratrol Carriers." *Journal of Materials Science* 49 (2014a): 4612–4619.

Cho, Ah Ra, Chun, Yong Gi, Kim, Bum Keun, and Park, Dong June. "Preparation of Chitosan–TPP Microspheres as Resveratrol Carriers." *Journal of Food Science* 79 (2014b): E568–E76.

Chuang, Yao-Chung, Chen, Shang-Der, Hsu, Chung-Yao et al. "Resveratrol Promotes Mitochondrial Biogenesis and Protects against Seizure-Induced Neuronal Cell Damage in the Hippocampus Following Status Epilepticus by Activation of the Pgc-1α Signaling Pathway." *International Journal of Molecular Sciences* 20 (2019): 998.

Constantinides, Panayiotis P. "Lipid Microemulsions for Improving Drug Dissolution and Oral Absorption: Physical and Biopharmaceutical Aspects." *Pharmaceutical Research* 12 (1995): 1561–1572.

Dai, Lin, Liu, Rui, Hu, Li-Qiu, Zou, Zhu-Fan, and Si, Chuan-Ling. "Lignin Nanoparticle as a Novel Green Carrier for the Efficient Delivery of Resveratrol." *ACS Sustainable Chemistry & Engineering* 5 (2017): 8241–8249.

Das, Surajit, Ng, Wai Kiong, and Tan, Reginald B. H. "Are Nanostructured Lipid Carriers (NLCs) Better Than Solid Lipid Nanoparticles (SLNs): Development, Characterizations and Comparative Evaluations of Clotrimazole-Loaded SLNs and NLCs?" *European Journal of Pharmaceutical Sciences* 47 (2012): 139–151.

Deng, Zhengyu, and Liu, Shiyong. "Inflammation-Responsive Delivery Systems for the Treatment of Chronic Inflammatory Diseases." *DrugDelivery and Translational Research* 11 (2021): 1475–1497.

de Sá Coutinho, Diego, Pacheco, Maria, Frozza, Rudimar, and Bernardi, Andressa. "Anti-Inflammatory Effects of Resveratrol: Mechanistic Insights." *International Journal of Molecular Sciences* 19 (2018): 1812.

Desai, Preshita P., Date, Abhijit A., and Patravale, Vandana B. "Overcoming Poor Oral Bioavailability Using Nanoparticle Formulations—Opportunities and Limitations." *Drug Discovery Today Technologies* 9 (2012): e87–e95.

de Vries, Konrad, Strydom, Morné, and Steenkamp, Vanessa. "Bioavailability of Resveratrol: Possibilities for Enhancement." *Journal of Herbal Medicine* 11 (2018): 71–77.

Dong, Qiannian, Yuan, Hui-Ling, Qian, Jia-Jia, Zhang, Cai-Yun, and Chen, Wei-Dong. "Preparation and In Vitro–In Vivo Characterization of Trans-Resveratrol Nanosuspensions." *Bio-Medical Materials and Engineering* 29 (2018): 333–345.

Doppalapudi, Sindhu, Mahira, Shaheen, and Khan, Wahid. "Development and In Vitro Assessment of Psoralen and Resveratrol Co-Loaded Ultradeformable Liposomes for the Treatment of Vitiligo." *Journal of Photochemistry & Photobiology B Biology* 174 (2017): 44–57.

Douillet-Breuil, Anne-Céline, Jeandet, Philippe, Adrian, Marielle, and Bessis, Roger. "Changes in the Phytoalexin Content of Various Vitis Spp. In Response to Ultraviolet C Elicitation." *Journal of Agricultural & Food Chemistry* 47 (1999): 4456–4461.

El-Far, Shaymaa W., Helmy, Maged W., Khattab, Sherine N. et al. "Phytosomal Bilayer-Enveloped Casein Micelles for Codelivery of Monascus Yellow Pigments and Resveratrol to Breast Cancer." *Nanomedicine* 13 (2018): 481–499.

El-Ghazaly, Mona A., Fadel, Noha A., Abdel-Naby, Doaa H. et al. "Potential Anti-Inflammatory Action of Resveratrol and Piperine in Adjuvant-Induced Arthritis: Effect on Pro-Inflammatory Cytokines and Oxidative Stress Biomarkers." *Egyptian Rheumatologist* 42 (2020): 71–77.

Elmowafy, Mohammed, and Al-Sanea, Mohammad M. "Nanostructured Lipid Carriers (NLCs) as Drug Delivery Platform: Advances in Formulation and Delivery Strategies." *Saudi Pharmaceutical Journal* 29 (2021): 999–1012.

Ephritikhine, Michel. "Cheminform Abstract: A New Look at the Mcmurry Reaction." *ChemInform* 30 (1999): no–no.

Fornara, V., Onelli, E., Sparvoli, F. et al. "Localization of Stilbene Synthase in Vitis Vinifera L. During Berry Development." *Protoplasma* 233 (2008): 83–93.

Franciosoa, A., Mastromarino, P., Masci, A., Erme, M. D., and Mosca, L. "Chemistry, Stability and Bioavailability of Resveratrol." *Medicinal Chemistry* 10 (2014): 237.

Gal, Roland, Deres, Laszlo, Toth, Kalman, Halmosi, Robert, and Habon, Tamas. "The Effect of Resveratrol on the Cardiovascular System from Molecular Mechanisms to Clinical Results." *International Journal of Molecular Sciences* 22 (2021): 10152.

Gandhi, Himanshu, Rathore, Charul, Dua, Kamal et al. "Efficacy of Resveratrol Encapsulated Microsponges Delivered by Pectin Based Matrix Tablets in Rats with Acetic Acid-Induced Ulcerative Colitis." *Drug Development and Industrial Pharmacy* 46 (2020): 365–375.

García-Pinel, Beatriz, Porras-Alcalá, Cristina, Ortega-Rodríguez, Alicia et al. "Lipid-Based Nanoparticles: Application and Recent Advances in Cancer Treatment." *Nanomaterials* 9 (2019): 638.

Gartziandia, Oihane, Lasa, Arrate, Pedraz, Jose et al. "Preparation and Characterization of Resveratrol Loaded Pectin/Alginate Blend Gastro-Resistant Microparticles." *Biomacromolecules* 23 (2018): 1886.

Gomes, Bruno Alexandre Quadros, Silva, João Paulo Bastos, Romeiro, Camila Fernanda Rodrigues et al. "Neuroprotective Mechanisms of Resveratrol in Alzheimer's Disease: Role of Sirt1." *Oxidative Medicine and Cellular Longevity* 2018 (2018): 1–15.

Gu, Tongyu, Wang, Nianmin, Wu, Tong et al. "Antioxidative Stress Mechanisms Behind Resveratrol: A Multidimensional Analysis." *Journal of Food Quality* 2021 (2021): 1–12.

Han, Yang, Chu, Xiaoyang, Cui, Lin et al. "Neuronal Mitochondria-Targeted Therapy for Alzheimer's Disease by Systemic Delivery of Resveratrol Using Dual-Modified Novel Biomimetic Nanosystems." *Drug Delivery* 27 (2020): 502–518.

Hasan, Mohidul, and Bae, Hanhong. "An Overview of Stress-Induced Resveratrol Synthesis in Grapes: Perspectives for Resveratrol-Enriched Grape Products." *Biomacromolecules* 22 (2017): 294.

Heck, R. F., and Nolley, J. P. "Palladium-Catalyzed Vinylic Hydrogen Substitution Reactions with Aryl, Benzyl, and Styryl Halides." *The Journal of Organic Chemistry* 37 (1972): 2320–2322.

Hilt, Gerhard, and Hengst, Christoph. "A Concise Synthesis of Substituted Stilbenes and Styrenes from Propargylic Phosphonium Salts by a Cobalt-Catalyzed Diels–Alder/Wittig Olefination Reaction Sequence." *The Journal of Organic Chemistry* 72 (2007): 7337–7342.

Honari, Mohadese, Shafabakhsh, Rana, Reiter, Russel J., Mirzaei, Hamed, and Asemi, Zatollah. "Resveratrol Is a Promising Agent for Colorectal Cancer Prevention and Treatment: Focus on Molecular Mechanisms." *Cancer Cell International* 19 (2019): 180.

Hu, Yu, Wang, Zhehui, Qiu, Yunfeng et al. "Anti-miRNA21 and Resveratrol-Loaded Polysaccharide-Based Mesoporous Silica Nanoparticle for Synergistic Activity in Gastric Carcinoma." *Journal of Drug Targeting* 27 (2019): 1135–1143.

Huang, Meigui, Liang, Cuiping, Tan, Chen et al. "Liposome Co-Encapsulation as a Strategy for the Delivery of Curcumin and Resveratrol." *Food &Function* 10 (2019): 6447–6458.

Huang, Xiang-Tao, Li, Xi, Xie, Ming-ling et al. "Resveratrol: Review on Its Discovery, Anti-Leukemia Effects and Pharmacokinetics." *Chemico Biological Interactions* 306 (2019): 29–38.

Hyrsova, Lucie, Vanduchova, Alena, Dusek, Jan et al. "Trans-Resveratrol, But Not Other Natural Stilbenes Occurring in Food, Carries the Risk of Drug-Food Interaction Via Inhibition of Cytochrome P450 Enzymes or Interaction with Xenosensor Receptors." *Toxicology Letters* 300 (2019): 81–91.

Ianni, Alen, and Waldvogel, Siegfried R. "Reliable and Versatile Synthesis of 2-Aryl-Substituted Cinnamic Acid Esters." *Synthesis* 37 (2006): 2103–2112.

Iglesias, Nieves, Galbis, Elsa, Díaz-Blanco, M. et al. "Nanostructured Chitosan-Based Biomaterials for Sustained and Colon-Specific Resveratrol Release." *International Journal of Molecular Sciences* 20 (2019): 398.

Jeandet, Philippe, Clément, Christophe, and Cordelier, Sylvain "Regulation of Resveratrol Biosynthesis in Grapevine: New Approaches for Disease Resistance?" *Journal of Experimental Botany* 70 (2019): 375–378.

Ji, Seung-Bae, Park, So-Young, Bae, Subin et al. "Comprehensive Investigation of Stereoselective Food Drug Interaction Potential of Resveratrol on Nine P450 and Six UGT Isoforms in Human Liver Microsomes." *Pharmaceutics* 13 (2021): 1419.

Junghanns, Jens-Uwe A. H., and Müller, Rainer H. "Nanocrystal Technology, Drug Delivery and Clinical Applications." *International Journal of Nanomedicine* 3 (2008): 295.

Junqueira, Mariana Volpato, and Bruschi, Marcos Luciano. "A Review about the Drug Delivery from Microsponges." *AAPS PharmSciTech* 19 (2018): 1501–1511.

Juszkiewicz, Katarzyna, Sikorski, Aleksander F., and Czogalla, Aleksander. "Building Blocks to Design Liposomal Delivery Systems." *International Journal of Molecular Sciences* 21 (2020): 9559.

Kamal, Rozy, Chadha, Vijayta D., and Dhawan, D. K. "Physiological Uptake and Retention of Radiolabeled Resveratrol Loaded Gold Nanoparticles (^{99m}Tc-Res-AuNP) in Colon Cancer Tissue." *Nanomedicine: Nanotechnology,Biology and Medicine* 14 (2018): 1059–1071.

Karakucuk, Alptug, and Tort, Serdar. "Preparation, Characterization and Antimicrobial Activity Evaluation of Electrospun PCL Nanofiber Composites of Resveratrol Nanocrystals." *Pharmaceutical Development and Technology* 25 (2020): 1216–1225.

Karam, Heba M., and Radwan, Rasha R. "Low Dose γ Radiation Enhances Antidepressant Effect of Resveratrol: Behavioral and Neurochemical Studies." *Environmental Toxicology* 35 (2020): 1137–1145.

Kaur, Jaskiran, Mishra, Vijay, Singh, Sachin Kumar et al. "Harnessing Amphiphilic Polymeric Micelles for Diagnostic and Therapeutic Applications: Breakthroughs and Bottlenecks." *Journal of Controlled Release* 334 (2021): 64–95.

Kececiler-Emir, Ceren, Ilhan-Ayisigi, Esra, Celen-Erden, Cigdem, Nalbantsoy, Ayse, and Yesil-Celiktas, Ozlem. "Synthesis of Resveratrol Loaded Hybrid Silica-Pamam Dendrimer Nanoparticles with Emphases on Inducible Nitric Oxide Synthase and Cytotoxicity." *PlantFoods for Human Nutrition* 76 (2021): 219–225.

Khosa, Archana, Reddi, Satish, and Saha, Ranendra N. "Nanostructured Lipid Carriers for Site-Specific Drug Delivery." *Biomedicine &Pharmacotherapy* 103 (2018): 598–613.

Kim, Dokyoon, Kim, Jonghoon, Park, Yong Il, Lee, Nohyun, and Hyeon, Taeghwan. "Recent Development of Inorganic Nanoparticles for Biomedical Imaging." *ACS Central Science* 4 (2018): 324–336.

Kim, Gyeungyun, Piao, Chunxian, Oh, Jungju, and Lee, Minhyung. "Self-Assembled Polymeric Micelles for Combined Delivery of Anti-Inflammatory Gene and Drug to the Lungs by Inhalation." *Nanoscale* 10 (2018): 8503–8514.

Kisková, Terézia, and Kassayová, Monika. "Resveratrol Action on Lipid Metabolism in Cancer." *International Journal of Molecular Sciences* 20 (2019): 2704.

Kong, Danli, Yan, Yan, He, Xiao-Yi et al. "Effects of Resveratrol on the Mechanisms of Antioxidants and Estrogen in Alzheimer's Disease." *BioMed Research International* 2019 (2019): 1–8.

Kotta, Sabna, Mubarak Aldawsari, Hibah, Badr-Eldin, Shaimaa M., Alhakamy, Nabil A., and Md, Shadab. "Coconut Oil-Based Resveratrol Nanoemulsion: Optimization Using Response Surface Methodology, Stability Assessment and Pharmacokinetic Evaluation." *Food Chemistry* 357 (2021): 129721.

Kuk, Do-Hoon, Ha, Eun-Sol, Ha, Dong-Hyun et al. "Development of a Resveratrol Nanosuspension Using the Antisolvent Precipitation Method without Solvent Removal, Based on a Quality by Design (QbD) Approach." *Pharmaceutics* 11 (2019): 688.

Kumar, Senthil, Chinnaiyan, Thangam, Ramar, Mary, Soloman Agnes et al. "Targeted Delivery and Apoptosis Induction of Trans-Resveratrol-Ferulic Acid Loaded Chitosan Coated Folic Acid Conjugate Solid Lipid Nanoparticles in Colon Cancer Cells." *Carbohydrate Polymers* 231 (2020): 115682.

Lee, Dong Gun, Go, Eun Byeol, Lee, Mindong et al. "Gold Nanoparticles Conjugated with Resveratrol Induce Cell Cycle Arrest in Mcf-7 Cell Lines." *Applied Biological Chemistry* 62 (2019): 1–6.

Li, Jingru, Wang, Zengming, Zhang, Hui, Gao, Jing, and Zheng, Aiping. "Progress in the Development of Stabilization Strategies for Nanocrystal Preparations." *Drug Delivery* 28 (2021): 19–36.

Li, Mengshuang, Zhang, Ling, Li, Rong, and Yan, Meixing. "New Resveratrol Micelle Formulation for Ocular Delivery: Characterization and In Vitro/In Vivo Evaluation." *Drug Development and Industrial Pharmacy* 46 (2020): 1960–1970.

Li, Xin, Dai, Yingnan, Yan, Shujun et al. "Resveratrol Lowers Blood Pressure in Spontaneously Hypertensive Rats Via Calcium-Dependent Endothelial No Production." *Clinical &Experimental Hypertension* 38 (2016): 287–293.

Li, Zerong, Qiao, Wenmei, Wang, Chenghao et al. "DPPC-Coated Lipid Nanoparticles as an Inhalable Carrier for Accumulation of Resveratrol in the Pulmonary Vasculature, a New Strategy for Pulmonary Arterial Hypertension Treatment." *Drug Delivery* 27 (2020): 736–744.

Lin, Chih-Hung, Chen, Chun-Han, Lin, Zih-Chan, and Fang, Jia-You. "Recent Advances in Oral Delivery of Drugs and Bioactive Natural Products Using Solid Lipid Nanoparticles as the Carriers." *Journal of Food & Drug Analysis* 25 (2017): 219–234.

Lin, Kai-Lieh, Lin, Kai-Jung, Wang, Pei-Wen et al. "Resveratrol Provides Neuroprotective Effects through Modulation of Mitochondrial Dynamics and ERK1/2 Regulated Autophagy." *Free Radical Research* 52 (2018): 1371–1386.

Lin, Meihua, Dai, Yu, Xia, Fan, and Zhang, Xiaojin. "Advances in Non-Covalent Crosslinked Polymer Micelles for Biomedical Applications." *Materials Science and Engineering: C* 119 (2021): 111626.

Lin, Yu-hua, Zhu, Lin-yun, Yang, Yan-qin et al. "Resveratrol Inhibits MUC5AC Expression by Regulating SPDEF in Lung Cancer Cells." *Phytomedicine* (2021): 153601.

Liswidowati, Melchior F., Hohmann, F., Schwer, B., and Kindl, H. "Induction of Stilbene Synthase by Botrytis Cinerea in Cultured Grapevine Cells." *Planta* 183 (1991): 307–314.

Liu, Qianyuan, Chen, Jingjing, Qin, Yang, Jiang, Bo, and Zhang, Tao. "Encapsulation of Pterostilbene in Nanoemulsions: Influence of Lipid Composition on Physical Stability, In Vitro Digestion, Bioaccessibility, and Caco-2 Cell Monolayer Permeability." *Food &Function* 10 (2019): 6604–6614.

Liu, Qiaoyu, Guan, Jian, Sun, Zheng et al. "Influence of Stabilizer Type and Concentration on the Lung Deposition and Retention of Resveratrol Nanosuspension-in-Microparticles." *International Journal of Pharmaceutics* 569 (2019): 118562.

Liu, Ye, Liang, Xiao, Zou, Yan et al. "Resveratrol-Loaded Biopolymer Core–Shell Nanoparticles: Bioavailability and Anti-Inflammatory Effects." *Food &Function* 11 (2020): 4014–4025.

Lomholt, S., Mellemkjaer, A., Iversen, M. B., Pedersen, S. B., and Kragstrup, T. W. "Resveratrol Displays Anti-Inflammatory Properties in an Ex Vivo Model of Immune Mediated Inflammatory Arthritis." *BMCRheumatology* 2 (2018): 27.

Lopes, Talíria Silva, Alves, Gutemberg Gomes, Pereira, Mariana Rodrigues, Granjeiro, Jose Mauro, and Leite, Paulo Emílio Corrêa. "Advances and Potential Application of Gold Nanoparticles in Nanomedicine." *Journal Of Cellular Biochemistry* 120 (2019): 16370–16378.

Lu, Ying, and Park, Kinam. "Polymeric Micelles and Alternative Nanonized Delivery Vehicles for Poorly Soluble Drugs." *International Journal of Pharmaceutics* 453 (2013): 198–214.

Luther, David C., Huang, Rui, Jeon, Taewon et al. "Delivery of Drugs, Proteins, and Nucleic Acids Using Inorganic Nanoparticles." *Advanced Drug Delivery Reviews* 156 (2020): 188–213.

Luzardo-Álvarez, Asteria, Lamela-Gómez, Iván, Otero-Espinar, Francisco, and Blanco-Méndez, José. "Development, Characterization, and In Vitro Evaluation of Resveratrol-Loaded Poly-(ε-Caprolactone) Microcapsules Prepared by Ultrasonic Atomization for Intra-Articular Administration." *Pharmaceutics* 11 (2019): 249.

Macedo, Luiza De O., Barbosa, Eduardo J., Löbenberg, Raimar, and Bou-Chacra, Nadia A. "Anti-Inflammatory Drug Nanocrystals: State of Art and Regulatory Perspective." *European Journal of Pharmaceutical Sciences* 158 (2021): 105654.

Machado, Noelia D., Fernández, Mariana A., Häring, Marleen, Saldías, César, and Díaz, David Díaz. "Niosomes Encapsulated in Biohydrogels for Tunable Delivery of Phytoalexin Resveratrol." *RSC Advances* 9 (2019): 7601–7609.

Machado, Noelia D., Gutiérrez, Gemma, Matos, María, and Fernández, Mariana A. "Preservation of the Antioxidant Capacity of Resveratrol Via Encapsulation in Niosomes." *Foods* 10 (2021): 988.

Maja, Leitgeb, Željko, Knez, and Mateja, Primožič. "Sustainable Technologies for Liposome Preparation." *The Journal of Supercritical Fluids* 165 (2020): 104984.

Marinheiro, Diogo, Ferreira, Bárbara J. M. L., Oskoei, Párástu, Oliveira, Helena, and Daniel-da-Silva, Ana L. "Encapsulation and Enhanced Release of Resveratrol from Mesoporous Silica Nanoparticles for Melanoma Therapy." *Materials* 14 (2021): 1382.

Meng, Tiantian, Xiao, Dingfu, Muhammed, Arowolo et al. "Anti-Inflammatory Action and Mechanisms of Resveratrol." *Molecules* 26 (2021): 229.

Miao, Liu, Daozhou, Liu, Ying, Cheng, Qibing, Mei, and Siyuan, Zhou. "A Resveratrol-Loaded Nanostructured Lipid Carrier Hydrogel to Enhance the Anti-UV Irradiation and Anti-Oxidant Efficacy." *Colloids & Surfaces B:Biointerfaces* 204 (2021): 111786.

Mocanu, Georgeta, Nichifor, Marieta, and Sacarescu, Liviu. "Dextran Based Polymeric Micelles as Carriers for Delivery of Hydrophobic Drugs." *Current Drug Delivery* 14 (2017): 406.

Moshawih, Said, Mydin, S. M. N., Basria, Rabiatul, Kalakotla, Shanker, and Jarrar, Qais B. "Potential Application of Resveratrol in Nanocarriers against Cancer: Overview and Future Trends." *Journal of Drug Delivery Science and Technology* 53 (2019): 101187.

Negi, Poonam, Aggarwal, Madhur, Sharma, Gajanand et al. "Niosome-Based Hydrogel of Resveratrol for Topical Applications: An Effective Therapy for Pain Related Disorder(s)." *Biomed Pharmacother* 882017 (2017): 480–487.

Neves, A. R., Lucio, M., Lima, J. L. C., and Reis, S. "Resveratrol in Medicinal Chemistry: A Critical Review of Its Pharmacokinetics, Drug-Delivery, and Membrane Interactions." *Current Medicinal Chemistry* 19 (2012): 1663.

Nie, Hongfei, Zhang, Yarong, Yu, Haiyang et al. "Oral Delivery of Carrier-Free Dual-Drug Nanocrystal Self-Assembled Microspheres Improved NAD^+ Bioavailability and Attenuated Cardiac Ischemia/Reperfusion Injury in Mice." *Drug Delivery* 28 (2021): 433–444.

Pandita, Deepti, Kumar, Sandeep, Poonia, Neelam, and Lather, Vincy. "Solid Lipid Nanoparticles Enhance Oral Bioavailability of Resveratrol, a Natural Polyphenol." *Food Research International* 62 (2014): 1165–1174.

Patel, Ketan R., Andreadi, Catherine, Britton, Robert G. et al. "Sulfate Metabolites Provide an Intracellular Pool for Resveratrol Generation and Induce Autophagy with Senescence." *Science Translational Medicine* 5 (2013): 205ra133.

Peng, Ru-Min, Lin, Guan-Ru, Ting, Yuwen, and Hu, Jing-Yu. "Oral Delivery System Enhanced the Bioavailability of Stilbenes: Resveratrol and Pterostilbene." *Biofactors* 44 (2018): 5–15.

Perkin, W. H. "XXIII.—On the Hydride of Aceto-Salicyl." *Journal of the Chemical Society* 21 (1868): 181–186.

Perris, Angela, Bhattacharya, Sanchari, Jawed, Junaid, and Hoda, Muddasarul. "Oncotherapeutic Application of Resveratrol-Based Inorganic Nanoparticles." *Pharmaceutical Nanotechnology* 9 (2021): 271–280.

Pignatello, Rosario, Pecora, Tiziana M. G., Cutuli, Giuseppa G. et al. "Antioxidant Activity and Photostability Assessment of Trans-Resveratrol Acrylate Microspheres." *Pharmaceutical Development and Technology* 24 (2019): 222–234.

Pouton, Colin W. "Lipid Formulations for Oral Administration of Drugs: Non-Emulsifying, Self-Emulsifying and 'Self-Microemulsifying' drug Delivery Systems." *European Journal of Pharmaceutical Sciences* 11 (2000): S93–98.

Zia, Aliabbas, Farkhondeh, Tahereh, Pourbagher-Shahri, Ali Mohammad, Samargh, and Ian, Saeed. "The Roles of Mitochondrial Dysfunction and Reactive Oxygen Species in Aging and Senescence." *Current Molecular Medicine* 22 (2022): 37–49.

Zuccari, Guendalina, Alfei, Silvana, Zorzoli, Alessia et al. "Increased Water-Solubility and Maintained Antioxidant Power of Resveratrol by Its Encapsulation in Vitamin E TPGS Micelles: A Potential Nutritional Supplement for Chronic Liver Disease." *Pharmaceutics* 13 (2021): 1128.

3 Novel Drug Delivery System for Gallic Acid

Afonso H.S. Júnior, Domingos L.P.Macuvele, and António B. Mapossa

3.1 INTRODUCTION

Gallic acid, 3,4,5-trihydroxybenzoic acid, is a plant-based phenolic compound that hasvarious medical and biological properties. Previous works reported many potential applications, such as antioxidant, antimicrobial, and anticancer uses and treatment of gastrointestinal diseases, metabolic diseases, and neuropsychological diseases, among others (Badhani et al., 2015; Kahkeshani et al., 2019). However, drawbacks, such as poor solubility, bioavailability, and stability, limit itspotential applications. Additionally, limitations were noted when gallic acid (GA) was applied in mice.

To address constraints on the application of GA, chemical modification (AL Zahrani et al., 2020) and the development of delivery systems are reported in the literature (Mitsou et al., 2020). Chemical modification, including the hybridization of gallic acid, aiming to improve their biological activity, solubility, and bioavailability, was extensively reviewed (AL Zahrani et al., 2020). The results of several studies are extremely promising. As a result, patents were applied, and clinical trials were performed with some hybrids and GA derivates. Despite these advances, the chemical modification of GA in some cases requires expensive reagents, leading to a greater cost. Thus, there are still many challenges to overcome regarding the use of GA delivery systems.

Conventional drug delivery systems, mainly injectables, pills, and tablets, havedrawbacks, such as uncontrolled drug delivery, poor bioavailability, use of higher concentrations of drugs, and intensified adverse effects. Inthe advent of nanoscience and nanotechnology, novel approaches aim to overcome the limitations of conventional drug delivery systems (Dorniani et al., 2016; Radwan et al., 2020; Sani Usman et al., 2017). Various nano-based drug delivery systems were developed for GA (Acevedo et al., 2018; Alfei et al., 2020; Dorniani et al., 2012, 2016; Guendouzi et al., 2020; Zhang et al., 2022a). These systems, including inorganic and organic nanoparticles, liposomes, dendrimers, and hybrid materials, showed promising results when applied as GA delivery systems. Despite these advances, many issues need to be addressed before real-life applications can be made. For example, the toxicity of some nanoparticles used in drug delivery systems was reported. Additionally, nanoparticles' efficacy depend on their size and shape, so more tests are still needed to determine the proper forms of some nanoparticles so that drug delivery systems can handle high drug load and release.

Therefore, this chapter aims to provide a general outlook of drug delivery systems for GA. Special attention is given to recent trends and novel approaches for GA delivery. Finally, this chapter will provide the main challenges and advances in the field.

3.2 TRENDS ON DRUG DELIVERY SYSTEM FOR GALLIC ACID

Gallic acid (GA) is a widely investigated bioactive molecule. GA is applied for numerous treatments, *e.g.*, anti-inflammatory, anti-diabetic, and anti-free radicals (Mitsou et al., 2020). Figure 3.1 shows the beneficial effects of using GA. Also, GA is an effective substance against several kinds of tumor cells (Patil & Killedar, 2021b). Unfortunately, cancer is one of the leading causes of death

DOI: 10.1201/9781003300588-3

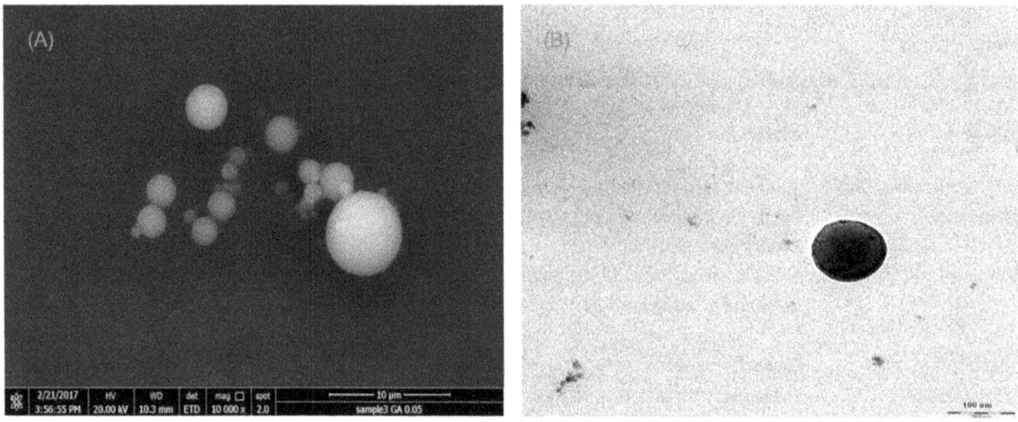

FIGURE 3.2 Gallic acid in the poly(D,L-lactide-co-glycolide) (PLGA) nanoparticles system (A) SEM and (B) TEM (reprinted under the terms of the Creative Commons CC BY license) (Mahboob et al., 2020).

protozoan, about 62% (25 µg/mL). SEM and TEM revealed that PLGA nanoparticles loaded with gallic acid had a spherical shape and about 100 nm. Figure 3.2 shows the morphological characteristics of the system through microscopic analysis. Gallic acid releasedfrom the system was performed *in vitro* as a function of time. The release rate of gallic acid was 98.24% in 48 hours. Several factors, including temperature, nanoparticle type, and pH, affect the release rate of chemicals from the matrices. Mahboob et al. (2020) reported that drug release was slow in the first 24 hours. Polymer content and reduction of surface area available for drug release may have been variables that influenced the performance of the system. In addition, strategies for controlling the release rate of gallic acid were suggested, for example, by adjusting the molecular parameters of PLGA. Furthermore, gallic acid is considered as a strong antioxidant, has cell-death-inducing effects on tumor cells, and protects healthy cells from the effects of reactive oxygen species. This feature is more fundamental for an effective treatment, as it can help maintain the lifespan of lymphocytes in the blood (important in the body's defense system against infections) (Acevedo et al., 2018).

Nowadays, the use of hydrogels in drug delivery is a topic widely explored by researchers in different field of science. Additionally, there is a rising in smart drug delivery systems (SDDS) (Jiang & Kobayashi, 2017). The main advantage of this class of material is the efficiency of bioactive release. Also, SDDS has "more" controllable properties than conventional systems, as there are external stimuli during the activity. SDDS have three relevant components—(1) the carrier, (2) the drug, and (3) the external stimulus. Polymeric drug delivery systems are quite effective for producing smart bioactive nano vehicles. External stimuli ideal for application in these systems must be noninvasive, safe, and painless. The main variables used for this purpose (stimulating the release of the compound) are pH, temperature, light, ultrasound, etc. The production of an intelligent gallic acid delivery system was reported using an ultrasound-stimulated chitin hydrogel (Jiang & Kobayashi, 2017). The ultrasound power, the experimental conditions, and the amount of gallic acid and chitin in the hydrogel matrix were varied to investigate the behavior of gallic acid release. Gallic acid released was improved by exposure to 43 kHz (ultrasound) with a release rate of 0.74 µg/mL·min. In addition, the bioactive release efficiency increased with the increase in ultrasound power, in the amount of gallic acid, and with the decrease in chitin concentration in the polymer matrix. Chitosan nanoparticles are widely used to produce support for the release of bioactive (Thanyacharoen et al., 2018). Encapsulation of gallic acid in chitosan nanoparticles generally shows high efficiency. However, some strategies for optimizing the gallic acid encapsulation process are required, for example, response surface methodology (RSM). In other words, an experimental design allows,

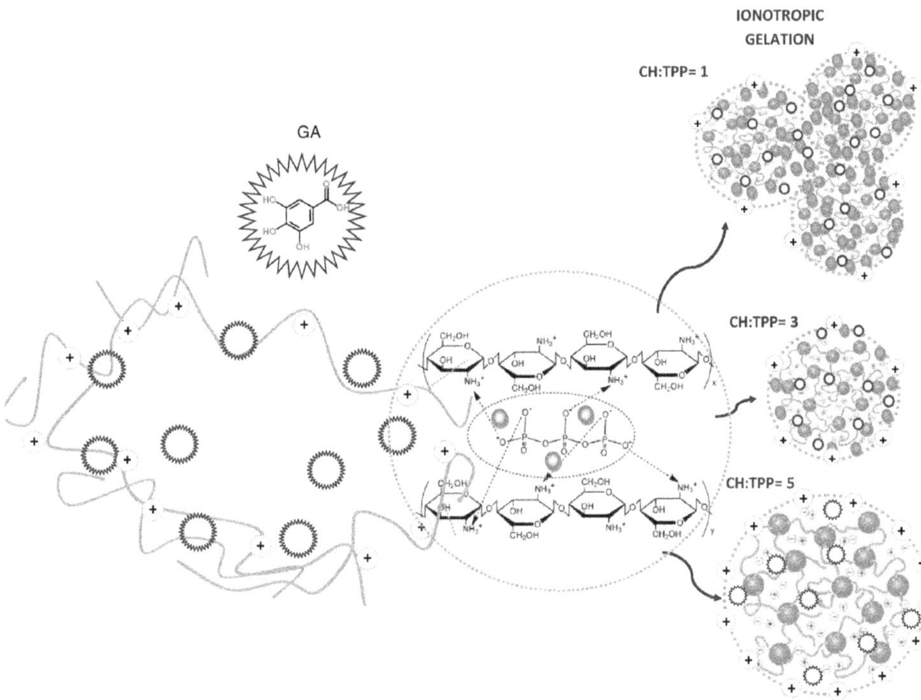

FIGURE 3.3 Illustration of the cross-linking and encapsulation process of gallic acid in a biopolymer matrix (copyright 2016, Elsevier; reproduced with permission) (Lamarra et al., 2016).

from a set of experiments, all variables to be evaluated simultaneously and in a systematic way. The application of RSM for the synthesis of gallic acid transport systems can be an interesting alternative for the subject. Gallic-acid-loaded chitosan-based nanoparticles were successfully synthesized using a statistical design (Lamarra et al., 2016). Responses such as zeta potential and encapsulation efficiency allowed optimizing the system as a function of chitosan concentration, chitosan (CH)/tripolyphosphate (TPP) ratio, and gallic acid concentration. Figure 3.3 shows a schematic illustration of CH/TPP cross-linking and gallic acid encapsulation. The methodology applied allowed a maximum encapsulation percentage of 82%. Therefore, it can be an efficient alternative to produce a biomaterial with the slow release of gallic acid in the disease treatment. Furthermore, the chemical investigation of the synthesized biomaterial revealed that the occurrence of hydrogen bonds and the ionic interactions of the biopolymer/TPP allowed an excellent result in the encapsulation of gallic acid. Likewise, these interactions made it possible to improve the stability of the active agent in the biomaterial.

Mechanism of complexation and the ability of β-cyclodextrin to encapsulate gallic acid to produce a slow-release system was investigatedby Guendouzi et al. (2020). According tothe published data, β-cyclodextrin demonstrated excellent ability to encapsulate and protect gallic acid. In addition, the lower energy structure presented two strong intramolecular hydrogen bonds that could contribute to a better stability result of the complex. Gallic acid encapsulation using gum arabic nanoparticles were evaluated for their antioxidant and antitumor effects through *in vitro* assays (Hassani et al., 2020). In addition, Hassani et al. (2020) reported that the synergistic effects of gum Arabic on encapsulation (1) contributed to increased selectivity toward cancer cells and (2) potentiated the antioxidant properties of the bioactive. Moreover, the formulated nanoparticles showed low toxicity in normal cells.

The application of magnetic nanoparticles provides drug delivery systems with numerous functions and can also act as a disease diagnosis material. The advancements in the production of nanomaterials applied to medicine, specifically for the manufacture of drug delivery systems, have grown exponentially in recent years. Progress in synthesis pathways, a better understanding of interactions of biomolecules with inorganic nanoparticles, and intensive *in vivo* assessments of toxicity to living organisms are important factors that have contributed to improvements in the field. Although there are not many products being researched commercially, the trend is that in the long term, many of these studiesthat are still at the laboratory level will be implemented in to treat the diseases of real-life patients.

3.2.3 Liposomes

Liposomes are small vesicles formed by phospholipids. Liposomes can be classified according to the number of lipid layers: (1) unilamellar or (2) multilamellar (Tan et al., 2021). Liposomes are mainly applied as a drug vehicle for the disease treatment. Therefore, each liposome type (unilamellar or multilamellar) can be more suitable for a particular kind of bioactive (water-soluble, lipid, etc.). The main advantages of using liposomes are the solubility and permeability increase of molecules with different chemical aspects (Giordani et al., 2019). However, conventional liposomes have disadvantages that preclude the wide application of these vesicles in targeting bioactives, such as easy degradation at acidic pH and by enzymes (*i.e.*, lipase). Currently, strategies to overcome these disadvantages involve the use of polymers, forming a liposome-polymer complex (Tan et al., 2021). In addition, the complexation of liposomes with polymers prevents the binding or adsorption of vesicles to some proteins during circulation. This interaction of the vesicles with the polymer may confer selectivity in the disease treatment. There are four types of interaction between the polymer and the liposome: (1) surface, (2) internal, (3) combination, and (4) incorporation into lamellar chains (Sriwidodo et al., 2022). Figure 3.6 shows the different types of polymers and liposome hybrid applied in the delivery of bioactive.

Gallic acid liposomes (GA-LIP) and lactoferrin-decorated gallic acid liposomes (LF-GA-LIP) were fabricated to assess their antibacterial potential (Zhang et al., 2019). *In vitro* digestion and

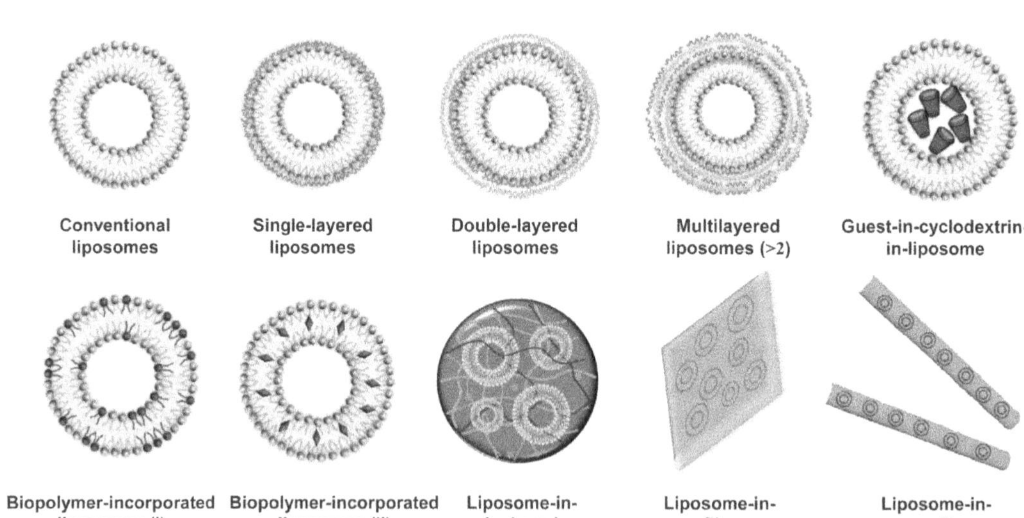

FIGURE 3.6 Types of polymer and liposome hybrids used in the production of drug delivery systems (copyright 2021, Elsevier; reproduced with permission) (Tan et al., 2021).

antibacterial activity were analyzed for the two complexes. LF-GA-LIP showed higher encapsulation efficiency and storage stability compared to GA-LIP. In addition, there was an electrostatic interaction between lactoferrin and the phospholipid bilayer of the liposome. The presence of gallic acid in the two complexes was through the amorphous form. Moreover, LF-GA-LIP showed a slower release effect compared to GA-LIP in the digestion simulation. Finally, the complex of gallic acid, lactoferrin, and liposome exerted greater antibacterial properties against *Escherichia coli* and *Staphylococcus aureus* than GA-LIP. It is suggested that the application of this hybrid in the release of gallic acid presents an enormous potential for science. The development of a new liposome for the simultaneous delivery of gallic acid and quercetin in the treatment of fungal disease was reported (Giordani et al., 2019). The fungal infection evaluated causes itching, pain, and local inflammation. Therefore, the production of this quercetin and gallic acid delivery system aimed to induce a synergy between the polyphenols to eradicate the infection while alleviating the symptoms of the disease. The synthesized liposome showed an excellent percentage of encapsulation of the bioactive. In addition, the biomaterial did not show cytotoxicity and presented a stronger anti-inflammatory effect than free polyphenols.

The use of liposomes and mixed micelles to deliver gallic acid was reported (Martí et al., 2014). The application of this gallic acid vehicle aims to create a biofunctional textile anti-aging drugs release (direct contact with the skin). The liposomes with encapsulated gallic acid were applied in numerous fibers: (1) cotton, (2) polyamide, (3) polyester, (4) acrylic, and (5) wool. The application of liposomes to different textiles showed a suitable aspect for most fibers. However, the high desorption in synthetic fibers (acrylic and polyester) showed that the application should be carried out preferentially in the other fibers. The use of textile fibers in the release of gallic acid can be an interesting strategy in the elaboration of smart textiles in the treatment of the skin in the fight against aging. However, the evaluation of the desorption of the active compound must be tested in-depth because a faster release than estimated may occur. Lipid nanocapsules loaded with reverse micelles were designed to deliver gallic acid (GA-RMLNC) in the liver fibrosis treatment (Radwan et al., 2020). GA-RMLNC presented size of approximately 30 nm. Furthermore, it demonstrated a high percentage of gallic acid encapsulation (63.95%) and excellent physical stability (9 months). The system showed burst gallic acid release in the first 2 hours followed by a sustained release profile. Finally, GA-RMLNC suppressed the activation, proliferation, and migration of activated liver cells. Manufacture of new biomaterials using liposomes should grow in the coming years due to the numerous advantages that present, such as (1) biocompatibility, (2) easy projection control, (3) excellent adaptability to drugs, (3) structural flexibility, and (5) size. However, it still requires a deeper understanding of the interactions that liposomes and possible stabilizers can interfere with the release of the drug. In addition, liposomes are easily degraded in the presence of acids and digestive enzymes. Therefore, before the development of drug delivery systems in living organisms can become successful, the challenges of using these lipid vesicles should still be overcome. But this category of material is promising for the area and especially in cancer treatment.

3.2.4 Hybrid-Based Materials

Drug delivery through nanoparticulate systems is a relevant aspect of science in the development of nanoformulations. However, some factors influence the release rate of the bioactive from the nanoparticles, for example, such as diffusion through the drug nanoparticles and the desorption phenomenon (Thanyacharoen et al., 2018). Thus, drug release by the system depends on numerous variables, mainly diffusion, and biodegradation. Therefore, outlining strategies to overcome these limitations is necessary for the current context. The production of hybrid drug delivery systems is a hot topic in medicine that aims to optimize these biomaterials in terms of drug diffusion and selectivity in the disease treatment (Hu et al., 2014; Tan et al., 2021; Zhang et al., 2022b). The use of single-wall carbon nanotubes (SWCNT) and polyhedral oligomeric silsesquioxanes (POSS) has been recently reported in the optimization of gallic acid delivery systems (Zhang et al., 2022b).

Dorniani, D., Saifullah, B., Barahuie, F., Arulselvan, P., Bin Hussein, M. Z., Fakurazi, S., & Twyman, L. J. (2016). Graphene oxide-gallic acid nanodeliverysystem for cancer therapy. *Nanoscale Research Letters*, *11*(1), 491. https://doi.org/10.1186/s11671-016-1712-2.

Giordani, B., Basnet, P., Mishchenko, E., Luppi, B., & Škalko-Basnet, N. (2019). Utilizing liposomal quercetin and gallic acid in ocalizedtreatment of vaginal candida infections. *Pharmaceutics*, *12*(1), 9. https://doi.org/10.3390/pharmaceutics12010009.

Guendouzi, O., Guendouzi, A., Ouici, H. B., Brahim, H., Boumediene, M., & Elkeurti, M. (2020). A quantum chemical study of encapsulation and stabilization of gallic acid in β-cyclodextrin as a drug delivery system. *Canadian Journal of Chemistry*, *98*(4), 204–214. https://doi.org/10.1139/cjc-2019-0464.

Hassani, A., Azarian, M. M. S., Ibrahim, W. N., & Hussain, S. A. (2020). Preparation, characterization and therapeutic properties of gum arabic-stabilized gallic acid nanoparticles. *Scientific Reports*, *10*(1), 17808. https://doi.org/10.1038/s41598-020-71175-8.

Hu, X., Liu, S., Zhou, G., Huang, Y., Xie, Z., & Jing, X. (2014). Electrospinning of polymeric nanofibers for drug delivery applications. *Journal of Controlled Release*, *185*(1), 12–21. https://doi.org/10.1016/j.jconrel.2014.04.018.

Jiang, H., & Kobayashi, T. (2017). Ultrasound stimulated release of gallic acid from chitin hydrogel matrix. *Materials Science and Engineering: C*, *75*, 478–486. https://doi.org/10.1016/j.msec.2017.02.082.

Kahkeshani, N., Farzaei, F., Fotouhi, M., Alavi, S. S., Bahramsoltani, R., Naseri, R., Momtaz, S., Abbasabadi, Z., Rahimi, R., Farzaei, M. H., & Bishayee, A. (2019). Pharmacological effects of gallic acid in health and disease: A mechanistic review. *Iranian Journal of Basic Medical Sciences*, *22*(3), 225–237. https://doi.org/10.22038/ijbms.2019.32806.7897.

Lai, J.Y., & Luo, L.J. (2015). Antioxidant gallic acid-functionalized biodegradable in situ gelling copolymers for cytoprotective antiglaucoma drug delivery systems. *Biomacromolecules*, *16*(9), 2950–2963. https://doi.org/10.1021/acs.biomac.5b00854.

Lamarra, J., Rivero, S., & Pinotti, A. (2016). Design of chitosan-based nanoparticles functionalized with gallic acid. *Materials Science and Engineering: C*, *67*, 717–726. https://doi.org/10.1016/j.msec.2016.05.072.

Liu, S., Zhou, G., Liu, D., Xie, Z., Huang, Y., Wang, X., Wu, W., & Jing, X. (2013). Inhibition of orthotopic secondary hepatic carcinoma in mice by doxorubicin-loaded electrospun polylactide nanofibers. *Journal of Materials ChemistryB*, *1*(1), 101–109. https://doi.org/10.1039/C2TB00121G.

Lu, S., Liu, L., Wang, H., Zhao, W., Li, Z., Qu, Z., Li, J., Sun, T., Wang, T., & Sui, G. (2019). Synthesis of dual functional gallic-acid-based carbon dots for bioimaging and antitumor therapy. *Biomaterials Science*, *7*(8), 3258–3265. https://doi.org/10.1039/C9BM00570F.

Mahboob, T., Nawaz, M., de Lourdes Pereira, M., Tian-Chye, T., Samudi, C., Sekaran, S. D., Wiart, C., & Nissapatorn, V. (2020). PLGA nanoparticles loaded with Gallic acid – a constituent of Leea indica against Acanthamoeba triangularis. *Scientific Reports*, *10*(1), 8954. https://doi.org/10.1038/s41598-020-65728-0.

Martí, M., Martínez, V., Lis, M. J., Valldeperas, J., de la Maza, A., Parra, J. L., & Coderch, L. (2014). Gallic acid vehiculized through liposomes or mixed micelles in biofunctional textiles. *The Journal of the Textile Institute*, *105*(2), 175–186. https://doi.org/10.1080/00405000.2013.833690.

Mitsou, E., Pletsa, V., Sotiroudis, G. T., Panine, P., Zoumpanioti, M., & Xenakis, A. (2020). Development of a microemulsion for encapsulation and delivery of gallic acid. The role of chitosan. *Colloids and Surfaces B: Biointerfaces*, *190*(March), 110974. https://doi.org/10.1016/j.colsurfb.2020.110974.

Patil, P., & Killedar, S. (2021a). Chitosan and glyceryl monooleate nanostructures containing gallic acid isolated from amla fruit: Targeted delivery system. *Heliyon*, *7*(3), e06526. https://doi.org/10.1016/j.heliyon.2021.e06526.

Patil, P., & Killedar, S. (2021b). Formulation and characterization of gallic acid and quercetin chitosan nanoparticles for sustained release in treating colorectal cancer. *Journal of Drug Delivery Science and Technology*, *63*(March), 102523. https://doi.org/10.1016/j.jddst.2021.102523.

Radwan, S. A. A., El-Maadawy, W. H., ElMeshad, A. N., Shoukri, R. A., & Yousry, C. (2020). Impact of reverse micelle loaded lipid nanocapsules on the delivery of gallic acid into activated hepatic stellate cells: A promising therapeutic approach for hepatic fibrosis. *Pharmaceutical Research*, *37*(9), 180. https://doi.org/10.1007/s11095-020-02891-z.

Sani Usman, M., Hussein, M., Fakurazi, S., Masarudin, M., & Ahmad Saad, F. (2017). Gadolinium-doped gallic acid-zinc/aluminium-layered double hydroxide/gold theranosticnanoparticles for a bimodal magnetic resonance imaging and drug delivery system. *Nanomaterials*, *7*(9), 244. https://doi.org/10.3390/nano7090244.

Sharma, S., Mittal, Di., Verma, A. K., & Roy, I. (2019). Copper-gallic acid nanoscale metal–organic framework for combined drug delivery and photodynamic therapy. *ACS Applied Bio Materials*, *2*(5), 2092–2101. https://doi.org/10.1021/acsabm.9b00116.

Sriwidodo, Umar, A. K., Wathoni, N., Zothantluanga, J. H., Das, S., & Luckanagul, J. A. (2022). Liposome-polymer complex for drug delivery system and vaccine stabilization. *Heliyon*, *8*(2), e08934. https://doi.org/10.1016/j.heliyon.2022.e08934.

Tan, C., Wang, J., & Sun, B. (2021). Biopolymer-liposome hybrid systems for controlled delivery of bioactive compounds: Recent advances. *Biotechnology Advances*, *48*(February), 107727. https://doi.org/10.1016/j.biotechadv.2021.107727.

Thanyacharoen, T., Chuysinuan, P., Techasakul, S., Nooeaid, P., & Ummartyotin, S. (2018). Development of a gallic acid-loaded chitosan and polyvinyl alcohol hydrogel composite: Release characteristics and antioxidant activity. *International Journal of Biological Macromolecules*, *107*(PartA), 363–370. https://doi.org/10.1016/j.ijbiomac.2017.09.002.

Wu, J., Chen, G., Jia, Y., Ji, C., Wang, Y., Zhou, Y., Leblanc, R. M., & Peng, Z. (2022). Carbon dot composites for bioapplications: A review. *Journal of Materials Chemistry B*, *10*(6), 843–869. https://doi.org/10.1039/D1TB02446A.

Zhang, X., An, D., Zhang, R., Huang, Y., & Liu, Z. (2022a). Preparation of carbon nanotubes and polyhedral oligomeric-reinforced molecularly imprinted polymer composites for drug delivery of gallic acid. *International Journal of Pharmaceutics*, *615*(January), 121476. https://doi.org/10.1016/j.ijpharm.2022.121476.

Zhang, X., An, D., Zhang, R., Huang, Y., & Liu, Z. (2022b). Preparation of carbonnanotubes and polyhedral oligomeric-reinforced molecularly imprinted polymer composites for drug delivery of gallic acid. *International Journal ofPharmaceutics*, *615*(August 2021), 121476. https://doi.org/10.1016/j.ijpharm.2022.121476.

Zhang, Y., Pu, C., Tang, W., Wang, S., & Sun, Q. (2019). Gallic acid liposomes decorated with lactoferrin: Characterization, in vitro digestion and antibacterial activity. *Food Chemistry*, *293*(March), 315–322. https://doi.org/10.1016/j.foodchem.2019.04.116.

TABLE 4.1
NEs with Herbal Active Substances

S. No.	NE with Herbal Bioactives	Biological Origins	Loaded Bioactives' Outcomes	References
1.	Basil oil NE	*Ocimum basilicum*	Bacterial resistance	(Ghosh *et al.*, 2013)
2.	Cumin-tincture-loaded NE	*Cuminum cyminum*	Antioxidant, anti-angiogenic, effective in removing free radicals in a dose-dependent manner and antibacterial activity	(Asgari *et al.*, 2021)
3.	Elemene oil NE	*Curcuma* species	Sprague Dawley rats had better oral bioavailability and stability compared to a commercial element emulsion.	(Zeng *et al.*, 2010)
4.	Essential oil NE	*Alhagi maurorum*	Increased bacterial and biofilm destroying the capacity	(Hassanshahian *et al.*, 2020)
5.	Essential oil NE	*Thymus vulgaris*	Act as treatment of cutaneous mycoses	(Moazeni *et al.*, 2021)
6.	Essential oil NE	*Myristica fragrans* or *Lavandula dentata*	Enhanced stability over a range of temperatures and storage settings	(Cossetin *et al.*, 2021)
7.	Linseed oil NE	*Linum usitatissimum*	Improved stability and physicochemical qualities for atopic dermatitis	(Kildaci *et al.*, 2021)
8.	*Nelumbo nucifera* crude extracts	*Nelumbo nucifera*	Improved antibacterial and stability properties	(Pengon *et al.*, 2020)
9.	*Nigella sativa* L. NE	*Nigella sativa*	Increased and dosages free radical—scavenging capability	(Arazmjoo *et al.*, 2021)
10.	Oregano oil NE	*Origanum vulgare*	Food-borne bacteria development in fresh lettuce has been minimized and suppressed	(Vickers *et al.*, 2017)
11.	Peppermint and rosemary essential oils NE	*Mentha piperita*	Enhanced antioxidant capacity and reduces arthritic pain	(Mohammadifar *et al.*, 2021)
12.	Quercetin NE	Many plants	With less skin retention, increased cutaneous permeability reached the systemic circulation	(Bonifácio *et al.*, 2014)

4.2.2 Liposomes

In 1960, Alec Bangham invented liposomes. These spherical polar lipid vesicles have a size range of 50 to 450 nm. They may surround an "aqueous core" that freely diffuses in one or more lipid bilayers that can be either natural or artificial (Ganesan *et al.*, 2017). They share similarities with cells in their membrane structure. The materials that makeup liposomes include both hydrophilic and lipophilic groups, enabling them to enclose both classes of pharmacological compounds in a single structure (Robson *et al.*, 2018). Due to its special feature of having phospholipid bilayers, they can improve drug solubility, bioavailability of the entrapped drug, drug delivery, absorption of drug within a cell, and distribution of drug throughout the body (Song *et al.*, 2017). To prepare vaccines, nutraceuticals, and cosmetics, it is necessary to modify the absorption, distribution, metabolism, and excretion (ADME) profiles of bioactives like herbs, enzymes, and proteins. Additionally, some unique qualities, such as the environmental shield of the medicinal molecule contained inside, catastrophic destruction of the loaded bioactive, prompt diagnosis, cost-effectiveness, and the least incidence of systemic disease, greatly increased their inclusion in biomedical formulations (Aqil *et al.*, 2013). On the other hand, liposomes can be enhanced by antibodies or ligands to improve target specificity. Thangapazham et al. fabricated PSMA-coated liposomes that contain curcumin and were tested on C4–2B and LNCaP human prostate cancer cell lines. As a result, a 10-fold dose advantage, enhanced targeted administration, and a 70–80% reduction in cell proliferation have been attained (Thangapazham *et al.*, 2008). Other examples of herbal substances that have been liposomal encapsulated are shown in Table 4.2.

TABLE 4.2
List of Liposomes with Herbal Bioactives

S. No.	Liposomes with Herbal Bioactives	Biological Origins	Loaded Bioactives' Outcomes	References
1.	Asparagus racemosus liposomes	*Asparagus racemosus*	Enhanced anti-inflammatory qualities	(Plangsombat *et al.*, 2016)
2.	Baicalin-loaded liposomes	*Scutellaria baicalensis*	Better solubility, prolonged release	(Li *et al.*, 2018)
3.	Brucine liposomes	*Nux vomica*	Reduced brucine side effects, such as severe convulsions	(Qin *et al.*, 2007)
4.	Capsaicin liposomes	*Capsicum*	Increasing bioavailability for the treatment of neuropathic pain	(Zhu *et al.*, 2015)
5.	Colchicine-loaded liposomes	*Colchicum autumnal*	Enhanced drug delivery in anti-gout treatment	(Singh *et al.*, 2009)
6.	Curcumin-loaded liposomes	*Curcuma longa*	Anti-inflammatory and anticancer properties	(Jadia *et al.*, 2018)
7.	Curcumin loaded liposomes	*Curcuma longa*	Curcumin's enhanced therapeutic index	(Kongkaneramit *et al.*, 2016)
8.	Guggul liposomes	*Commiphora* (mukul)	Enhanced anti-inflammatory qualities	(Paliwal *et al.*, 2011)
9.	Liposomal neem gel	*Azadirachta indica*	Increased antibacterial effects	(Singh *et al.*, 2014)
10.	Naringenin-loaded liposomes	Immature orange fruit and peels of grape fruits	Enhanced tissue dispersion, solubility, absorption, and stability	(Wang *et al.*, 2017)
11.	Paclitaxel-loaded liposomes	The bark of *Taxus brevifolia*	Enhanced biodistribution, bioavailability, solubility, and absorption inside cells.	(Hong *et al.*, 2016)
12.	Polydatin-loaded liposomes	*Polygonum cuspidatum*	Enhanced solubility, prolonged release, and increased oral bioavailability.	(Wang *et al.*, 2015)
13.	Sterols-loaded liposomes	*Flammulina velutipes*	Enhance tissue dispersion, oral absorption, and water solubility	(Yi *et al.*, 2013)

TABLE 4.5
Herbal-Medicine-Containing Phytosomes

S. No.	Phytosomes with Herbal Bioactives	Biological Origins	Results and Findings	References
1.	*Aloe vera* extract–loaded phytosome	*Aloe Vera*	Increased oral administration of aloe vera, inhibition of cancer cell line development	(Murugesan *et al.*, 2021)
2.	*Butea monosperma* extract–loaded phytosome	*Butea monosperma*	Enhanced solubility, bioavailability, stability, release dissolving pattern and demonstrated strong scavenging of free radical activity	(Gahandule *et al.*, 2016)
3.	Epigallocatechin gallate–loaded phytosome	*Camellia sinensis*	Enhanced solubility, bioavailability and stability of physicochemical reactions at varying temperatures	(Anwar *et al.*, 2018)
4.	Gingerol-loaded phytosome	*Zingiber officinale*	Better stability, bioavailability, oral absorption, sustained release, antibacterial activity against *S. aureus* and *E. coli*, and improved anti-inflammatory properties.	(Singh *et al.*, 2018)
5.	*Morinda lucida* extract–loaded phytosome	*Morinda lucida*	Increased antimalarial impact	(Calister *et al.*, 2020)
6.	Rutin-loaded phytosome	Citrus fruits	Better solubility, stability, release dynamics, and bioavailability; effective antioxidant agent	(Hooresfand *et al.*, 2015)
7.	Soybean seed phytosome–based thermogel	*Glycine max*	Reduced adipose tissue weight, body weight, and improved absorption, stability, and release	(El-Menshawe *et al.*, 2018)
8.	*Swertia perennis* extract–loaded phytosome	*Swertia perennis*	Enhanced drug release and effectiveness of implanted phytomedicine.	(Chouhan *et al.*, 2021)

a lipid-soluble state (Soni *et al.*, 2018). Numerous studies to evaluate their benefits and effectiveness in contrast to other conventional delivery methods have been conducted. Different researchers used the several flavonoids, including apigenin, kaempferol, and quercetin, to develop a phytosome known as a flavonosome. This phytosome were utilized as a hepatoprotective agent, antioxidant, and heat supplement. Herbal-medicine-containing phytosomes are enlisted in Table 4.5.

4.2.6 Ethosomes

Ethosomes are phospholipid-based elastic soft vesicular systems, having vehicles like water, glycolipids, phosphatidylserine, phosphatidylcholine, and phosphatidyl, as well as 30–45% ethanol and isopropyl alcohol. For both hydrophilic and lipophilic medicines, the ethosomes' proportion composition enhances their transdermal transport, topical drug delivery, and trapping capabilities. In addition to promoting blood circulation, they carry substances into deeper tissue. Flexible lecithin bilayers are what give ethosomes a physical stability advantage over liposomes. Ethosomes, on the

TABLE 4.6
Ethosomes Used with Herbal Drugs

S. No.	Ethosomes with Herbal Bioactives	Biological Origins	Results and Findings	References
1.	Apigenin-loaded ethosomes	Many fruits and vegetables	Potent anti-inflammatory effects of topical application	(Khaled *et al.*, 2014)
2.	*Berberis aristata* extract–loaded ethosomal gel	*Berberis aristata*	Transdermal delivery and improved permeability profiles	(Fatima *et al.*, 2017)
4.	Colchicine trans ethosomal gel	*Colchicum*	Enhanced skin diffusion, prolonged release, solubility, and stability	(Abdulbaqi *et al.*, 2018)
3.	Cryptotanshinone-loaded ethosomal gel	*Salvia miltiorrhiza*	Skin permeability, increased transdermal flux, increased anti-acne effectiveness, and decreased skin irritation	(Yu *et al.*, 2016)
5.	Piperine-loaded ethosomes	*Piper nigrum*	New drug carrier for atopic dermatitis therapy	(Kumar *et al.*, 2021, 2022)

other hand, have certain drawbacks, including less stability, a tendency to increase from nano- to micrometer in size as a result of alcohol evaporation, and the eventual leaking of entrapped material. The combination of alcohol with propylene glycol and trehalose was used for reducing this limitation. The development of "curcumin-encapsulated PEGylated and conventional liposomes and ethosomes" and evaluation in a rat model to see how well they could transport molecules that were entrapped to the skin was performed. PEGylated liposomes were subsequently demonstrated to be the most promising transdermal drug delivery technique, more effectively decreasing paw edema in the rat model (Abdulbaqi *et al.*, 2018; Zhao *et al.*, 2013). Ethosomes used with herbal drugs are enlisted in Table 4.6.

4.2.7 Cubosomes

These are thermodynamically highly stable self-assembled nanoparticles developed from lipid cubic phase and stabilized by a polymer-based outer corona, considered as biocompatible nanocarriers for drug delivery (Garg *et al.*, 2021). They have the ability to entrap both entrap both water-soluble and non-soluble molecules due to large internal surface area and a 3D structure with hydrophilic and hydrophobic domains. Because of their broad interfacial surface, biodegradable, bio-adhesive, and digestible lipid components, and release medication molecules that have been confined over time, they offer a range of diffusion paths (Tian *et al.*, 2017). In the liquid phase, they are usually produced by dispersing or breaking up the cubic stages of gel. The cubosomes have been used to successfully encapsulate drugs, such as indomethacin, somatostatin, insulin, and rifampicin. Furthermore, cubosomes have been examined for their potential use in peptides, antimuscarinic properties, enzymatic effects, antibiotics, and analgesic delivery (Azmi *et al.*, 2015). Cubosomes can easily release the bioactive that has been contained inside them into the epidermis since their structure is nearly matching to that of the stratum corneum. Additionally, cubosomes' increased adherence and penetration suggest their potential value in the treatment of skin cancer. In a work, polymer-free cubosomes were developed for photodynamic

TABLE 4.7
Cubosomes Encapsulating Herbal Bioactives

S. No.	Cubosomes with Herbal Bioactives	Biological origin	Limitations of free drugs	Results and findings	References
1.	*Achyranthes bidentata* extract–loaded cubosomes	*Achyranthes bidentata*	Low stability	Enhancement of stability and immunomodulatory impact	(Ou *et al.*, 2018)
2.	Capsaicin-loaded cubosomes	Plants of the *Capsicum* genus	Skin irritation	Improved stability in the presence of heat and light, reduced skin irritation,	(Peng *et al.*, 2015)
3.	Curcumin-loaded cubosomes	*Curcuma longa*	Low stability	Increased antibacterial capabilities, nanoscale vesicle formation, and improved stability	(Archana *et al.*, 2015)
4.	Piperine-loaded cubosomes	Fruits of the Piperaceae family	Low stability	Better anti-inflammatory, anti-apoptotic, and antioxidant	(Elnaggar *et al.*, 2015)

skin therapy and malignant skin cancers that can be bioimaging with very little damage to the cutaneous system (Bazylińska *et al.*, 2018). Examples of cubosomes that include herbal bioactives are provided in Table 4.7.

4.2.8 Metallic Nanoparticles

Metallic nanoparticles (MNPs) are the most popular inorganic nanoparticles and considered as an effective way to combat antibiotic resistance. They also target a wide range of biomolecules, which puts the development of resistant strains in danger (Slavin *et al.*, 2017). MNPs produced utilizing green techniques have gained appeal due to their physicochemical characteristics and applications in biotechnology. Due to their low toxicity in human body and minimally hazardous impact on the environment, green-synthesizing nanoparticles derived from plant extracts have recently attracted the attention of many researchers. Without the use of chemical agents, plant extracts have been used to reduce and stabilize the toxicity of nanoparticles in both environment and human body (Yadi *et al.*, 2022). The formulation of herbal-medicine-based nanoparticles in dentistry is a good example of significant advancement due to the ability of these plant-mediated biogenic metal nanoparticles to overcome the limitations of herbal treatments. The formulation of MNPs, including copper, silver, and gold nanoparticles, using medicinal plant extracts is more successful than traditional materials at treating a range of oral/dental illnesses. These can also be used in mouthwash, toothpaste, and other dental care products. More study is required since there is a knowledge gap regarding the safety of nanomaterials. Nanoparticles and a combination of plant extracts could be used to address a wide range of issues, including drug resistance. The safety, naturalness, and affordability of herbal extracts make them valuable in dental care. In this case, spreading knowledge about the benefits of using herbal remedies to prevent oral disorders will be very helpful. As these MNPs efficiently overcome herbal medicine's drawbacks, such as low water solubility and absorption, poor bioavailability, physical instability, and sluggish and harmful pharmacological activities, nanotechnology is anticipated to be applied in all dental applications (Minelli *et al.*, 2016). Examples of magnetic nanoparticles that include herbal bioactives are provided in Table 4.8.

TABLE 4.8
Magnetic Nanoparticles Encapsulating Natural Bioactives

S. No.	Nanoparticles with Herbal Bioactives	Biological Origins	Results and Findings	References
1.	Ag nanoparticles	*Camellia sinensis*	Ag-SiO_2 nanoparticles with potent antibacterial properties	(Rodrigues *et al.*, 2020)
2.	Ag nanoparticles	*Acacia senegal*	Ag nanoparticles' potent antibacterial action against *S. mutans* opens the way for novel dental care products to be developed.	(Al-Ansari *et al.*, 2021)
3.	Au Nanoparticles	*Anogeissus latifolia*	Bone-inducing and effective analgesics for implantation treatment	(Wang *et al.*, 2020)
4.	Au nanoparticles	*Salacia chinensis*	Using Au nanoparticles as a powerful bone-inducing agent during dental implant treatment	(Jadhav *et al.*, 2018)
5.	Cu nanoparticles	*Zingiber officinale*	Strong scavenging of free radicals' action, several pharmaceutical and dental applications, and copper nanoparticles created with *Zingiber* are all possible.	(Acosta *et al.*, 2012)
6.	Fe nanoparticles	Rose, *Azadirachta indica*, *Syzygium aromaticum*	Polyvinylpyrrolidone (PVP) is included to preserve its crystalline structure.	(Afsheen *et al.*, 2018)

4.2.9 Dendrimers

The majority of bioactive phytochemical components, including curcumin, carotenoids, numerous phenolic compounds, etc., are hydrophobic; it is impossible to synthesize them freely at the necessary dosage (Akhavan *et al.*, 2018). Furthermore, because of their poor solubility, roughly 15% of the phytochemical bioactive substances used perform sub-optimally. Reduced solubility and permeability of phytochemicals, resulting in low bioavailability (Assadpour *et al.*, 2017). The use of organic solvents and issues with their stability restricts their use. Common techniques to increase the bioavailability of these substances include altering their surface and shrinking their size. Phytochemical bioactive components' bioavailability has recently been effectively increased by the use of delivery methods (Abaee *et al.*, 2017). Dendrimers provide a novel method for dissolving these components due to its specific polyvalency and nanoencapsulation feature (Kulhari *et al.*, 2011). Dendrimers govern the release of loaded phytochemicals while shielding their contents from adverse conditions like enzyme breakdown, pH changes, etc. The use of dendrimers to nanoencapsulate various phytoconstituents is explored in the paragraphs that follow. Examples of dendrimers that include herbal bioactives are provided in Table 4.9.

4.2.10 Solid Lipid Nanoparticles

Solid lipid nanoparticles (SLNs) are nanosized (50 to 1,000 nm) spherical solid-core-containing lipid carriers that can incorporate both kind of drug (hydrophilic and hydrophobic). They are composed of biocompatible materials which make them a preferred type of drug delivery (Paliwal *et al.*, 2020). SLNs offered many advantages over other drug delivery such as low toxicity, higher bioavailability, and feasibility for large-scale production (Bayón-Cordero *et al.*, 2019). The main components of SLNs include solid lipid, emulsifier, drug, and solvent as required. Further surface modifiers are added for the formulation of a mucoadhesive and target specific drug delivery system

TABLE 4.9
Some Examples of Phytochemical-Loaded Dendrimers

S. No.	Phytochemical Bioactive Ingredient	Dendrimer Type	Dendrimer generation	Reference
1.	Capsaicin	PAMAM	—	(Yamauchi *et al.*, 2010)
2.	Crocetin	PAMAM and PPI	G4	(Soltani *et al.*, 2017)
3.	Folic acid	PAMAM	G5	(Dongen *et al.*, 2014)
4.	Glucosamine	PAMAM	G3.5	(Barata *et al.*, 2011)
5.	Isothiocyanate	PAMAM	G3	(Rezvani *et al.*, 2014)
6.	N-Acetylcysteine	PAMAM	G4	(Sharma *et al.*, 2018)
7.	Peptide amphiphiles	Unique dendrimer	G5	(Lin *et al.*, 2011)

(Duan *et al.*, 2020). For the formulation of SLNs, many interesting methods/techniques have been developed, such as high-pressure homogenization method, o/w microemulsion breaking technique, solvent injection method, solvent-emulsification diffusion technique, ultrasonication, w/o/w double emulsion method, membrane contractor technique, electrospray technique, and super critical fluid technique (Duan *et al.*, 2020; Ganesan & Narayanasamy, 2017). Furthermore, the types of SLNs depend on the nature of the solid and liquid lipids used. Based on that, SLNs are three types: imperfect type, structureless type, and multiple "oil in solid fat in water" types (Xu *et al.*, 2022). By SLNs, improved drug delivery at the target site can be achieved by three mechanisms: the first one is through a passive mechanism that takes the support of the tumor microenvironment, the second one is through an active mechanism that involves surface modification of SLNs, and the third one is through codelivery mechanisms in which more than one drug can be incorporated (Bayón-Cordero *et al.*, 2019). As SLNs offers so numerous advantages over other delivery system, the incorporation of phytomedicines in SLNs leads to the development of an innovative approach for the improvement of solubility, absorption, and permeation. These properties enhance the bioavailability and therapeutic properties of the drug (Thangadurai *et al.*, 2020). Kim *et al.* have formulated *Houttuynia cordata* extract SLNs for oral delivery by using Poloxamer188 and 407 through hot homogenization and ultrasonication method. They found that SLNs provide a sustained release pattern of quercitrin and 70% cell viability in Cacp-2 cells was observed (Kim *et al.*, 2017). Andrographolide, a diterpenoid obtained from *Andrographis paniculata* has been reported to have several activities such as oxidative stress-mediated neurotoxicity, cerebral ischemia, and inflammation-mediated neurodegeneration. But it has some limitations such as low hydrophobicity, bioavailability, and stability. SLNs of *Andrographis* were formulated by using Compritol 888 ATO and Brij 78 and evaluations showed that the developed SLNs have improved permeation capacity across the blood brain barrier as well as good physical and chemical stability (Graverini *et al.*, 2018). In another study, He *et al.* formulated curcumin- and dexanabinol-loaded SLNs by using low-temperature solidification and emulsification method. *In vitro* evaluations in PC 12 cells showed that the developed formulation induces greater dopamine release with reduced corticosterone-induced apoptotic cell death (He *et al.*, 2017). Furthermore, a study was conducted to determine the effect of SLNs on the chemical and biological efficacy of the *Eucalyptus globulus*. The evaluations showed that the formulation has decreased the minimum bactericidal concentration (MBC), minimum fungicidal concentration, and minimum inhibitory concentration (MIC) significantly (Azadmanesh *et al.*, 2021). An investigation was done for the development of dual layered SLNs of curcumin by using pectin and skimmed milk powder. The developed formulation showed a significantly higher entrapment efficacy with better stability and sustained release pattern of curcumin. The formulation could also improve oral bioavailability (up to 72 hours) on colon-targeted release (Mohamed *et al.*, 2021). Current research on phytomedicines bearing SLNs was enlisted in Table 4.10.

TABLE 4.10
Various Types of Phytomedicines Containing SLNs and Nanostructured Lipid Carriers

Phytomedicines	Ingredients	Methods of Preparation	Results	References
		SLNs		
Carvacrol	Cocoa butter, 3,5-Di-tert 4-butylhydroxytoluene, polysorbate 80	Fusion-emulsification method	Minimized oxidative stress and histological damages generated from smoke	(Carvalho *et al.*, 2020)
Curcumin	Myrj52, stearic acid, lecithin	Emulsification-evaporation method	Activation of Bcl-2 Family and P38 MAPK pathways	(Huang *et al.*, 2020)
Curcumin and piperine	TocopherylPEG succinate (TPGS) and Brij 78	Emulsification-evaporation low-temperature solidification method	Drug delivery to drug-resistant A2780/ Taxol cells.	(Tang *et al.*, 2017)
Genistein	Phosphatidylcholine, Tween 80	Hot homogenization process	Good stability and high drug-loading efficiency	(Obinu *et al.*, 2021)
β-sitosterol	Compritol 888 ATO, Phospholipid 90 G	Double-emulsion solvent-displacement method	Antiarthritic activity by suppression of NF-kB and activation of HO-1/ Nrf-2 pathway	(Zhang *et al.*, 2020)
		Nanostructured Lipid Carriers		
Berberine	Glyceryl behenate, olive oil	High-pressure homogenization	Inhibits NF-κB pathway in ulcerative colitis	(Deng *et al.*, 2020)
Epigallocatechin gallate	Arginyl-glycyl-aspartic acid	Hot homogenization technique	IC50 value on MDA-MB-231 cells was 38.2 μM	(Hajipour *et al.*, 2018)
Hesperidin and clarithromycin	Chinese wax, oleic acid, soybean lecithin	Hot homogenization and ultrasonication method	The most effective inhibition rate against H. pylori was 25 μM	(Sharaf *et al.*, 2021)
Silymarin	Cetyl palmitate, Lauroglycol 90, Brij S20	Emulsion/evaporation/ solidifying methodemulsion/ evaporation/solidifying method	Enhanced permeation and internalization with significant downregulation of blood glucose and triglyceride levels	(Piazzini *et al.*, 2019)

4.2.11 Poly-Lactic-Co-Glycolic Acid Nanoparticles

In the field of personalized nanomedicines, poly-lactic-co-glycolic acid (PLGA) nanocarriers are the most versatile drug delivery system. PLGA, an FDA-approved copolymer, is extensively used for the design and fabrication of drug delivery systems because of its biodegradability biocompatibility, biosafety, and versatility in the formulation. PLGA, when used in nanocarrier systems, such as nanoparticles of PLGA, protects drugs from premature degradation, provides a sustained release pattern with targeted delivery, and facilitates intracellular penetration (Ghitman *et al.*, 2020). Synthesis of PLGA occurs through a ring-opening copolymerization reaction. It is composed of two types of monomer units, they are poly lactic acid and poly glycolic acid (Hashemi *et al.*, 2020).

techniques (Konoshenko *et al.*, 2018). Plant-derived EVs are emerging as novel therapeutic tools. Cui et al. have isolated *Momordica charantia*–derived EVs by using the density gradient centrifugation technique and tested its protective action against radiation-induced heart disease. They found that isolated EVs can promote cell proliferation, suppress cell apoptosis, and also alleviate the DNA damage in irradiated H9C2 cells (Cui *et al.*, 2022). Another study on *Momordica charantia*–derived EVs has shown that the vesicles inhibit glioma proliferation, migration, and invasion by regulating PI3K/AKT signaling pathway in U251 glioma cells (Wang *et al.*, 2022). In another research, *Kaempferia parviflora* EVs were isolated through the sucrose density gradient centrifugation technique. Their studies showed that the isolated vesicles were able to internalize the adenocarcinoma gastric cells and also reduce cell viability in a dose-dependent manner (Nemidkanam & Chaichanawongsaroj, 2022). Furthermore, Jang et al. have studied the effect of *Codium fragile* and *Sargassum fusiform* EVs on MNT-1 human melanoma cells and they found that the EVs have significantly decreased α-MSH-mediated melanin synthesis as well as downregulated TRP1 (tyrosinase-related proteins 1), MITF (microphthalmia-associated transcription factor), and tyrosinase (Jang *et al.*, 2021). Research done on *Panax ginseng*–derived extracellular vesicles has revealed that the vesicles facilitate the anti-senescence effect by downregulating molecules associated with senescence and/or melanogenesis-related proteins in human skin cells (Cho *et al.*, 2021). A study was done by Kim *et al.* to evaluate the cytotoxic effects of EVs derived from plant sap namely *Dendropanax morbifera, Pinus densiflora, Thuja occidentalis,* and *Chamaecyparis obtuse*. They found that the EVs of *Dendropanax morbifera* and *Pinus densiflora* sap have a strong cytotoxic effect on tumor cells, but EVs of *Thuja occidentalis* and *Chamaecyparis obtuse* had no significant effect (Kim *et al.*, 2020). Aloe vera peels are reported to have antioxidant activity; therefore, Kim et al. isolated the EVs by ultracentrifugation and tangential flow filtration and studied their effect on wound healing. They found that isolated vesicles reduce the intracellular ROS level in HaCaT cells and ultimately activate the antioxidant defense mechanism and wound healing process through Nrf2 activation (Kim *et al.*, 2021).

4.2.14 Nano Polymersomes

Nano polymersomes are a self-assembled structure of amphiphilic block copolymers having polymeric shells of 5–50 nm thickness. Diblock (AB) and triblock (ABA, BAB, ABC) copolymers are generally employed for polymersome synthesis, in which A and C could be hydrophilic polymers and B could be hydrophobic ones. The block length of polymer and hydrophilic-to-hydrophobic block ratio governs the major properties of membranes, such as thickness, rigidity, permeability, and morphology (Lo & Zeng, 2022). The synthesis technique of polymersome is dependent on desired size, drug encapsulation efficiency, and reproducibility. Various synthesis techniques used are film rehydration, direct hydration, nanoprecipitation, double emulsification technique, microfluidic technique, and supercritical fluid technique (Kotha *et al.*, 2022). Furthermore, various types of stimulus-responsive polymersomes were developed, which are categorized as responsive to temperature, redox, pH, enzyme, photo, magnetic field, voltage, or electric field (Singh *et al.*, 2022). For the targeted delivery of phytomedicines, polymersomes are the best suitable carriers, as an investigation done by Goyal et al. showed that bacoside-loaded targeted lactoferrin conjugated tri-block PEG-PLA-PCL-OH polymersomes were more efficient in restoring scopolamine-induced impaired memory than its natural form (Goyal *et al.*, 2021). In another study, curcumin-loaded polymersomes showed affinity toward neurons with neuroprotective activity and improved cognitive impairment (Jia *et al.*, 2016). The phytomedicine docetaxel is widely used for the treatment of different types of cancer, but it has dose-limiting side effects, such as neurotoxicity and myelosuppression. Therefore, Zou and their team have formulated cyclic cNGQGEQc-peptide-directed polymersome docetaxel as targeted therapy for non-small-cell lung cancer. The developed formulation has better antitumor activity in α3β1 integrin overexpressing A549 human lung cancer cells than free and nontargeted docetaxel (Zou *et al.*, 2020).

4.2.15 Polymeric Nanoparticles

Polymeric nanoparticles (PNPs) have gained a large interest among researchers in the last few decades owing to their unique small size and properties. They have a vast range of applications both in the treatment and diagnosis of disease. PNPs offer certain advantages, such as the controlled release of the drug, and they can combine both in targeting and in imaging and protect the drug from degradation (Pandey *et al.*, 2022). Targeting through PNPs facilitates improvement in the therapeutic index (Haider T. *et al.*, 2020; Crucho & Barros, 2017). PNPs are collectively used for many types of nanoparticles but mainly include nanospheres and nanocapsules. Nanospheres are a solid mass of matrix polymers in which drug is entrapped. Nanocapsules are drug reservoirs in which drug retains in an aqueous or non-aqueous core of a vesicle cavity, enclosed by the solid polymeric shell (Khalid & El-Sawy, 2017). Polymers used for the synthesis of PNPs are categorized into two types based on their source: natural and synthetic polymers. Examples of natural polymers are silk fibroin, chitosan, alginate, gelatin, and albumin, while synthetic polymers include polylactides, polyglycolide, polyglutamic acid, polyethylene glycol, etc. (Nunes *et al.*, 2022; Baghel & Bhattacharya, 2021; Pandey *et al.*, 2020). The methods of preparation for PNPs depend upon the nature of the drug to be loaded and route of administration. Two main strategies are generally used which are either dispersion of preformulated polymers or polymerization of monomers. Different techniques for the preparation of nanoparticles are solvent evaporation method, emulsification/solvent diffusion method, emulsification/reverse salting-out method, and nanoprecipitation methods (Zielińska *et al.*, 2020; Haider & Soni, 2022). PNPs are extensively investigated by researchers for the targeted and safer delivery of phytomedicines. Tanshinone IIA has been reported to have anticancer activity, but it suffers from poor oral bioavailability and extensive first-pass metabolism. Therefore, Ashour et al. have encapsulated tanshinone IIA into liquid nanocapsules and evaluated its *in vitro* activity as well as oral pharmacokinetic property in rats. They found a 3.6-fold increase in the $AUC_{0\text{-inf}}$ value of the developed formulation (Ashour *et al.*, 2020). Oregano-essential-oil-based polycaprolactone nanocapsules were formulated by using polycaprolactone as a polymer through nanoprecipitation and double emulsion methods. The developed nanocapsules were found to have good stability over long-term storage conditions (Fraj *et al.*, 2019). Various phytomedicinal polymeric nanoparticles are enlisted in Table 4.11.

TABLE 4.11
Phytomedicines Containing Polymeric Nanocarriers (Nanocapsules)

Phytomedicine	Formulation	Polymer Used	Activity	Reference
Achyrocline satureioides essential oil	Nanocapsules	Polycaprolactone	Reduces cytotoxic damage in the liver of rats	(Ritter *et al.*, 2017)
Curcumin-polyallyhydrocarbon	Nanocapsules	Polyallyhydrocarbon	Inhibits Wnt/β-catenin pathway in colorectal cancer	(Al Moubarak *et al.*, 2021)
Docetaxel and thymoquinone	Nanocapsules	Chitosan	Improves uptake and endosomal escape effect	(Fraj *et al.*, 2019)
Linalool	Nanocapsules	Chitosan	Sustains release pattern obtain	(Xiao *et al.*, 2017)
Lychnopholide	Nanocapsules	Poly(d,l-lactide)-block-polyethylene glycol	Better result in cure of Chagas disease	(Branquinho *et al.*, 2020)
Matricaria chamomilla essential oil	Nanocapsules	Chitosan	IC_{50} values against promastigotes and amastigotes are 3.33 µg/mL and 14.56 µg/mL respectively.	(Karam *et al.*, 2020)
Origanum vulgare essential oil	Nanocapsules/ microspheres	Polycaprolactone	Stable over a wide range of temperature	(Fraj *et al.*, 2019)

Calister U, Lydia U, Edith D, Chioma N, Ben A, Andrew E, Nwokedi U. Formulation and evaluation of Morinda lucida based-phytosome complexes for malaria treatment. Technology. 2020;6(9).

Cao C, Wang Q, Liu Y. Lung cancer combination therapy: Doxorubicin and β-elemene co-loaded, pH-sensitive nanostructured lipid carriers. Drug Design, Development and Therapy. 2019;13:1087.

Caruso S, Poon IK. Apoptotic cell-derived extracellular vesicles: More than just debris. Frontiers in Immunology. 2018;9:1486.

Carvalho FO, Silva ER, Nunes PS, Felipe FA, Ramos KP, Ferreira LAS, Lima VN, Shanmugam S, Oliveira AS, Guterres SS. Effects of the solid lipid nanoparticle of carvacrol on rodents with lung injury from smoke inhalation. Naunyn-Schmiedeberg's Archives of Pharmacology. 2020;393(3):445–455.

Casula L, Lai F, Pini E, Valenti D, Sinico C, Cardia MC, Marceddu S, Ailuno G, Fadda AM. Pulmonary delivery of curcumin and beclomethasone dipropionate in a multicomponent nanosuspension for the treatment of bronchial asthma. Pharmaceutics. 2021;13(8):1300.

Chatterjee M, Chanda N. Formulation of PLGA nano-carrier: Specialized modification for cancer therapeutic applications. Materials Advances. 2022;3:837–858.

Chauhan I, Yasir M, Verma M, Singh AP. Nanostructured lipid carriers: A groundbreaking approach for transdermal drug delivery. Advanced Pharmaceutical Bulletin. 2020;10(2):150.

Chen L, Nasab EM, Athari SS. Effect of loaded glycyrrhizic acid on PLGA nano-particle on treatment of allergic asthma. Iranian Journal of Allergy, Asthma and Immunology. 2022;21(1):65.

Cho EG, Choi SY, Kim H, Choi EJ, Lee EJ, Park PJ, Ko J, Kim KP, Baek HS. Panax ginseng-derived extracellular vesicles facilitate anti-senescence effects in human skin cells: An eco-friendly and sustainable way to use ginseng substances. Cells. 2021;10(3):486.

Chouhan A, Shah SK, Tyagi C, Budholiya P, Wazid A, Khan F. Formulation, development and evaluation of phytosomes of swertia perennis. Asian Journal of Pharmaceutical Education and Research. 2021;10:55–56.

Cossetin LF, Garlet QI, Velho MC, Gündel S, Ourique AF, Heinzmann BM, Monteiro SG. Development of nanoemulsions containing Lavandula dentata or Myristica fragrans essential oils: Influence of temperature and storage period on physical-chemical properties and chemical stability. Industrial Crops and Products. 2021;160:113115.

Crucho CI, Barros MT. Polymeric nanoparticles: A study on the preparation variables and characterization methods. Materials Science and Engineering: C. 2017;80:771–784.

Cui WW, Ye C, Wang KX, Yang X, Zhu PY, Hu K, Lan T, Huang LY, Wang W, Gu B. Momordica. charantia-derived extracellular vesicles-like nanovesicles protect cardiomyocytes against radiation injury via attenuating DNA damage and mitochondria dysfunction. Frontiers in Cardiovascular Medicine. 2022;9:864188.

Deng J, Wu Z, Zhao Z, Wu C, Yuan M, Su Z, Wang Y, Wang Z. Berberine-loaded nanostructured lipid carriers enhance the treatment of ulcerative colitis. International Journal of Nanomedicine. 2020;15:3937.

Dong Z, Iqbal S, Zhao Z. Preparation of ergosterol-loaded nanostructured lipid carriers for enhancing oral bioavailability and antidiabetic nephropathy effects. AAPS Pharm Sci Tech. 2020;21(2):1–11.

Du J, Li X, Zhao H, Zhou Y, Wang L, Tian S, Wang Y. Nanosuspensions of poorly water-soluble drugs prepared by bottom-up technologies. International Journal of Pharmaceutics. 2015;495(2):738–749.

Duan Y, Dhar A, Patel C, Khimani M, Neogi S, Sharma P, Kumar NS, Vekariya RL. A brief review on solid lipid nanoparticles: Part and parcel of contemporary drug delivery systems. RSC Advances. 2020;10(45):26777–26791.

Ebrahimian M, Shahgordi S, Yazdian-Robati R, Etemad L, Hashemi M, Salmasi Z. Targeted delivery of galbanic acid to colon cancer cells by PLGA nanoparticles incorporated into human mesenchymal stem cells. Avicenna Journal of Phytomedicine. 2022;12(3).

El-Batal AI, Elmenshawi SF, Ali AM, Goodha E. Preparation and characterization of silymarin nanocrystals and phytosomes with investigation of their stability using gamma irradiation. Indian Journal of Pharmaceutical Education and Research. 2018;52:S174–183.

El-Menshawe SF, Ali AA, Rabeh MA, Khalil NM. Nanosized soy phytosome-based thermogel as topical anti-obesity formulation: An approach for acceptable level of evidence of an effective novel herbal weight loss product. International Journal of Nanomedicine. 2018;13:307.

Elnaggar YS, Etman SM, Abdelmonsif DA, Abdallah OY. Novel piperine-loaded Tween-integrated monoolein cubosomes as brain-targeted oral nanomedicine in Alzheimer's disease: Pharmaceutical, biological, and toxicological studies. International Journal of Nanomedicine. 2015;10:5459.

Elsharkasy OM, Nordin JZ, Hagey DW, de Jong OG, Schiffelers RM, Andaloussi SE, Vader P. Extracellular vesicles as drug delivery systems: Why and how? Advanced Drug Delivery Reviews. 2020;159:332–343.

Farahpour MR, Hamishehkar H. Effectiveness of topical caraway essential oil loaded into nanostructured lipid carrier as a promising platform for the treatment of infected wounds. Colloids and Surfaces A: Physicochemical and Engineering Aspects. 2021;610:125748.

Fatima Z. Formulation and performance evaluation of berberis aristata extract loaded ethosomal gel. Asian Journal of Pharmaceutics (AJP). 2017;11(03).

Fraj A, Jaâfar F, Marti M, Coderch L, Ladhari N. A comparative study of oregano (Origanum vulgare L.) essential oil-based polycaprolactone nanocapsules/microspheres: Preparation, physicochemical characterization, and storage stability. Industrial Crops and Products. 2019;140:111669.

Gahandule MB, Jadhav SJ, Gadhave MV, Gaikwad DD. Formulation and development of hepato-protective butea monosperma-phytosome. International Journal of Research Pharmacy and Pharmaceutical Sciences. 2016;1(4):21–27.

Ganesan P, Narayanasamy D. Lipid nanoparticles: Different preparation techniques, characterization, hurdles, and strategies for the production of solid lipid nanoparticles and nanostructured lipid carriers for oral drug delivery. Sustainable Chemistry and Pharmacy. 2017;6:37–56.

Garg M, Goyal A, Kumari S. An update on the recent advances in cubosome: A novel drug delivery system. Current Drug Metabolism. 2021;22(6):441–450.

Ghitman J, Biru EI, Stan R, Iovu H. Review of hybrid PLGA nanoparticles: Future of smart drug delivery and theranostics medicine. Materials & Design. 2020;193:108805.

Ghosh V, Mukherjee A, Chandrasekaran N. Ultrasonic emulsification of food-grade nanoemulsion formulation and evaluation of its bactericidal activity. Ultrasonics Sonochemistry. 2013;20(1):338–344.

Goyal K, Kumar A, Sharma S, Konar A, Koul V. Evaluation of bacosides and lauric acid nanoformulations in regulation of neuronal and epigenetic markers in primary culture of mouse cortical neurons: A comparative study to rivastigmine, a synthetic analog. Trends in Biomaterials & Artificial Organs. 2021;35(1): 48–56.

Graverini G, Piazzini V, Landucci E, Pantano D, Nardiello P, Casamenti F, Pellegrini-Giampietro DE, Bilia AR, Bergonzi MC. Solid lipid nanoparticles for delivery of andrographolide across the blood-brain barrier: In vitro and in vivo evaluation. Colloids and Surfaces B: Biointerfaces. 2018;161:302–313.

Gupta V, Ramakanth D, Verma C, Maji PK, Gaikwad KK. Isolation and characterization of cellulose nanocrystals from amla (Phyllanthus emblica) pomace. Biomass Conversion and Biorefinery. 2021:1–2.

Haider M, Abdin SM, Kamal L, Orive G. Nanostructured lipid carriers for delivery of chemotherapeutics: A review. Pharmaceutics. 2020;12(3):288.

Haider T, Pandey V, Behera C, Kumar P, Gupta PN, Soni V. Spectrin conjugated PLGA nanoparticles for potential membrane phospholipid interactions: Development, optimization and in vitro studies. Journal of Drug Delivery Science and Technology. 2020;60:102087.

Haider T, Soni V. Response surface methodology and artificial neural network-based modeling and optimization of phosphatidylserine targeted nanocarriers for effective treatment of cancer: In vitro and in silico studies. Journal of Drug Delivery Science and Technology. 2022;75:103663.

Hajipour H, Hamishehkar H, Nazari Soltan Ahmad S, Barghi S, Maroufi NF, Taheri RA. Improved anticancer effects of epigallocatechin gallate using RGD-containing nanostructured lipid carriers. Artificial Cells, Nanomedicine, and Biotechnology. 2018;46(sup1):283–292.

Hashemi M, Shamshiri A, Saeedi M, Tayebi L, Yazdian-Robati R. Aptamer-conjugated PLGA nanoparticles for delivery and imaging of cancer therapeutic drugs. Archives of Biochemistry and Biophysics. 2020;691:108485.

Hassanshahian M, Saadatfar A, Masoumipour F. Formulation and characterization of nanoemulsion from Alhagi maurorum essential oil and study of its antimicrobial, antibiofilm, and plasmid curing activity against antibiotic-resistant pathogenic bacteria. Journal of Environmental Health Science and Engineering. 2020;18(2):1015–1027.

He X, Yang L, Wang M, Zhuang X, Huang R, Zhu R, Wang S. Targeting the endocannabinoid/CB1 receptor system for treating major depression through antidepressant activities of curcumin and dexanabinol-loaded solid lipid nanoparticles. Cellular Physiology and Biochemistry. 2017;42(6):2281–2294.

Hessvik NP, Llorente A. Current knowledge on exosome biogenesis and release. Cellular and Molecular Life Sciences. 2018;75(2):193–208.

Hong SS, Choi JY, Kim JO, Lee MK, Kim SH, Lim SJ. Development of paclitaxel-loaded liposomal nanocarrier stabilized by triglyceride incorporation. International Journal of Nanomedicine. 2016;11:4465.

Hooresfand Z, Ghanbarzadeh S, Hamishehkar H. Preparation and characterization of rutin-loaded nanophytosomes. Pharmaceutical Sciences. 2015;21(3):145–151.

Huang R, Zhu Y, Lin L, Song S, Cheng L, Zhu R. Solid lipid nanoparticles enhanced the neuroprotective role of curcumin against epilepsy through activation of Bcl-2 family and P38 MAPK pathways. ACS Chemical Neuroscience. 2020;11(13):1985–1995.

Iadnut A, Mamoon K, Thammasit P, Pawichai S, Tima S, Preechasuth K, Kaewkod T, Tragoolpua Y, Tragoolpua K. In vitro antifungal and antivirulence activities of biologically synthesized ethanolic extract of propolis-loaded PLGA nanoparticles against Candida albicans. Evidence-Based Complementary and Alternative Medicine. 2019;2019.

Jadhav K, Rajeshwari HR, Deshpande S, Jagwani S, Dhamecha D, Jalalpure S, Subburayan K, Baheti D. Phytosynthesis of gold nanoparticles: Characterization, biocompatibility, and evaluation of its osteo-inductive potential for application in implant dentistry. Materials Science and Engineering: C. 2018;93:664–670.

Jadia R, Kydd J, Piel B, Rai P. Liposomes aid curcumin's combat with cancer in a breast tumor model. Oncomedicine. 2018;3:94–109.

Jang B, Chung H, Jung H, Song HK, Park E, Choi HS, Jung K, Choe H, Yang S, Oh ES. Extracellular vesicles from Korean Codium fragile and Sargassum fusiforme negatively regulate melanin synthesis. Molecules and Cells. 2021;44(10):736.

Jia T, Sun Z, Lu Y, Gao J, Zou H, Xie F, Zhang G, Xu H, Sun D, Yu Y. A dual brain-targeting curcumin-loaded polymersomes ameliorated cognitive dysfunction in intrahippocampal amyloid-β1–42-injected mice. International Journal of Nanomedicine. 2016;11:3765.

Junyaprasert VB, Morakul B. Nanocrystals for enhancement of oral bioavailability of poorly water-soluble drugs. Asian Journal of Pharmaceutical Sciences. 2015;10(1):13–23.

Kalra H, Drummen GP, Mathivanan S. Focus on extracellular vesicles: Introducing the next small big thing. International Journal of Molecular Sciences. 2016;17(2):170.

Kao CY, Papoutsakis ET. Extracellular vesicles: Exosomes, microparticles, their parts, and their targets to enable their biomanufacturing and clinical applications. Current Opinion in Biotechnology. 2019;60:89–98.

Karakucuk A, Tort S. Preparation, characterization and antimicrobial activity evaluation of electrospun PCL nanofiber composites of resveratrol nanocrystals. Pharmaceutical Development and Technology. 2020;25(10):1216–1225.

Karam TK, Ortega S, Nakamura TU, Auzély-Velty R, Nakamura CV. Development of chitosan nanocapsules containing essential oil of Matricaria chamomilla L. for the treatment of cutaneous leishmaniasis. International Journal of Biological Macromolecules. 2020;162:199–208.

Karimi N, Ghanbarzadeh B, Hamishehkar H, Mehramuz B, Kafil HS. Antioxidant, antimicrobial and physicochemical properties of turmeric extract-loaded nanostructured lipid carrier (NLC). Colloid and Interface Science Communications. 2018;22:18–24.

Katopodi A, Detsi A. Solid lipid nanoparticles and nanostructured lipid carriers of natural products as promising systems for their bioactivity enhancement: The case of essential oils and flavonoids. Colloids and Surfaces A: Physicochemical and Engineering Aspects. 2021;630:127529.

Kaur H, Mohanta GC, Gupta V, Kukkar D, Tyagi S. Synthesis and characterization of ZIF-8 nanoparticles for controlled release of 6-mercaptopurine drug. Journal of Drug Delivery Science and Technology. 2017;41:106–112.

Khaled SA, Burley JC, Alexander MR, Roberts CJ. Desktop 3D printing of controlled release pharmaceutical bilayer tablets. International Journal of Pharmaceutics. 2014;461(1–2):105–111.

Khalid M, El-Sawy HS. Polymeric nanoparticles: Promising platform for drug delivery. International Journal of Pharmaceutics. 2017;528(1–2):675–691.

Khani S, Keyhanfar F, Amani A. Design and evaluation of oral nanoemulsion drug delivery system of mebudipine. Drug Delivery. 2016 July 23;23(6):2035–2043.

Khosa A, Reddi S, Saha RN. Nanostructured lipid carriers for site-specific drug delivery. Biomedicine & Pharmacotherapy. 2018;103:598–613.

Kildaci I, Budama-Kilinc Y, Kecel-Gunduz S, Altuntas E. Linseed oil nanoemulsions for treatment of atopic dermatitis disease: Formulation, characterization, in vitro and in silico evaluations. Journal of Drug Delivery Science and Technology. 2021;64:102652.

Kim JH, Baek JS, Park JK, Lee BJ, Kim MS, Hwang SJ, Lee JY, Cho CW. Development of houttuynia cordata extract-loaded solid lipid nanoparticles for oral delivery: High drug loading efficiency and controlled release. Molecules. 2017;22(12):2215.

Kim K, Yoo HJ, Jung JH, Lee R, Hyun JK, Park JH, Na D, Yeon JH. Cytotoxic effects of plant sap-derived extracellular vesicles on various tumor cell types. Journal of Functional Biomaterials. 2020;11(2):22.

Kim MK, Choi YC, Cho SH, Choi JS, Cho YW. The antioxidant effect of small extracellular vesicles derived from aloe vera peels for wound healing. Tissue Engineering and Regenerative Medicine. 2021; 18(4):561–571.

Kongkaneramit L, Aiemsum-ang P, Kewsuwan P. Development of curcumin liposome formulations using polyol dilution method. Songklanakarin Journal of Science & Technology. 2016;38(6).

Konoshenko MY, Lekchnov EA, Vlassov AV, Laktionov PP. Isolation of extracellular vesicles: General methodologies and latest trends. BioMed Research International. 2018;2018.

Kotha R, Kara DD, Roychowdhury R, Tanvi K, Rathnanand M. Polymersomes based versatile nanoplatforms for controlled drug delivery and imaging. Advanced Pharmaceutical Bulletin. 2022; 13(2):218–232.

Kulhari H, Pooja D, Prajapati SK, Chauhan AS. Performance evaluation of PAMAM dendrimer based simvastatin formulations. International Journal of Pharmaceutics. 2011;405(1–2):203–209.

Kumar P, Sharma DK, Ashawat MS. Development of phospholipids vesicular nanocarrier for topical delivery of tea tree oil in management of atopic dermatitis using BALB/c mice model. European Journal of Lipid Science and Technology. 2021;123(10):2100002.

Kumar P, Sharma DK, Ashawat MS. Topical creams of piperine loaded lipid nanocarriers for management of atopic dermatitis: Development, characterization, and in vivo investigation using BALB/c mice model. Journal of Liposome Research. 2022;32(1):62–73.

Li N, Feng L, Tan Y, Xiang Y, Zhang R, Yang M. Preparation, characterization, pharmacokinetics and biodistribution of baicalin-loaded liposome on cerebral ischemia-reperfusion after iv administration in rats. Molecules. 2018;23(7):1747.

Lin BF, Marullo RS, Robb MJ, Krogstad DV, Antoni P, Hawker CJ, Campos LM, Tirrell MV. De novo design of bioactive protein-resembling nanospheres via dendrimer-templated peptide amphiphile assembly. Nano Letters. 2011;11(9):3946–3950.

Lo CH, Zeng J. Application of polymersomes in membrane protein study and drug discovery: Progress, strategies, and perspectives. Bioengineering & Translational Medicine. 2022:e10350.

Lovelyn C, Attama AA. Current state of nanoemulsions in drug delivery. Journal of Biomaterials and Nanobiotechnology. 2011;2(05):626.

Mahato R. Nanoemulsion as targeted drug delivery system for cancer therapeutics. Journal of Pharmaceutical Sciences and Pharmacology. 2017;3(2):83–97.

Majidzadeh H, Araj-Khodaei M, Ghaffari M, Torbati M, Dolatabadi JEN, Hamblin MR. Nano-based delivery systems for berberine: A modern anti-cancer herbal medicine. Colloids and Surfaces B: Biointerfaces. 2020;194:111188.

Minelli C. Talk on 'Measuring nanoparticle properties: Are we high and dry or all at sea?' at 'Nanoparticle characterisation–challenges for the community' Event–IOP (Institute of Physics), book of abstracts, July 2016, London.

Moazeni M, Davari A, Shabanzadeh S, Akhtari J, Saeedi M, Mortyeza-Semnani K, Abastabar M, Nabili M, Moghadam FH, Roohi B, Kelidari H. In vitro antifungal activity of Thymus vulgaris essential oil nanoemulsion. Journal of Herbal Medicine. 2021;28:100452.

Mohamed JM, Alqahtani A, Ahmad F, Krishnaraju V, Kalpana K. Pectin co-functionalized dual layered solid lipid nanoparticle made by soluble curcumin for the targeted potential treatment of colorectal cancer. Carbohydrate Polymers. 2021;252:117180.

Mohammadifar M, Aarabi MH, Aghighi F, Kazemi M, Vakili Z, Memarzadeh MR, Talaei SA. Anti-osteoarthritis potential of peppermint and rosemary essential oils in a nanoemulsion form: Behavioral, biochemical, and histopathological evidence. BMC Complementary Medicine and Therapies. 2021;21(1):1–2.

Murugesan MP, Ratnam MV, Mengitsu Y, Kandasamy K. Evaluation of anti-cancer activity of phytosomes formulated from aloe vera extract. Materials Today: Proceedings. 2021;42:631–636.

Nagalakshmi S, Krishnaraj K, Jothy AM, Chaudhari PS, Pushpalatha HB, Shanmuganthan S. Fabrication and characterization of herbal drug-loaded nonionic surfactant based niosomal topical gel. Journal of Pharmaceutical Sciences and Research. 2016;8(11):1271.

Nazeer AA, Rajan HV, Vijaykumar SD, Saravanan M. Evaluation of larvicidal and repellent activity of nanocrystal emulsion synthesized from F. glomerata and neem oil against mosquitoes. Journal of Cluster Science. 2019;30(6):1649–1661.

Nemidkanam V, Chaichanawongsaroj N. Characterizing Kaempferia parviflora extracellular vesicles, a nanomedicine candidate. Plos One. 2022;17(1):e0262884.

Ni R, Zhao J, Liu Q, Liang Z, Muenster U, Mao S. Nanocrystals embedded in chitosan-based respirable swellable microparticles as dry powder for sustained pulmonary drug delivery. European Journal of Pharmaceutical Sciences. 2017;99:137–146.

Noor NM, Sheikh K, Somavarapu S, Taylor KM. Preparation and characterization of dutasteride-loaded nanostructured lipid carriers coated with stearic acid-chitosan oligomer for topical delivery. European Journal of Pharmaceutics and Biopharmaceutics. 2017;117:372–384.

Nunes D, Andrade S, Ramalho MJ, Loureiro JA, Pereira MC. Polymeric nanoparticles-loaded hydrogels for biomedical applications: A systematic review on in vivo findings. Polymers. 2022;14(5):1010.

Obinu A, Burrai GP, Cavalli R, Galleri G, Migheli R, Antuofermo E, Rassu G, Gavini E, Giunchedi P. Transmucosal solid lipid nanoparticles to improve genistein absorption via intestinal lymphatic transport. Pharmaceutics. 2021;13(2):267.

Ou N, Sun Y, Zhou S, Gu P, Liu Z, Bo R, Hu Y, Liu J, Wang D. Evaluation of optimum conditions for Achyranthes bidentata polysaccharides encapsulated in cubosomes and immunological activity in vitro. International Journal of Biological Macromolecules. 2018;109:748–760.

Paliwal R, Paliwal SR, Kenwat R, Kurmi BD, Sahu MK. Solid lipid nanoparticles: A review on recent perspectives and patents. Expert Opinion on Therapeutic Patents. 2020;30(3):179–194.

Paliwal S, Sharma J, Tilak A, Sohgaura A, Gupta C, Dave V. Synergistic effect of guggul liposomes lipid nanovesicles for the treatment of inflammation. International Journal of Life Sciences and Biotechnology. 2011;6:1–3.

Pandey V, Haider T, Agrawal P, Soni S, Soni V. Advances in natural polymeric nanoparticles for the drug delivery. In Advanced Drug Delivery Systems, 2022 Oct 31. IntechOpen.

Pandey V, Haider T, Jain P, Gupta PN, Soni V. Silk as a leading-edge biological macromolecule for improved drug delivery. Journal of Drug Delivery Science and Technology. 2020;55:101294.

Pardhi VP, Verma T, Flora SJ, Chandasana H, Shukla R. Nanocrystals: An overview of fabrication, characterization and therapeutic applications in drug delivery. Current Pharmaceutical Design. 2018;24(43):5129–5146.

Peng X, Zhou Y, Han K, Qin L, Dian L, Li G, Pan X, Wu C. Characterization of cubosomes as a targeted and sustained transdermal delivery system for capsaicin. Drug Design, Development and Therapy. 2015;9:4209.

Pengon S, Chinatangkul N, Limmatvapirat C, Limmatvapirat S. Development of Antimicrobial Nanoemulsions Containing Nelumbo nucifera Extract. In Key Engineering Materials, 2020 (Vol. 859, pp. 226–231). Trans Tech Publications Ltd.

Piazzini V, Micheli L, Luceri C, D'Ambrosio M, Cinci L, Ghelardini C, Bilia AR, Mannelli LDC, Bergonzi MC. Nanostructured lipid carriers for oral delivery of silymarin: Improving its absorption and in vivo efficacy in type 2 diabetes and metabolic syndrome model. International Journal of Pharmaceutics. 2019;572:118838.

Plangsombat N, Rungsardthong K, Kongkaneramit L, Waranuch N, Sarisuta N. Anti-inflammatory activity of liposomes of Asparagus racemosus root extracts prepared by various methods. Experimental and Therapeutic Medicine. 2016;12(4):2790–2796.

Qin XQ, Yuan Y, Liu CS, Wang QY, Shen X, Yang BC. Preparation of liposomal brucine and its pharmaceutical/pharmacodynamic characterization. Acta Pharmacologica Sinica. 2007;28(11):1851–1858.

Rameshk M, Sharififar F, Mehrabani M, Pardakhty A, Farsinejad A, Mehrabani M. Proliferation and in vitro wound healing effects of the microniosomes containing Narcissus tazetta L. bulb extract on primary human fibroblasts (HDFs). DARU Journal of Pharmaceutical Sciences. 2018;26(1):31–42.

Rezvani Alanagh H, Khosroshahi ME, Tajabadi M, Keshvari H. The effect of pH and magnetic field on the fluorescence spectra of fluorescein isothiocyanate conjugated SPION-dendrimer nanocomposites. Journal of Superconductivity and Novel Magnetism. 2014;27(10):2337–2345.

Ritter CS, Baldissera MD, Grando TH, Souza CF, Sagrillo MR, da Silva APT, Moresco RN, Guarda NS, da Silva AS, Stefani LM. Achyrocline satureioides essential oil-loaded in nanocapsules reduces cytotoxic damage in liver of rats infected by Trypanosoma evansi. Microbial Pathogenesis. 2017;103:149–154.

Robson AL, Dastoor PC, Flynn J, Palmer W, Martin A, Smith DW, Woldu A, Hua S. Advantages and limitations of current imaging techniques for characterizing liposome morphology. Frontiers in Pharmacology. 2018;9:80.

Rodrigues MC, Rolim WR, Viana MM, Souza TR, Gonçalves F, Tanaka CJ, Bueno-Silva B, Seabra AB. Biogenic synthesis and antimicrobial activity of silica-coated silver nanoparticles for esthetic dental applications. Journal of Dentistry. 2020;96:103327.

Rohilla R, Garg T, Goyal AK, Rath G. Herbal and polymeric approaches for liver-targeting drug delivery: Novel strategies and their significance. Drug Delivery. 2016;23(5):1645–1661.

Scognamiglio I, De Stefano D, Campani V, Mayol L, Carnuccio R, Fabbrocini G, Ayala F, La Rotonda MI, De Rosa G. Nanocarriers for topical administration of resveratrol: A comparative study. International Journal of Pharmaceutics. 2013;440(2):179–187.

Sharaf M, Arif M, Khan S, Abdalla M, Shabana S, Chi Z, Liu C. Co-delivery of hesperidin and clarithromycin in a nanostructured lipid carrier for the eradication of Helicobacter pylori in vitro. Bioorganic Chemistry. 2021;112:104896.

Sharma R, Sharma A, Kambhampati SP, Reddy RR, Zhang Z, Cleland JL, Kannan S, Kannan RM. Scalable synthesis and validation of PAMAM dendrimer-N-acetyl cysteine conjugate for potential translation. Bioengineering & Translational Medicine. 2018;3(2):87–101.
Singh AS, Vengurlekar PR, Rathod SU. Design, development and characterization of liposomal neem gel. International Journal of Pharmaceutical Sciences and Research. 2014;5:140–148.
Singh HP, Utreja P, Tiwary AK, Jain S. Elastic liposomal formulation for sustained delivery of colchicine: In vitro characterization and in vivo evaluation of anti-gout activity. The AAPS Journal. 2009;11(1):54–64.
Singh RP, Gangadharappa HV, Mruthunjaya K. Phytosome complexed with chitosan for gingerol delivery in the treatment of respiratory infection: In vitro and in vivo evaluation. European Journal of Pharmaceutical Sciences. 2018;122:214–229.
Singh V, Md S, Alhakamy NA, Kesharwani P. Taxanes loaded polymersomes as an emerging polymeric nanocarrier for cancer therapy. European Polymer Journal. 2022;162:110883.
Slavin YN, Asnis J, Häfeli UO, Bach H. Metal nanoparticles: Understanding the mechanisms behind antibacterial activity. Journal of Nanobiotechnology. 2017;15(1):1–20.
Soltani F, Ramezani M, AmelFarzad S, Mokhtarzadeh A, Hashemi M. Comparison study of the effect of alkyl-modified and unmodified PAMAM and PPI dendrimers on solubility and antitumor activity of crocetin. Artificial Cells, Nanomedicine, and Biotechnology. 2017;45(7):1356–1362.
Song Z, Lin Y, Zhang X, Feng C, Lu Y, Gao Y, Dong C. Cyclic RGD peptide-modified liposomal drug delivery system for targeted oral apatinib administration: Enhanced cellular uptake and improved therapeutic effects. International Journal of Nanomedicine. 2017;12:1941.
Soni DK, Jain P, Pandey V, Dixit VK, Soni V. Development and characterization of novel vesicular system using purpurin-phosphatidyl choline as a skin antiaging agent. Asian Journal of Pharmaceutics. 2018;12(04).
Souto EB, Baldim I, Oliveira WP, Rao R, Yadav N, Gama FM, Mahant S. SLN and NLC for topical, dermal, and transdermal drug delivery. Expert Opinion on Drug Delivery. 2020;17(3):357–377.
Tang J, Ji H, Ren J, Li M, Zheng N, Wu L. Solid lipid nanoparticles with TPGS and Brij 78: A co-delivery vehicle of curcumin and piperine for reversing P-glycoprotein-mediated multidrug resistance in vitro. Oncology Letters. 2017;13(1):389–395.
Thangadurai D, Sangeetha J, Prasad R. Functional Bionanomaterials, 2020. Springer.
Thangapazham RL, Puri A, Tele S, Blumenthal R, Maheshwari RK. Evaluation of a nanotechnology-based carrier for delivery of curcumin in prostate cancer cells. International Journal of Oncology. 2008;32(5):1119–1123.
Tian Y, Li JC, Zhu JX, Zhu N, Zhang HM, Liang L, Sun L. Folic acid-targeted etoposide cubosomes for theranostic application of cancer cell imaging and therapy. Medical Science Monitor: International Medical Journal of Experimental and Clinical Research. 2017;23:2426.
Van Dongen MA, Silpe JE, Dougherty CA, Kanduluru AK, Choi SK, Orr BG, Low PS, Banaszak Holl MM. Avidity mechanism of dendrimer–folic acid conjugates. Molecular Pharmaceutics. 2014;11(5):1696–1706.
Vickers NJ. Animal communication: When i'm calling you, will you answer too? Current Biology. 2017;27(14):R713–715.
Wang B, Guo XJ, Cai H, Zhu YH, Huang LY, Wang W, Luo L, Qi SH. Momordica charantia-derived extracellular vesicles-like nanovesicles inhibited glioma proliferation, migration, and invasion by regulating the PI3K/AKT signaling pathway. Journal of Functional Foods. 2022;90:104968.
Wang F, Ye X, Zhai D, Dai W, Wu Y, Chen J, Chen W. Curcumin-loaded nanostructured lipid carrier induced apoptosis in human HepG2 cells through activation of DR5/caspases-mediated extrinsic apoptosis pathway. Acta Pharmaceutica. 2020;70(2):227–237.
Wang X, Guan Q, Chen W, Hu X, Li L. Novel nanoliposomal delivery system for polydatin: Preparation, characterization, and in vivo evaluation. Drug Design, Development and Therapy. 2015;9:1805.
Wang Y, Wang S, Firempong CK, Zhang H, Wang M, Zhang Y, Zhu Y, Yu J, Xu X. Enhanced solubility and bioavailability of naringenin via liposomal nanoformulation: Preparation and in vitro and in vivo evaluations. AAPS Pharmscitech. 2017;18(3):586–594.
Xiao Z, Xu Z, Zhu G. Production and characterization of nanocapsules encapsulated linalool by ionic gelation method using chitosan as wall material. Food Science and Technology. 2017;37:613–619.
Xiong S, Liu W, Li D, Chen X, Liu F, Yuan D, Pan H, Wang Q, Fang S, Chen T. Oral delivery of puerarin nanocrystals to improve brain accumulation and anti-parkinsonian efficacy. Molecular Pharmaceutics. 2019;16(4):1444–1455.
Xu L, Wang X, Liu Y, Yang G, Falconer RJ, Zhao CX. Lipid nanoparticles for drug delivery. Advanced NanoBiomed Research. 2022;2(2):2100109.
Yadi M, Azizi M, Dianat-Moghadam H, Akbarzadeh A, Abyadeh M, Milani M. Antibacterial activity of green gold and silver nanoparticles using ginger root extract. Bioprocess and Biosystems Engineering. 2022;45(12):1905–1917.

Yamauchi T, Saitoh T, Shirai K, Fujiki K, Tsubokawa N. Immobilization of capsaicin onto silica nanoparticle surface and stimulus properties of the capsaicin-immobilized silica. Journal of Polymer Science Part A: Polymer Chemistry. 2010;48(8):1800–1805.

Yi C, Fu M, Cao X, Tong S, Zheng Q, Firempong CK, Jiang X, Xu X, Yu J. Enhanced oral bioavailability and tissue distribution of a new potential anticancer agent, Flammulina velutipes sterols, through liposomal encapsulation. Journal of Agricultural and Food Chemistry. 2013;61(25):5961–5971.

Yin J, Hou Y, Yin Y, Song X. Selenium-coated nanostructured lipid carriers used for oral delivery of berberine to accomplish a synergic hypoglycemic effect. International Journal of Nanomedicine. 2017;12:8671.

Yu Z, Lv H, Han G, Ma K. Ethosomes loaded with cryptotanshinone for acne treatment through topical gel formulation. PLoS One. 2016;11(7):e0159967.

Zeng Z, Zhou G, Wang X, Huang EZ, Zhan X, Liu J, Wang S, Wang A, Li H, Pei X, Xie T. Preparation, characterization and relative bioavailability of oral elemene o/w microemulsion. International Journal of Nanomedicine. 2010;5:567.

Zhang F, Liu Z, He X, Li Z, Shi B, Cai F. β-Sitosterol-loaded solid lipid nanoparticles ameliorate complete Freund's adjuvant-induced arthritis in rats: Involvement of NF-κB and HO-1/Nrf-2 pathway. Drug Delivery. 2020;27(1):1329–1341.

Zhao YZ, Lu CT, Zhang Y, Xiao J, Zhao YP, Tian JL, Xu YY, Feng ZG, Xu CY. Selection of high efficient transdermal lipid vesicle for curcumin skin delivery. International Journal of Pharmaceutics. 2013;454(1):302–309.

Zhu Y, Wang M, Zhang J, Peng W, Firempong CK, Deng W, Wang Q, Wang S, Shi F, Yu J, Xu X. Improved oral bioavailability of capsaicin via liposomal nanoformulation: Preparation, in vitro drug release and pharmacokinetics in rats. Archives of Pharmacal Research. 2015;38(4):512–521.

Zielińska A, Carreiró F, Oliveira AM, Neves A, Pires B, Venkatesh DN, Durazzo A, Lucarini M, Eder P, Silva AM. Polymeric nanoparticles: Production, characterization, toxicology and ecotoxicology. Molecules. 2020;25(16):3731.

Zou Y, Sun Y, Guo B, Wei Y, Xia Y, Huangfu Z, Meng F, van Hest JC, Yuan J, Zhong Z. α3β1 integrin-targeting polymersomal docetaxel as an advanced nanotherapeutic for nonsmall cell lung cancer treatment. ACS Applied Materials & Interfaces. 2020;12(13):14905–14913.

5 Liposomes and Nanoliposomes as Potential Nano-Enabled Platforms for Enhanced Bioavailability of Phytoconstituents

Petra Obioma Nnamani, Ozioma Blessing Onokala, Kruga Owodeha-Ashaka, Ibilola Mary Cardoso-Daodu, Margaret Ilomuanya, Christian Chukwuemeka Mbah, and Anthony Amaechi Attama

5.1 INTRODUCTION

Historically, folks have made use of naturally occurring materials sourced from plants and animal as medicaments to treat their ailments. This dates as far back as 600 years ago. Application of these materials or medicaments was not without its detriments as consumption of toxic plants mistaken as therapeutic medicaments may have occurred. Over time the human race has been able to acquire knowledgeable information on use of safe therapeutic materials sourced from plants through the practice of trial and error.

Presently, therapeutically active materials sourced from plants are termed phytoconstituents. The first phytoconstituent to be extracted was an alkaloid called morphine from the plant opium in 1805 by a European pharmacist named Friedrich Serturner. So far many more phytoconstituents have been extracted and used therapeutically to treat several conditions (Yuan et al, 2016). Phytoconstituents comprise a significant part of the pharmaceutical market. Alkaloids, for example, have produced a variety of active pharmaceutical ingredients, such as quinine (an antimalarial), reserpine (an antihypertensive), and tubocurarine (a muscle relaxant).

Phytoconstituents can also be described as chemical constituents produced from plants. They are also responsible for the organoleptic properties of plants. Phytoconstituents also serve as prototype drug molecules in the development of modern medicines. Phytoconstituents provide an unrivalled diversity in presenting a platform for the discoveries of new molecules as lead compounds (Dias et al, 2012). The plague of communicable and non-communicable diseases comes with obstacles in identifying ideal pharmacologically active compounds that have adequate therapeutic properties for treating these ailments. Presently, a novel approach is to go back to nature through phytoconstituents for solutions as it has provided answers in the field of drug discovery and disease treatment in the past (Thomford et al, 2018). Phytoconstituents extracted from plants usually display an inherent setback of poor oral bioavailability and chemical instability due to their hydrophilicity. These water-soluble phytoconstituents have low lipophilicity and large molecular weights, which may hinder cellular permeability. There is, therefore, a need for a proper delivery system to help deliver these phytoconstituents in active forms to the target site for pharmacological activity. Recently scientific interest has been geared toward the use of novel carriers as a viable option. Novel carriers, such

DOI: 10.1201/9781003300588-5

as microparticles, nanoparticles, liposomes, and niosomes, have been considered for the safe and targeted delivery of phytoconstituents. Ideally, a novel carrier should clearly tackle two important factors in the delivery of phytoconstituents. Firstly, it should be able to cargo the phytoconstituents without changing their pharmacological properties. Secondly, it should deliver the active phytoconstituents safely, with minimal environmental interaction, to the site of action. Liposomes fit this description in a unique way; they are made up of phospholipid that rarely interact with phytoconstituents. Liposomes can accommodate phytoconstituents in their inner aqueous core or their outer bilipid layer and transport them safely to the target site.

5.1.1 General Introduction

Phytoconstituents are produced naturally in plants. The term "phyto" is a Greek word for "plant." There are at least 10,000 varieties of phytoconstituents that have been assessed for the treatment of various diseases, such as bacterial/fungal/viral infections, cancer, hypertension, and diabetes. The use of nano-delivery to deliver phytoconstituents is becoming of popular interest, specifically within the area of pharmaceutical research. The major goal of an ideal drug delivery system is to improve the solubility, pharmacokinetics, pharmacodynamics, bioavailability, and efficacy of phytoconstituents (Singh, 2020).

Liposomes and nanoliposomes represent one of the most widely used lipid drug delivery systems. The major reasons for their popularity being that they are nontoxic and biocompatible and can encapsulate a wide variety of compounds (including phytoconstituents) regardless of their unique properties. Presently there are several successfully developed stable liposomes loaded with phytoconstituents (Savaghebi et al, 2019). In a study by Shariare et al, *Aphonamixis polystachya* leaf extract was encapsulated in liposomes for the treatment of neurodegenerative diseases such as Alzheimer's and Parkinson's disease. Gas chromatography mass spectrometry revealed that the main compounds in the extract were octadic 9-enoic acid, 5-hydoxypipeloic, and 2-pentone. Liposomes were stable. A clear enhancement of memory, locomotion, and ambulatory performance were observed in mice with induced dementia when treated with a liposome-encapsulated extract. This indicates successful liposomal delivery of the phytoconstituent extract through the blood-brain barrier to the target site within the brain (Shariare et al, 2020). Another example is a study by Karimi et al, where turmeric, which is rich in phenols and flavonoids, was encapsulated in liposomes prepared by thin-film hydration method (Yata et al, 2021). The antioxidant and antimicrobial properties of turmeric extract were higher in liposome-loaded extracts than in the free turmeric extracts (Karimi et al, 2019).

The deployment of liposomes as a drug delivery system has proven a necessity as it ensures localized delivery of phytoconstituents through passive targeting, reduces the possibilities of side effects, allows for uniformity of dosage, and prevents early degradation of phytoconstituents during circulation (Karimi et al, 2019).

5.1.2 Liposomes and Nanoliposomes

Liposomes were first discovered in University of Cambridge by Alec Bangham in 1961 at Babarham Institute. It was observed through an electron microscope as the self-assembling of phosphatidyl choline mixed with water. The name "liposome" comes from "lipos," which means "fat," and "soma," which means "body" (Allahou et al, 2021). Liposomes are spherically shaped vesicles that comprise of one or more bilipid membrane while nano-liposomes are nanosized aqueous phospholipid vesicles composed of two concentric spherical lipid bilayers, entrapping a distinct aqueous compartment (Tekade, 2019), (Worsham et al, 2018). Liposomes may be classified as small unilamellar vesicles (20–80 nm), large unilamellar vesicles (100–1,000 nm) and multilamellar vesicles (0.1–0.5 μm). Small unilamellar vesicles, as the name implies, are small. Due to their small size, they are only able to encapsulate a limited volume of aqueous fluid but are more stable with longer time for blood circulation. They can either be prepared by sonication through reduction of multilamellar vesicles and large vesicles by the process of agitation or ethanol/ether solvent injection method. Large unilamellar vesicles can encapsulate more aqueous volume and therefore more of the drug. The upside to large unilamellar vesicles is that the drug can be encapsulated with minimal lipid content. Unless attached covalently to polyethylene glycol, large unilamellar vesicles are easily detected by phagocytic bodies and the reticuloendothelial system. Multilamellar vesicles consist of multiple bilayers of phospholipid and cholesterol. They are also easily identified and cleared by phagocytic bodies and reticuloendothelial system. Due to their multilayered phospholipid bilayers, there is more volume to entrap hydrophobic drugs, while for hydrophilic drugs, their aqueous compartment is of moderate volume (Memvanga et al, 2021; Labouta et al, 2018). Nanoliposomes generally give better surface area in comparison to liposomes; as a result, they are more stable and soluble and have better bioavailability, efficacy, and targeting ability. The major difference between nanoliposomes and liposomes is primarily in their size; liposomes are usually microscopic, while nanoliposome are nanosized. Liposomes and nanoliposomes have exceptional features that make them stand out as drug delivery systems. They are biologically compatible, nontoxic, and biodegradable. Other features, such as good encapsulation efficiency, invisibility to the body's immune system, controlled release, and the possibility of enhanced targeting, depend on the parameters and techniques involved in liposome preparation (Savaghebi et al, 2019). Liposomes are particularly advantageous because they can act as nanocarriers for a wide variety of phytoconstituents.

Liposomes are formed spontaneously when phospholipids are hydrated excessively by an aqueous medium. Once phospholipids are hydrated, there must be an introduction of external energy input in the form of sonication, shaking, homogenization, or heating. This leads to lipid-lipid, lipid-water hydrophilic, and hydrophobic interactions, which result in formation of bilayer vesicles to achieve thermodynamic equilibrium. Liposomes can be prepared using several means, such as mechanical dispersion, solvent dispersion, electro formation, and supercritical fluidic and microfluidic methods. The most widely adopted is the mechanical dispersion method, mainly because of its cost effectiveness and simplicity.

Over the years, liposomes have evolved through innovations and smart scientific modifications. The very first set of liposomes known as the conventional liposomes are formed mainly by phospholipids and cholesterol self-assembled in an aqueous environment. However, they are easily recognized by the reticuloendothelial system during systemic circulation. The unique advantages and disadvantages of conventional liposomes are listed in Table 5.1.

The problem of easy recognition by the reticuloendothelial system led to modifications by the attachment of polyethylene glycol (PEG) on the surface of the liposome. PEG acts as a shield preventing clearance by the reticuloendothelial system, allowing longer circulation time. PEG decreases

TABLE 5.1
Advantages and Disadvantages of Conventional Liposomes

Advantages of Conventional Liposomes	Disadvantages of Conventional Liposomes
Increase therapeutic efficacy of the drug they carry	Low stability
Increase bioavailability to the site of action through targeting	Short circulation time, easily detected by the reticuloendothelial system
Decrease systemic toxicity of the drug by shielding the drug from normal cells	Amount of drug carried is limited to the volume of aqueous compartment for hydrophilic drug
Allow for controlled and timely release of the drug	Amount of hydrophobic drug is limited to the thickness of the phospholipid layer
Can carry hydrophilic and hydrophobic drugs	

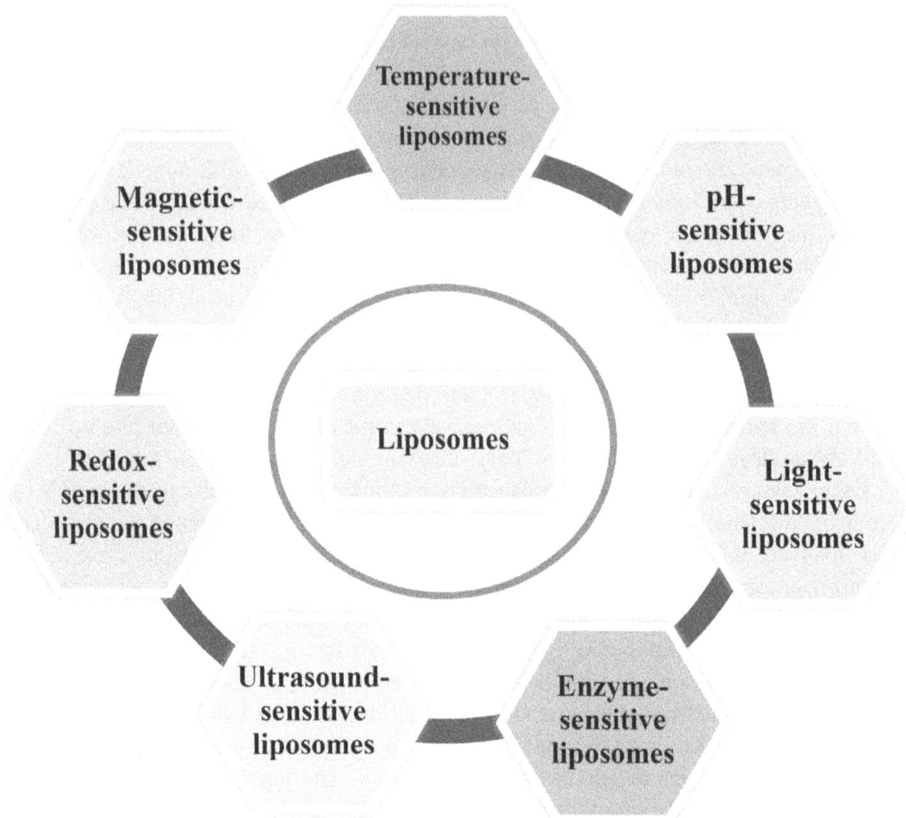

FIGURE 5.1 Flow chart showing a variety of stimulus responsive liposomes.

opsonization, thereby reducing the tendency of phagocytes to identify and target liposomes during circulation. Liposomes coated with PEG are usually referred to as stealth liposomes. Although the challenge of short liposome circulation time has been addressed through the attachment of PEG at the liposome's lipid bilayer surface, other major challenges with conventional liposomes, such as specific targeting and total release of cargo at the site of pharmacological action, remain unresolved. In order to address the problem of specific targeting, proteins, folates, and peptides whose receptors are excessively expressed at the targeted site of action have been deployed. To ensure complete release of loaded cargo, in this case phytoconstituents, the engagement of stimulus response as a property in the formulation of liposomes has been utilized. The development of temperature-sensitive liposomes that release its cargo in high temperatures has recorded many successes. pH-sensitive liposomes, which respond to pH variations in their environment, qualify as another clinically useful innovation. Other smart ideas, such as liposomes sensitive to magnetic fields, redox, light, ultrasound, and enzymes, have been investigated in the field of lipid drug delivery with the intent of ensuring proper release drug content or phytoconstituents from liposome formulations (Heidarli et al, 2017).

5.1.3 Phytoconstituents

Phytoconstituents are chemical components that are produced by plants through metabolic pathways. Phytoconstituents may be bioactive, pharmaceutically active, or inert (Chhikara et al, 2019).

One fundamental advantage of phytoconstituents is their low toxicity (Mahomoodally et al, 2019). Phytoconstituents have a very promising potential for sustaining health and maintaining well-being (Wang et al, 2014). A broad variety of phytoconstituents have been proven to possess pharmacological properties that will be useful in the treatment of serious pathological conditions. Currently, one major setback is the development of an ideal mode of delivery of these phytoconstituents. Challenges such as achieving a standard and metric dose and dosing frequency, consistent pharmacokinetics profiles, safety (drug irritation), ease of compliance, and patient variation have been encountered in drug development (Wen et al, 2015; Shariare et al, 2021b).

Phytoconstituents have a wide array of pharmacological actions and medicinal uses as seen in Table 5.2; it is, therefore, pertinent we take advantage of this attribute. An ideal drug delivery system for the transportation of phytoconstituents from its point of administration to site of action is vital. Liposomes and nanoliposomes are lipid vesicles that can encapsulate hydrophilic and hydrophilic large molecular weight compounds. Phytoconstituents are therefore excellent candidates for encapsulation by liposomes. Liposomes have been vastly used for delivering drugs, active pharmaceutical ingredients, extracts, and phytoconstituents as a way of enhancing stability, controlled release and improving bioavailability. Liposomes present multiple advantages, such as extension of half-life of phytoconstituents, reduced interaction with environment before reaching the target site,

TABLE 5.2
Phytoconstituents, Sources, Pharmacological Actions, and Derived Pharmaceutically Active Compounds

S/N	Phytoconstituents	Sources	Pharmacological Actions	References
1	Tanins	Seagrass, seaweeds, cranberry, grapes, and walnuts	Used for diarrhea, dysentery, and cancer; astringent	(Cuong et al, 2019; Singh et al, 2019b)
2	Steroids	Fenugreek, safed musli, creatine	Reduces inflammation, used in the treatment of asthma and eczema	(Harmata, 2021)
3	Flavonoids	Onion, berries, grapes, tomatoes, apples	Anti-inflammatory and antioxidant properties	(Panche et al, 2016; Xiao et al, 2016)
4	Saponins	Tomato, beans, peas, lentils	Antidote for lead poisoning, treatment of hypercalciuria, anti-inflammatory and antibacterial	(Ashour et al, 2019)
5	Glycosides	Purple foxglove, woolly foxglove, ouabain, yellow oleander	Used in the treatment of heart failure, also used as purgative and laxative	(Triana-Martínez et al, 2019)
6	Terpenoids	Mosses, mushrooms	Anticancer, anti-inflammatory, antiviral, antifungal, antiparasitic, antihyperglyceamic	(Nirmal et al, 2019)
7	Glucosinolates	Cruciferous vegetables	Anticancer and anti-inflammatory	(Egbuna et al, 2021)
8	Carotenoids	Carrots, parsley, spinach	Antioxidant	(Bhanu, 2020)
9	Limonoids	Citrus fruits	Anticancer, antibacterial, antifungal, antimalarial, antiviral, reduction of blood cholesterol	(Panche et al, 2016; Tundis et al, 2013)
10	Alkaloids	Opium poppy, buttercups, nightshades	Narcotic (pain relief), muscle relaxant, treatment of bronchial asthma, hay fever, sinusitis, and common colds	(Kurek et al, 2019)

and predetermined release of phytoconstituents at target sites. In a study by Olatunde et al, ethanolic coconut husk extract was encapsulated in liposomes. The extract is a wealthy source of phytoconstituents, such as tannin, flavonoids, and tannic acid, which are known to have great antioxidant and antibacterial properties, inhibiting the growth of a myriad of bacterial microorganisms. On investigation it was revealed that liposomal encapsulation of the extract improved or enhanced the antibacterial properties of the extract by improving stability, targeting, and overall bioavailability (Olatunde et al, 2019).

In another study by Dag et al, liposomes were developed and loaded with green tea extract using microfluidic method. Green tea extract is rich in the phytoconstituent polyphenol and is known to protect against cancer. The liposomes were then incorporated with biopolymers, gum arabic, and whey protein on their surface. Results from the experiment showed that liposomes incorporated with biopolymers were more stable than control liposomes. A similar study was carried out by Pavaloiei et al, where liposomes were developed and loaded with phenol-rich *Lucium barbarum* leaf extract. Entrapment efficiency was above 70%. Liposomes had a narrow polydispersity index and were stable for over 12 weeks (Dag et al, 2019; Păvăloiu et al, 2021). The current panorama in the development of liposomes as a delivery system for phytoconstituents is exceedingly encouraging. Phytoconstituents have the potential to be an ideal, safe, cost-effective, and efficient option for the treatment of a wide variety of disease conditions with the right choice of delivery system. Liposomes are an excellent choice for the delivery of liposomes orally, topically, or systemically.

5.2 LIPOSOME- AND NANOLIPOSOME-BASED DRUG DELIVERY SYSTEMS

Liposomes are spherical vesicles that can be formulated using cholesterol and natural phospholipids that can exhibit nontoxic *in vivo* characteristics. The hydrophilic, biocompatible, and hydrophobic characteristic accompanied with the size makes liposomes an ideal drug system. Liposomes have their properties altered in tandem with various factors, which include lipid composition, surface charge, size, and the formulation method (Tang et al, 2017). Liposomes are advantageous in drug delivery stems because they can increase efficacy and therapeutic index of drug and increase the stability of varying API due to the encapsulation of the API within the lipid vesicle. The choice of bilayer components determines the rigidity or fluidity and the charge of the bilayer. For instance, unsaturated phosphatidylcholine species from natural sources (egg or soybean phosphatidylcholine) give much more permeable and less stable bilayers, whereas the saturated phospholipids with long acyl chains (for example, dipalmitoylphos phatidylcholine) form a rigid, rather impermeable bilayer structure (Justo et al, 2005).

Due to the intrinsic amphiphilicity of their lipid bilayer shells, liposomes exhibit both lipophilic and hydrophilic properties that facilitate encapsulation of a broad range of pharmaceuticals. Diverse liposomal drug carriers and formulations have been intensively investigated, patented, and developed. Liposome formulations are extensively investigated for many different routes of administration, such as parenteral, oral, transdermal, nasal, pulmonary, and ophthalmic. Compared to conventional parenteral drug solutions, liposomal drug systems combining drugs into sub-micron lipid carriers as aqueous parenteral dispersions require additional considerations for regulatory approval, producing a more complicated development and regulatory process with higher costs and risks. Nanoliposomes may contain, in addition to phospholipids, other molecules, such as cholesterol, which is an important component of most natural membranes (Tang et al, 2017). The incorporation of cholesterol can increase stability by modulating the fluidity of the lipid bilayer preventing crystallization of the phospholipid acyl chains and providing steric hindrance to their movement. Further advances in liposome research found that surface modification with polyethylene glycol (PEG), which is inert in the body, generally reduces the clearance of liposome by RES and therefore allows longer circulatory life of the drug delivery system in the blood (Mezzomo et al, 2019). PEGylated liposomal doxorubicin has shown great prolonged circulation and substantial efficacy in breast cancer treatment. The net charge of nanoliposomes is also an important factor and generally anionic

TABLE 5.3
Examples of Various Liposomal and Nanoliposomal Formulations Detailing Their Bioavailability and Stability Properties

	API Encapsulated for a Specific Disease Condition	Type of Lipid Carrier	Bioavailability	Stability	Reference
1	Methyl 6-methoxy-3-(4-methoxyphenyl)-1H-indole-2-carboxylate evaluated for the in vitro cell growth inhibition on breast adenocarcinoma, melanoma, and non-small-cell lung cancer	Nanoliposome	—	Dynamic light scattering measurements showed that nanoliposomes with the encapsulated compound are generally monodisperse and with hydrodynamic diameters lower than 120 nm, hence exhibiting excellent stability.	(Abreu et al, 2011)
2	Kanamycin administered via inhalation for management of multidrug-resistant tuberculosis	Liposome	—	Storage at 5°C of the liposomes encapsulating kanamycin, monitored for 8 weeks, showed no aggregation or fusion. These vesicles can be considered as adequately stable.	(Justo et al, 2005)
3	Creatine nanoliposomes developed to combat oxidative stress in the cerebral cortex of young rats chemically induced to have hyperphenylalaninemia	Nanoliposome	Creatine nanoliposomes had excellent bioavailability, which resulted in prevention of the inhibition of the activities of the complexes II + III, caused by hyperphenylalaninemia. The results suggests that these nanoliposomal vesicle protect against neurotoxicity caused by hyperphenylalaninemia	The encapsulated nanoliposome showed an average particle size of 236.44 ± 0.61 nm, PDI of 0.27 ± 0.005, ZP of −23.9 ± 0.63, and pH of 5.98 ± 0.09, which was stable over a period of 48 hours.	(Mezzomo et al, 2019)
4	Pyronaridine and atovaquone nanoliposomes developed for antimalaria therapy	Nanoliposomes (immunoliposomes)	The nanoliposomal drugs had activities significantly higher than their free forms when tested in in vitro *Plasmodium falciparum* cultures. Pyronaridine and atovaquone concentrations, when encapsulated in nanoliposomes, resulted in a 50% inhibition of parasite growth	The encapsulated pyronaridine and atovaquone, over a storage time of two weeks at 4°C, showed relative stability of about 90% and 80%, respectively. No drug leakage was observed over two weeks; thus, the pyronaridine and atovaquone uptake did not destabilize the nanoliposomal structure.	(Biosca et al, 2019)

(Continued)

TABLE 5.3 *(Continued)*
Examples of Various Liposomal and Nanoliposomal Formulations Detailing Their Bioavailability and Stability Properties

	API Encapsulated for a Specific Disease Condition	Type of Lipid Carrier	Bioavailability	Stability	Reference
5	Curcumin- and β-arteether-loaded lipid-based system developed for oral antimalarial therapy	Liposomal vesicles	Combinations of antimalarial-loaded lipid-based systems drastically decreased blood parasite; hence, the bioavailability of the formulation was optimized.	A preliminary study of stability was conducted at room temperature (20 ± 2°C) and protected from light. Over 180 days, the percentage of curcumin and β-arteether remaining in the formulations was greater than 98.5%. The formulations did not exhibit any change in color or appearance under storage conditions tested.	(Memvanga et al, 2013)
6	Bupivacaine extended-release liposome injection developed as a local anesthetic formulation	Liposomes	Subcutaneous administration of 20 ml of 2% bupivacaine liposomes vs. 20 ml of 0.5% plain bupivacaine in human volunteers was compared by Davidson et al. The Cmax values found between the two groups was 0.83 ± 0.34 and 0.87 ± 0.45 μg/mL in liposomal and plain groups, respectively, showing no difference despite a 4-fold higher plain bupivacaine dose. This was attributed to the sustained release of liposomal bupivacaine	Exparel®, a bupivacaine extended-release liposome injection, is currently marketed with excellent stability indices and an expiration date of 24 months.	(Bulbake et al, 2017; Davidson et al, 2010)
7	Liposomal amphotericin B (AmBisome®) for treatment of opportunistic fungal pathogens	Liposomes	Studies in mice, rats, and rabbits all suggest that the plasma pharmacokinetics of liposomal amphotericin are linear and the drug bioavailable reaching therapeutic concentrations in all the animal models studied. In human studies, renal clearance is approximately 1–2 L/h and the volume of the central compartment is approximately 20 L, which is significantly less than estimates for members of the triazole class of antifungals. Generally, there is a dose-dependent increase in tissue concentrations.	Marketed formulations such as AmBisome® have been shown to be stable at 4°C for 48 months.	(Stone et al, 2016)

	API Encapsulated for a Specific Disease Condition	Type of Lipid Carrier	Bioavailability	Stability	Reference
8	Interferon-alpha-2b (IFN-alpha-2b) liposomes via injection for dermal wound healing	Liposomes	Bioavailability was evaluated via northern analysis. The expression of mRNAs for type I and type III collagens in response to IFN-alpha-2b showed a marked reduction in abundance of the transcripts for the pro-alpha1(I) chain of type 1 collagen on days 11 and 14 after wounding in the animal model, reflective of wound healing due to systemic bioavailability of the liposomal formulation.	—	(Ghahary et al, 2008; Wang et al, 2021)
9	Rapidly dissolvable oral Pediatric formulation for mefloquine liposomes for malaria management	Liposomes	In mice, orally administered lyophilized mefloquine liposomes showed comparable absorption with bioavailability of 81–86%, while the absorption of the standard mefloquine suspension was significantly lower with bioavailability of 70% and reflected almost a 20% decrease in maximum plasma concentration.	Long-term room temperature storage was improved via lyophilization of the mefloquine liposomes. This ensured the formulation was completely dissolvable in water within ten seconds. The lyophilized mefloquine liposomes was stable at room temperature for 3 months.	(Tang et al, 2017)
10	Mitoxantrone hydrochloride liposome injection for management of advanced breast cancer	Liposomes	Liposomal mitoxantrone was safe and effective at 12 to 24 mg/m^2. At this dose bioavailability was optimized with minimal systemic toxicity when compared to other dosage forms of mitoxantrone. Liposomal mitoxantrone provided a different toxicity profile, which might be associated with the altered distribution of the drug, hence increasing its safety profile.	Mitoxantrone hydrochloride liposome injection utilized in Chinese patients with advanced breast cancer during phase II clinical trial had a stability profile of 2 years.	(Li et al, 2008; Wang et al, 2019)

and neutral liposomes survive longer than cationic liposomes in the blood circulation after intravenous injection (Jakoby et al, 2015; Niu et al, 2010). The main chemical ingredients of nanoliposomes are lipid and/or phospholipid molecules (Mozafari, 2010; Abreu et al, 2011). Lipids are fatty acid derivatives with various head group moieties. When taken orally, lipids are subjected to conversion by gastrointestinal lipases to their constituent fatty acids and head groups. Triglycerides are lipids made from three fatty acids and a glycerol molecule (a three-carbon alcohol with a hydroxyl group [OH] on each carbon atom). The amount of cholesterol to be used in the nanoliposome formulations largely depends on the intended application (Mozafari, 2010). For liposomes in the form of multilamellar vesicles, both anionic (with dicetylphosphate, DCP, as a negatively charged ingredient) and neutral (without DCP) liposomes containing PC can interact with model membrane systems when containing a 10% molar ratio of cholesterol. Studies have shown that lipid composition and cholesterol content are among the major parameters in the research and development of nanoliposome formulations.

5.3 FORMULATION OF LIPOSOMES AND NANOLIPOSOMES

5.3.1 Materials Used in the Preparation of Liposomes and Nanoliposomes

As lipid-based delivery systems, liposomes and nanoliposomes are fundamentally composed of two phases, a lipid and an aqueous phase. The lipid phase would typically contain phospholipids or lipids as the main component of the bilayer in addition to any other lipid soluble component of the vesicle, while the aqueous phase would house the components of suitable solubility. Structurally, phospholipids being amphiphilic possess a polar head and two hydrophobic alkyl tails, the natures of which affect important properties of the phospholipids like their phase transition temperatures (T_c) and charge under physiological conditions (Fan et al, 2021). Phospholipids are biocompatible materials and are able to self-assemble, forming liposomes with an aqueous core and outer lipid membranes (Mozafari et al, 2010).

As the properties of liposomes (LP) and nanoliposomes (NLP) have expanded over the years, the range and nature of materials used in their formulation has broadened as well, going from simple natural lipids and phospholipids to modified natural, to semi-synthetic, to synthetic phospholipids. Phopsphatidylcholine, commonly from eggs or soy, the most commonly used natural phospholipid (Large et al, 2021). A range of semi-synthetic and synthetic phospholipids that have been used include 1,2-dipalmitoyl-sn-glycero-3-phosphocholine (DPPC) (Abrishami et al. 2020; Dai et al, 2020); 1,2-distearoyl-sn-glycero-3-phosphocholine (DSPC) (Haeri et al, 2011); 1,2-distearoyl-sn-glycero-3-phospho-rac-(1-glycerol) sodium salt (DSPG) (Assadpour et al, 2021); 1,2-dipalmitoyl-sn-glycero-3-phosphoethanolamine (DPPE) (Ramezani et al, 2011); 1-palmitoyl-2-oleoyl-sn-glycero-3-phosphocholine (POPC) and 1,2-dioleoyl-sn-glycero-3-phosphocholine (DOPC) (Maherani et al, 2012); and monostearoyl phosphatidylcholine (MSPC) (Wang et al, 2016). In some formulations, phospholipids conjugated with polyethylene glycol are used to produce stealth vesicles, improving the biological stability of the formulation. For example, 1,2-distearoyl-sn-glycero-3-phosphoethanolamine-N-[methoxy (polyethylene glycol)-2000] (mPEG2000-DSPE) is often used in LP and NLP formulations (Wang et al, 2016; Wang et al, 2012; Barati et al, 2021).

Specialized liposomes and nanoliposomes contain other lipids with characteristics that fine tune vesicular properties to fit the intended use. 1,2-dioleoyl-3-trimethylammonium-propane (DOTAP) is often used in preparing liposomes or nanoliposomes carrying biologics like genes, etc. (Barati et al, 2021; Baradaran et al, 2022; Hamed et al, 2019). DOTAP is a cationic lipid that confers a positive charge on the liposomes, improving their stability as well as their intracellular uptake by interactions with negatively charged cell membranes (Baradaran et al, 2022). Anionic 1,2-dipalmitoyl-sn-glycero-3-phosphorylglycerol sodium salt (DPPG) is another charged lipid used in the delivery of bioactives (Carvalho et al, 2013). Negatively charged phospholipids 1,2-dimyristoyl-sn-glycero-3-phosphoglycerol sodium salt (DMPG), 1,2-dimyristoyl-sn-glycero-3-phopho-L-serine sodium salt (DMPS) and 1,2-dimyristoylsn-glycero-3-phosphate monosodium salt (DMPA) have been used to improve liposome interaction with bacterium in delivery of antibiotics (Ma et al, 2013). 1,2-dioleoyl-sn-glycero-3-phosphoethanolamine (DOPE), a pH-sensitive lipid, is another phospholipid often used in formulation of NLP and LP for facilitating delivery of various bio-based therapeutic agents due to its sensitivity to intracellular pH and the structural changes it induces (Rastakhiz et al, 2019).

Sterols like cholesterol and phytosterols (less frequently) are also frequently added to LP and NLP formulations, having a strong influence on membrane fluidity, permeability, and stability and acting as a membrane stabilizer (Panahi et al, 2017; Daraee et al, 2016; Naziris et al, 2020; Amiri et al, 2019). Some formulations use pure fatty acids like oleic acid (Pinilla et al, 2019), palmitic acid (Molaveisi et al, 2021) or other fatty acid sources (Gorjian et al, 2021) alone or together with cholesterol and have been seen to improve stability, particles size and polydispersity index properties.

Liposomes and nanoliposomes often utilize in their formulation processes organic solvents (methanol, ethanol, chloroform, diethyl ether, etc.) for dissolving lipids (Wang et al, 2012; Pinilla et al, 2019; Shakeel et al, 2019) and buffers like phosphate, chloride, or citric acid buffers for lipid hydration (Li et al, 2013; Gosangari et al, 2012). Other additives, like charge inducers, such as diacetyl phosphate (DCP) or stearylamine (SA), are sometimes added to configure the surface charge of vesicles, either negatively or positively respectively to give the best properties suited to their intended function (Abrishami et al, 2020). Surfactants have been added to nanoliposomes as stabilizing agents, as edge activators, to improve drug solubility in liposomes or, generally, the properties of the vesicles. Some surfactants that have been used are Tween 80 (commonly), Birj 35, sodium deoxycholate (SDC), dodecyltrimethylammonium bromide (DTAB), Pluronic P85 (Zhou et al, 2014;Lujan et al, 2019; Sarabandi et al, 2019; Ghassemi et al, 2018). LP and NLP are sometimes formulated with unconjugated polyethylene glycol or biopolymers, like chitosan, as coating material to facilitate drug delivery through physiological membranes, as well as increase circulation time (Tamaddon et al, 2021; Gao et al, 2012; Najlah et al, 2019).

5.3.2 Synthesis/Preparation of Liposomes and Nanoliposomes

The preparation of liposomes has been achieved by a range of methods having the same base principles involving drying down lipids from an organic solvent, dispersing the lipid in aqueous media, purifying the resultant liposome, and analyzing the final product, but this process has evolved with various modifications over the years (Akbarzadeh et al, 2013; Khorasani et al, 2018). During preparation, active molecules can be passively or actively loaded into vesicles, of which passive loading has been further categorized according to technique: mechanical dispersion method, solvent dispersion method, and detergent removal method (Akbarzadeh et al, 2013). Different methods produce vesicles with characteristic properties like particle size, lamellarity, and entrapment efficiency. Advantages and disadvantages of various methods have been outlined by Maja et al (2020).

Under the mechanical dispersion method, there is the Bangham thin-film hydration (TFH), which is perhaps the most widely used method for LP preparation. The phospholipids and other suitably soluble vesicular components are dissolved in organic solvents and the resulting homogenous mixture is evaporated under vacuum, usually in a rotary evaporator. Nitrogen is applied to ensure complete removal of organic solvent and a thin-lipid film is formed. An appropriate hydrating medium (deionized water or buffer solution) maintained at temperatures above the phase transition temperatures of the phospholipids and sometimes containing other aqueous components of the vesicle is added to hydrate the solid film. Hydration is done under agitation for 1–2 hours, resulting in a liposomal suspension containing multilamellar vesicles (MLVs). Different researchers have used this method, some with slight modifications of different aspects of the process (Pinilla et al, 2019; Huang et al, 2021; Han et al, 2021; Natarajan et al, 2012; Manikkath et al, 2020).

Ethanol injection is a solvent dispersion method involving the rapid injection of an ethanolic solution of lipids into a large excess of aqueous buffer while mechanically agitating. This results in the immediate formation of liposomal vesicles and agitation is continued until complete removal of organic solvent (Barati et al, 2021; Gao et al, 2012). Ether injection is another solvent dispersion method where the lipids are dissolved in diethyl, either alone or in combination with other organic solvents. The lipid mixture is then injected slowly into the aqueous phase containing the active molecule and other aqueous-soluble vesicle components. As the organic solvent(s) is removed under vacuum or by elevated temperatures, liposomes, usually large unilamellar vesicles (LUVs) are formed (Shakeel et al, 2019; Akbarzadeh et al, 2013).

In revere-phase evaporation, lipid and amphiphilic materials are dissolved in organic solvent and the aqueous phase is added. Under reduced pressure, the organic solvent is evaporated resulting in an aqueous solution of a mix of LUVs and MLVs (Large et al, 2021; Sheoran et al, 2020).

Using the freeze-drying method, Heravi Shargh *et al* prepared temozolomide inclusion complexes loaded into nanoliposomes for glioma treatments (Heravi Shargh et al, 2021). A homogenous solution of lipids in appropriate buffer containing the drug was prepared and freeze-dried overnight. Using deionized water, the lyophilized powder was reconstituted, and the suspension was sonicated to give a homogenous nanoliposomal dispersion. Baradaran *et al* used a modified freeze-drying of a monophase solution method to develop plasmid DNA nanoliposomes (Baradaran et al, 2022). Lipids were dissolved in tert-butyl alcohol and mixed in equal ratios with a water-based solution containing sucrose and the pDNA to form a monophase solution. The resulting mixture was then snap-frozen and freeze-dried at a low condensing pressure and sub-zero temperature. Liposomes were obtained after reconstituting the dried material in distilled water and allowing the dispersion to stand.

Shreya *et al* described a one-pot spray drying method that they used to prepare modified proliposomes of tramadol (Shreya et al, 2021). The lipids were completely dispersed in ethanol via sonication-aided dissolution. The aqueous phase with the drug was mixed with the alcoholic lipid solution, maintaining the temperature at 60°C, after which the coating solution was added and stirred until completely solubilized. The mixture was spray-dried at set temperatures and liposomes were obtained by redispersing powders in distilled water.

Depressurization of an expanded liquid organic solution into aqueous solution (DELOS-susp) method was used in the preparation of liposomal α-galactosidase A (GLA) (Merlo-Mas et al, 2021). Lipids were dissolved in an alcoholic dimethyl sulfoxide solution which as loaded into a high-pressure vessel at atmospheric pressure. Compressed CO_2 was used to volumetrically expand the solution while stirring till a specific molar fraction and working temperature was reached, as well as achievement of complete homogenization and thermal equilibrium. Over an aqueous GLA solution of desired concentration, the lipid/organic phase was depressurized with a flow of compressed nitrogen applied to maintain constant pressure. The resulting mixture is a dispersion of nanoGLA liposomes.

Microfluidic method: A number of researchers have utilized the simil-microfluidic (SMF) method (De Simone et al, 2020; Bochicchio et al, 2020; Dalmoro et al, 2019). This method capable of directly producing nanosized unilamellar vesicles employs the use of a microfluidizer apparatus. As described by Fathi et al, it involves dividing a high-pressure stream into two parts (an alcoholic lipid solution and the aqueous phase, either buffer or water), passing through fine orifices and their flow directed at each other inside the chamber (Fathi et al, 2012). Mixing results in liposome formation. During their interaction, cavitation and shear reduce the particle sizes of the liposomes.

Some researchers have used a combination of methods in their experiments. For example, Yang *et al* prepared vitamin c nanoliposomes using Thin film evaporation followed by microfluidization (DHPM) (Yang et al, 2012) while Liu *et al* used thin layer dispersion method combined with dynamic high-pressure microfluidization (DHPM) to formulate coated nanoliposomes (Liu et al, 2013).

Most of the mentioned methods involve the use of one or more organic solvents, raising concerns of toxicity and safety. In response to this, some methods that circumvent the need for these solvents have been developed. Rasti *et al* successfully demonstrated the formulation of liposomes using the Mozafari method where lipids and fatty acids were directly hydrated using water and glycerol heated and maintained at 30°C for 1 hour while stirring (Rasti et al, 2012). The formed dispersion was left under nitrogen to anneal and stabilize for a minimum of an hour.

Some other methods that have been used for synthesizing LP include lyophilization and rehydration method by Wei *et al* (Wei et al, 2011) and a similar method used Bazrgari *et al* (Bazrgari et al, 2020) in nanoliposomes of rifampin and N-acetyl cysteine, both modelled after the dehydration-rehydration method but still including the use of organic solvents; fusion method (Golmohammadzadeh et al, 2011); emulsification-evaporation and low-temperature curing method (Wang et al, 2012); modified heating method (MHM) (Naziris et al, 2020); effervescence dispersion technique (Wei et al, 2014); dehydration-rehydration method; detergent removal method; and freeze-thawing method. These and other less commonly used methods have been extensively described in a number of reviews (Large et al, 2021; Akbarzadeh et al, 2013; Dua et al, 2012). In active loading, the transmembrane pH gradient method is often used to remotely encapsulate active molecules in solution into previously formed empty vesicles (Shalaby et al, 2020; Bajelan et al, 2012). It typically uses buffer or ammonium sulfate gradients and strongly influences active loading of a weak base (Maja et al, 2020). Haeri *et al* has also used a "remote film loading" method where a methanolic solution of sirolimus was evaporated after which empty liposomes formulated prior by lipid film hydration were added (Haeri et al, 2011). The resulting mixture was sonicated for the desired period and allowed to stand afterward.

Formulation of nanoliposomes typically involves at least one additional step for size reduction from liposomes. Several size reduction methods have been employed, high-pressure homogenization using a homogenizer (Li et al, 2015), extrusion (Assadpour et al 2021), and sonication (Han et al, 2021), and they are often considered as mechanical methods for formulations of smaller unilamellar LPs and NLPs from multilamellar vesicles (Maja et al, 2020). One or more methods can be used in a formulation like combing homogenization and sonication (Gorjian et al, 2021) or sonication and extrusion (Gheibi et al, 2020). Sonication using a bath or probe sonicator involves the use of ultrasonic energy to break down MLV and LUV in an inert atmosphere. A disadvantage of the process is that it generates heat and can sometimes lead to the destruction of the vesicles (Maja et al, 2020). In extrusion, multilamellar vesicles are forced by high pressure through polycarbonate filters of specific sizes to form small unilamellar vesicles.

5.3.3 Characterization of Liposomes and Nanoliposomes

5.3.3.1 Particle Size and Polydispersity Index (PDI)

Liposomes sizes range from very small (0.025 μm) to large (2.5 μm), vesicles while nanoliposomes have average sizes of 50–150 nm. Nanoliposomes have larger surface areas than liposomes due to their smaller size. One of the most common techniques that have been used is based on dynamic light scattering (DLS), also known as photon correlation spectroscopy. It uses time-dependent fluctuations of the scattered light from particles in Brownian motions to determine hydrodynamic diameter and is quick and reliable (Fan et al, 2021). Electron microscopy techniques that include transmission electron microscopy (TEM) and atomic force microscopy (AFM) have also been used in particle size determination (Gao et al, 2012). These, however, give accuracy to a lesser degree than light-scattering techniques (Fan et al, 2021). Other methods for size determination include flow cytometry, laser diffractometry, size exclusion chromatography, and a range of field flow fractionation methods (Kanásová et al, 2017).

PDI is a measure of homogeneity of a vesicular dispersion. It represents the particle size distribution, which could be influenced by storage conditions and time, thus being an important indication of stability. It is often determined using a zetasizer with the DLS technique. PDI values are range from 0 to 1 with values less than 0.05 seen to be highly homogenous. Values greater than 0.7 indicate a hetero-dispersed vesicle dispersion.

5.3.3.2 Morphology and Lamellarity

The shape of vesicles can be visualized using various microscopic techniques including the commonly used TEM and scanning electron microscopy (SEM). Confocal scanning fluorescent laser microscopy has also been used in identifying microstructure images of liposomal suspensions and the process involves applying a dye solution to sample before visualizing (Peng et al, 2018). Others are cryo-SEM, cryo-TEM (Merlo-Mas et al, 2021), AFM, freeze-fracture TEM (FF-TEM), bright field microscopy, negative staining microscopy, optical light microscopy, environmental scanning electron microscopy (ESEM), and field emission microscopy.

Lamellarity gives insight as to the number of lipid bilayers the formed LP possess. They could possess one (unilamellar), two to five (oligolamellar), or more than five (multilamellar) bilayer membranes. It is often determined by ^{31}P nuclear magnetic resonance (^{31}P NMR), as well as by small-angle X-ray scattering (SAXS) (Merlo-Mas et al, 2021; Kanásová et al, 2017). Electron microscopy techniques like cryo TEM have also been used (Cai et al, 2020).

5.3.3.3 Zeta Potential

Zeta potential expresses the surface charge potential of vesicles which is typically conferred by the phospholipid component, incorporating a positive or a negative charge (Fan et al, 2021; Nikkhah et al, 2020). This potential is also affected by other components of the vesicles such as the pH of the buffer, the charge of the active molecule, additives like charge inducers, etc. (Naziris et al, 2020). It impacts the stability of the vesicular dispersion significantly as it determines the magnitude of particle-particle interactions. The unit of measurement is electrophoretic mobility, which is the velocity of particles moving through an electric field and is obtained by determination of frequency change of laser light scattered as they move. Values <-30 mV or >30 mV are thought to give sufficient intra-particle repulsive force for preventing aggregation and maintaining the best stability (Fan et al, 2021; Wilson et al, 2001). Techniques that have been used to analyze zeta potential of liposomes and nanoliposomes in the last decade include phase analysis light scattering (PALS) (Haeri et al, 2011), electrophoretic light scattering (ELS) (Maherani et al, 2012; Risaliti et al, 2020; Ghosh et al, 2020), photon correlation spectroscopy (De Simone et al, 2020; Chaves et al, 2020; Praveen et al, 2019), and laser doppler velocimetry (Manikkath et al, 2020; Socaciu et al, 2019).

5.3.3.4 Entrapment Efficiency

This is a measure of the ability of the lipid bilayer to entrap and retain the active molecule(s) within its vesicle. It is represented by the percentage of active ingredient actually present within the vesicles at a given time compared to how much was introduced in the formulation process. In its determination, unentrapped drug is separated from the drug-containing vesicles in the colloidal dispersion via ultracentrifugation or the dialysis bag method (Abrishami et al, 2020; Shalaby et al, 2020). The vesicles are lysed, and the active molecule is quantified using techniques such as high performance chromatography (HPLC), UV-Vis spectroscopy, LC-MS/MS, spectrofluorimetric analysis, and other appropriate methods (Assadpour et al, 2021; Maherani et al, 2012).

5.3.3.5 Stability

The properties of formulated liposomes and nanoliposomes are affected by the length of storage and the conditions under which they are kept. These could effect changes in particle size changes leading to changes in size distribution, in the measure of surface charge, and even in the entrapment efficiency. An important quality of optimum LP and NLP is that they are able to maintain their physical and chemical stability physiologically and throughout the handling process from manufacture to final use. As such, formulated NLP and LP are often subjected to tests over a storage period and under various conditions of temperature and relative humidity according to stipulated guideline and their properties are evaluated.

5.3.3.6 *In Vitro* Release

For a formulation to be clinically useful, the vesicles must release the active agent in the effective quantities, at the appropriate site and over the desired duration. A simulation of the release kinetics of the vesicles in the intended physiological medium gives insight into the behavior of the vesicles in this regard.

5.3.3.7 Others

Other LP and NLP properties that are sometimes analyzed are thermal stability using techniques like differential scanning colorimetry (DSC), Fourier transform infrared spectroscopy (FTIR), attenuated total reflectance (ATR), or a combination of both—ATR-FTIR and x-ray diffraction analysis. These also give insight into the phase transmission behavior, drug stability during encapsulation, drug-excipients interaction, and general solid-state characteristics of the vesicles (Molaveisi et al, 2021; Suri et al, 2020; Li et al, 2019). Merlo-Mas *et al* used Turbiscan stability index (TSI) determined by static multiple light scattering (SMLS) as a measure of the physical stability of α-galactosidase A (GLA) liposomes in their experiment (Merlo-Mas et al, 2021). Phospholipid content in the bilayer membrane has been analyzed by Barlette phosphate assay (Gheibi et al, 2020) and phosphate assay (Rastakhiz et al, 2019). Zagana *et al* also used Stewart assay following size exclusion chromatography (SEC) and ultracentrifugation for quantification of lipids (Zagana et al, 2020). Membrane fluidity is another property that is sometimes evaluated is often determined by fluorescence anisotropy (Hasan et al, 2020).

For specialized liposomes or nanoliposomes that are designed to be carriers for specific active molecules, such as genes and some other bio-based molecules, their evaluation includes other tests like agarose gel retardation assay. In an experiment with DNA-loaded nanoliposomes for gene therapy, it was used to determine the efficacy of DNA encapsulation into the vesicles (Baradaran et al, 2022).

5.4 DEVELOPMENT CHALLENGES OF LIPOSOMES AND NANOLIPOSOMES AS DRUG DELIVERY VEHICLES

Drug delivery in form of liposomes have been successfully translated into the clinics for treatment of many diseases for several decades now (Allen et al, 2013; Pattni et al, 2015; Das et al, 2019; Mu et al, 2018). Obstacles have been identified in the use of liposomes for drug delivery purposes. These include drug leakage, opsonization, the reticuloendothelial system (RES), immunogenicity, scale-up issues, lyophilization, quality control, and safety and toxicological concerns.

5.4.1 Drug Leakage

Unwanted drug leakage can be premature and lead to decreased drug concentration at the target site as well as cause damage to healthy organs (Liu et al, 2016; Kim et al, 2019). The moment liposomes enter the body, opsonins, high-density lipoproteins (HDLs), and low-density lipoproteins (LDLs) meet with them as they circulate in the bloodstream. These opsonins (immunoglobulins and fibronectin) help RES to recognize and eliminate liposomes, whereas HDLs and LDLs interact with liposomes to decrease their stability (Patra et al, 2018). Liposomal accumulation at sites such as the liver and the spleen is advantageous as this favors high drug concentration/build-up of liposomes, which is helpful to treat certain disease, yet this can be disadvantageous because it can lead to leakage of drugs and hence lower the concentration of the drug reaching a target site as seen in cancers. Drug leakage results from disruption of the integrity of liposome bilayers by plasma proteins (lipoproteins and opsonins) as well as blood flow shear stress, leading to increased liposomal membrane deformation and permeability (Bernard et al, 2005; Yoshimoto et al, 2013; Kogan et al,

2014). This is responsible for the different types of liposomal formulations being developed, and despite the potential of liposomes, low physical stability and high sensitivity to temperature and pH variations need to be overcome when commercial use is intended (Zhong-Jian et al, 2020; Aguilar-Pérez et al, 2020).

5.4.2 Cost and Scale-Up Issues

Liposomal translation from laboratory to commercial scale is challenging in terms of reproducibility. A formulation is robust when it is scalable at the lab, pilot, and industrial levels. Despite advances in scale-up technology and quality by design (QbD) concepts, which have made significant progress in smoothing the commercialization process of liposomes, efforts are still being made to bring liposomal formulations to the market (Goranthla et al, 2020).

5.4.3 Lyophilization of Liposomes

Oxidation/hydrolysis of lipids, vesicle fusion or aggregation affect the in vivo biodistribution, efficacy and safety of liposomal products (Franzé et al, 2018). When design parameters such as use of saturated or high transition phase lipids, antioxidant agents, charged/functionalized particles ore storage conditions are insufficient, the only option is to overcome instability is to dry the liposomal product, to remove the aqueous environment precipitating the mentioned conditions. For liposomes intended for oral or pulmonary routes, it is unavoidable to dry the colloidal suspension (Misra et al, 2009; Wilkhu et al, 2017; Jensen et al, 2017). Excipient choice and process parameters to protect the membrane integrity from freezing and dehydration stresses is still challenging. Since liposomes contain water and spontaneously assemble is water, therefore, removal of water may cause significantly irreversible changes in their structure (Chen et al, 2010).

5.4.4 Quality Control of Liposomes

In pharmaceutical product development, quality and cost-effective production are important factors to consider. Stability and the process of manufacture define quality of liposomes. Liposomal formulations experience challenges in nanoparticle preparation, such as batch-to-batch reproducibility, dispersion stability, and safety issues (Zhong-Jian et al, 2020). Little variation in the process parameters may lead to changes in particle size distribution, encapsulation efficiency, drug release, and therapeutic efficacy (Goranthla et al, 2020). In-process quality control testing and finished product testing ensures safety and efficacy of a pharmaceutical product. Regulatory agencies (United States Food and Drug Administration and Indian Regulator bodies, Central Drug Standard Control Organization, Indian Council of Medical Research) have issued guidelines for product specification and characterization required for product approval processes. The regulatory bodies ensure that safety and product efficacy are stringently met for analytical method validation, stability studies, applicable comparative analysis of innovator product, preclinical data, clinical data, and sample testing protocols (Guidelines for Evaluation of Nanopharmaceuticals in India, 2019; https://www.drishtiias.com/daily-updates/daily-news-analysis/guidelines-for-evaluation-of-nanopharmaceuticals-in-india).

5.4.5 Safety and Toxicity Concerns

The main issues for clinical approval of liposomes are safety and toxicity, and because of their biocompatibility, they are safer for interaction with biological membranes which has accentuated their presence in the market (Beltrán-Gracia et al, 2019).

5.5 APPLICATIONS OF LIPOSOMES AND NANOLIPOSOMES AS DRUG DELIVERY VEHICLES FOR PHYTOCONSTITUENTS

5.5.1 ORAL

The bioavailability of therapeutic phytochemicals administered orally is constrained by a number of factors some of which are related to the physicochemical properties of the agent such as solubility, permeability, molecular size, (Kalepu et al, 2013; Raju et al, 2011), and others to the nature as well as the physiological conditions of the biological system, including first-pass metabolism (Kalepu et al, 2013; Renukuntla et al, 2013). Liposomes and nanoliposomes have been found very useful as delivery devices for efficient delivery of phytochemicals and other drugs (Singh et al, 2019b). The suitability of these delivery devices arises from their capability to entrap the phytochemicals of different natures, including both lyophilic and lyophobic compounds, and their ability to deliver them through biological barriers for absorption (Mbah et al, 2018). The major constituent of liposomes is the phospholipid, which occurs as a natural component of cell membranes (Karpuz et al, 2020b; Mbah et al, 2014). Phospholipid molecules have intrinsic property of self-assembly in aqueous medium, forming flexible vesicles and, in the process, entrapping particles of any incorporated drug. Being a natural component of the cell membrane, the phospholipid molecules easily interact and align with the like lipid components of cell membranes, thereby penetrating, releasing, or carrying through the entrapped drug particles (Mbah et al, 2014). This biocompatibility makes liposomes additionally advantageous in drug delivery. Oral delivery of phytochemicals via liposomes is achieved as a result of enhanced permeation through the lipid-rich biological membranes, the intestinal wall in this case, by different mechanisms (Mbah et al, 2014; Silindir et al, 2012).

Rodriguez et al (2019) encapsulated quercetin in nanoliposomes following thin film-sonication technique with the aim of enhancing its stability, delivery, and biological activity. The results of the in vitro studies showed improved stability, controlled release, and enhanced (1,004-fold antioxidant and 2-to-5-fold anti-angiogenic) bioactivities over the unencapsulated quercetin (Rodriguez et al, 2019).

Baicalin is an active flavonoid obtained from the dry roots of *Scutellaria baicalensis* G., which has folkloric use in China against ulcer, fever, inflammation, cancer, and infections (Wei et al, 2014; Kim et al, 2014; Srinivas et al, 2010). As a result of the poor aqueous solubility of baicalin, Wei et al (2014) worked on the development of a liposomal formulation using effervescent dispersion method for its improved bioavailability via oral administration in rats. The results obtained from the biodistribution and pharmacokinetic studies indicated significant increases in tissue biodistributions, maximum blood concentration and overall bioavailability of the phytoconstituent from the liposomes compared to its suspension preparation (Wei et al, 2014).

A liposomal formulation was prepared by solvent injection technique for the delivery of phytochemicals extracted from the leaves of *Aphanamixis polystachya* for therapeutic applications in Parkinson's and Alzheimer's diseases. The results of gas chromatography and mass spectrometry analyses of the ethanol extract showed content of some major phytoconstituents, including beta-elemene, 2-pentanone, and 5-hydoxypipecolic acid, among others, probably responsible for the observed significant improvements in locomotor activity, ambulatory performance, and memory function of mice with induced dementia by the liposomes relative to the to extract alone (Shariare et al, 2020).

Liposomes and other phospholipid-based formulations of some phytoconstituents were demonstrated to have more efficient in vitro and in vivo antineoplastic (epigallocatechin gallate) and pain-relief (thymoquinone) activities relative to the natural product isolates alone (Shariare et al, 2021a, 2021b). The milestones notwithstanding, liposomes, however, encounter delimiting conditions in the gastrointestinal tract that cause either disintegration of the vesicle or leakage of the cargo prior to reaching the target delivery site. However, formulation scientists have developed some techniques of overcoming the drawbacks, such as cross-linking of the vesicle membrane (Hu et al, 2013; Chen et al, 1996), using saturated phospholipids (Kokkona et al., 2000), coating the vesicle with polymers like chitosan and polyethylene glycol (Iwanaga et al, 1997; Takeuchi et al, 2005; Werle et al, 2009), or attaching suitable ligands (Mbah et al, 2014; Chavda et al, 2022). The last two approaches have gained popularity in the pharmaceutical industry. Many liposomal pharmaceuticals, including phytomedicines, have been developed through the application of the mentioned approaches and other conventional ones (Chavda et al, 2022).

5.5.2 Topical

Conventional liposomes are capable of delivering encapsulated drug particles onto the surface and into limited inner layers of the skin, owing to the barrier function of the *stratum corneum* outer layer. However, penetration into deeper layers is achieved through the inclusion of penetration enhancers in liposome formulation (Mbah et al, 2014; Neupane et al, 2021; Elsayed et al, 2007). Thus, liposomes could be used to encapsulate phytochemicals and other drugs for dermal and intradermal or transdermal delivery. Park (2013), in a study, demonstrated enhanced transdermal permeation of quercetin and rutin from the liposome-hydrogel complex formulation relative to the hydrogel-, liposome- or phosphate buffer (pH 7.4)-only phytochemical preparations (Park, 2013)

Paclitaxel is a diterpenoid pseudoalkaloid extracted from *Taxus brevifolia*, which possesses potent anticancer activity and is usually administered in parenteral form (Singla et al, 2002; Horwitz, 1992; Wani et al, 1971). The paclitaxel (and its derivative, docetaxel) formulations are popularly used in many malignancies, such as breast, ovarian, and lung cancers, among others. Utreja et al (2011) attempted the development of an elastic liposome formulation for topical delivery of paclitaxel in skin cancer (Utreja et al, 2011). The results showed maximum drug loading of 6.0 mg/ml in the elastic liposome relative to 2.0 mg/ml in the conventional liposomes due to the hydrophobic nature of paclitaxel and inclusion of a surfactant in the former. The elastic liposomes also showed the highest permeation and deposition in the skin compared to the conventional liposome formulation and the drug solution.

Also, a curcumin liposome formulation was reported to have exhibited improved antitumor activities via topical application on the skin (Lin et al, 2012). Chavda et al (2022), in a review, outlined a list of other plant-derived phytochemicals that have proven anticancer activities, with their plant sources and mechanisms of action (Chavda et al, 2022). The use of liposome technology in their formulations may enhance the therapeutic applications. The development of effective topical formulations for dermal, intradermal, or transdermal delivery of phytochemicals should involve, among other factors, taking cognizance of the physicochemical properties of the agent related to aqueous solubility, lipophilicity, permeability, molecular and dose sizes, as well as safety (Naik et al, 2000). Successful development of topically applied forms of medicinally potent phytoconstituents for various diseases is highly encouraged so as to avail the consumers of the numerous benefits, including dose and cost reduction, safety from gastrointestinal and hepatic first-pass effects and systemic toxicities, and convenience of application and termination of dosage (Wiedersberg et al, 2014; Bos et al, 2000).

5.5.3 Parenteral

The parenteral route of drug administration is usually utilized when immediate action is desired especially for water-soluble agents. Liposomal formulations are versatile, being able to encapsulate both hydrophilic and hydrophobic drug particles; thus, they can be applied for parenteral pharmaceutical forms to improve solubility and bioavailability of poorly soluble drugs, including phytochemicals.

A curcumin-cyclodextrin complex was encapsulated in liposomes for parenteral (intravenous) administration (Dhule et al, 2012; Lan et al, 2005), which showed significantly higher anticancer activities compared to the plain drug. Also, a cationic liposome containing curcumin showed higher anticancer efficiency and caused more apoptosis than the plain phytochemical and its other types of formulation (Saengkrit et al, 2014). Bomana et al (1995) prepared liposomes of vincristine for achieving prolonged circulation, retention of drug in the target malignant tissue and improved in vivo anticancer activity (Khan et al, 2018; Bomana et al, 1995). Yuan (2006) developed a liposomal formulation of quercetin, a potent antioxidant phytoconstituent, which exhibited higher dose- and time-dependent percent in vitro inhibition of CT26 cancer cells proliferation compared to the plain drug (Yuan, 2006). Rajamanicka et al (2016) formulated a D-α-tocopheryl polyethene glycol 1000 succinate-coated resveratrol liposome, which showed significantly higher anticancer action than the uncoated liposomes and the free drug, with potentials for application in brain cancer (Rajamanicka et al, 2016).

Although doxorubicin is derived from *Streptomyces peucetius* (Johnson-Arbor et al, 2022), a bacterium, it would be important to briefly discuss the liposomal formulation of the chemotherapeutic agent here. Doxorubicin was approved in the early 1960s as an anticancer agent due to its effectiveness against varieties of cancers but has serious adverse effects, including cardiotoxicity and cytotoxicity on noncancerous cells. It is usually administered parenterally as a dispersion, the commercial product being available as Doxil®, a liposome formulation (Mbah et al, 2018; Bulbake et al, 2017; Tacara et al, 2012; European Medicines Agency (EMEA), 2011; Alza Corporation, 2011; Bhatnagar et al, 2019). Due to the severe systemic and cytotoxic adverse effects, efforts have been in progress to develop transdermal delivery forms of the potent chemotherapeutic agent as alternatives. In line with the objective, some researchers have worked on the development of microneedle-based liposomal and other vesicular systems of doxorubicin for transdermal delivery (Yang et al, 2019; Nguyen et al, 2018; Ahmad et al, 2018). However, no oral or transdermal form of doxorubicin has yet been approved for clinical application (Ahmad et al, 2018).

Formulation of liposomal parenteral phytochemicals for treatment of various diseases should incorporate strategies that would offer potentials for targeting, protection of the agent from opsonization by the circulating phagocytic endothelial system, and prolongation of circulation time, as could be achieved by stealth liposomes.

5.5.4 Pulmonary

The pulmonary route of drug administration, which involves inhalation of the dosage form, usually fine powder spray, into the respiratory tract down to the lungs, is a potential strategy to be explored for phytomedicines delivery. The suitability of the physicochemical properties of the agents and their development to the point of standardization for approval for clinical use may pose some challenges. Moreover, the self-defense mechanism of the respiratory system could be a limiting factor to the successful development of pulmonary delivery systems for phytoconstituents. However, a lot of researches have been conducted in this regard. Chavda et al (2022) had provided a recent comprehensive review on liposomes formulations for phytomedicines in cancer (Elsayed et al, 2007). The review dealt with recent developments in the application of liposome technology for pulmonary delivery of relevant phytochemicals in lung cancer and many other types of cancers, including breast, ovarian, liver, colorectal, prostate, cervical, blood (leukemia), pancreatic, brain, skin, and so on. The authors provided lists of recently registered and patented liposome preparations containing phytochemicals and also that of already marketed phytochemicals-loaded liposomal products for anticancer treatment, with their clinical trial records and tradenames. More than 60% of anticancer medications are sourced from phytoconstituents (Khan et al, 2018).

5.6 MECHANISM OF ENHANCEMENT OF THE BIOAVAILABILITY OF PHYTOCONSTITUENTS BY LIPOSOMES AND NANOLIPOSOMES

Phytoconstituents possess therapeutic potentials in the management multiple diseases, including cancer and, recently, COVID-19. They have received enormous acceptance because they are naturally sourced. However, their therapeutic and medicinal uses are limited due to poor solubility, bioavailability, and stability (Ngwa et al, 2020; Mansouri et al, 2020). Liposomes and nanoliposomes have been reported to be efficient vehicles for delivery of phytoconstituents because they improve their bioavailability through various mechanisms by which they entrap and preserve phytoconstituents. In addition, the bioavailability of phytoconstituents can be further improved by modification of liposome and nanoliposomes formulation methods, formulation designs, and structural modification of their membranes (Karpuz et al, 2020b); Sogut et al, 2021). This section will highlight the mechanisms of enhancement of bioavailability of specific phytoconstituents by liposomes and nanoliposomes.

5.6.1 Liposomal Formulation of Phytoconstituents in *Salvadora Persica* L. (Salvadoraceae) by the Spraying Technique

Bioavailability enhancement of phytoconstituents can be achieved by post-processing liposomes using the spraying technique (Yu et al, 2021; Refaat et al, 2019). The spraying technique involves an interplay of formulation factors, including temperature, polymer type, and quantity, as well as the spray gas flow (Khatib et al, 2021; Altin et al, 2018). These factors are then modified to formulate liposomes with desirable physicochemical properties which enhance bioavailability (Yu et al, 2021). A liposomal formulation containing extracts from *Salvadora persica* L. was developed as a potential anti-SARS-CoV-2 agent (Ngwa et al, 2020). The liposomal formulation was post-processed by the spraying technique (Refaat et al, 2019). The efficacy of the formulation was compared with 11 flavonoids derived from the aqueous plant extract. Inhibition of *in vitro* viral replication was 38.09 ± 0.83 and 85.56 ± 1.12% for the aqueous plant extract and the liposomal formulation, respectively. The *in vitro* efficacy of the liposomal formulation was comparable with that of the FDA-approved anti-COVID-19 agent, remdesivir, which was 91.20 ± 1.71%. The study confirmed that formulation of phytoconstituents into liposomes increases their bioavailability and efficacy (Ngwa et al, 2020).

5.6.2 Curcumin Nanoliposome Formulation by the Micelle-to-Vesicle Transition Method

The bioavailability of phytoconstituents have been proven to improve when formulated into nanoliposomes by the micelle-to-vesicle transition (MVT) method. Curcumin is a phytoconstituent derived from turmeric, which possesses antioxidant potential but is limited by its low bioavailability and rapid degradation (Voronin et al, 2021; Chen et al, 2015; Cheng et al, 2017; Salem et al, 2014; De Leo et al, 2018). In order to overcome these limitations, curcumin was encapsulated into nanoliposomes by the MVT method without the use of harmful organic solvents and cationic materials (De Leo et al, 2018). Mixed micelles were prepared by formulating a lipid-cholesterol-curcumin complex, which was added to a high-critical micellar concentration detergent solution of sodium cholate. Thereafter, the detergent molecules were eliminated from the micelles via size exclusion chromatography. This led to formation of curcumin-loaded nanoliposomes, with curcumin trapped within the lipid bilayer. The entrapment of curcumin within the lipid layer resulted in nanoliposome formulations that were stable, with encapsulation efficiency of 98% and optimal curcumin release profiles. The nanoliposomes were further modified by coating them with Eudragit S100, a pH-responsive polymer. The Eudragit S100 polymer protected premature release of curcumin at different parts of the gastrointestinal system while ensuring optimal release in the target organ. The entire process confirmed the enhanced bioavailability of Eudragit S100-coated, curcumin-loaded nanoliposomes for potential targeted colonic delivery of curcumin in colon cancer (De Leo et al, 2018; Jain et al, 2005).

5.6.3 Encapsulation of Naringenin into Nanoliposomes by Microfluidization and pH-Driven Method

Naringenin (5,7-trihydroxyflavonone) is a hydrophobic phytoconstituent mainly found in citrus fruits and tomatoes (Arafah et al, 2020). It has potential therapeutic applications in the treatment of Parkinson's disease, non-alcoholic fatty liver disease, and various cancer types. However, it is rapidly degraded during hepatic first-pass metabolism, which results in low plasma concentrations (Arafah et al, 2020). The pH-driven method involves dissolution of hydrophobic phytoconstituents in alkaline conditions and then adjusting the pH to a suitable value after incorporating them in the formulation matrix. It is an energy-saving, cost-effective, and easy procedure that eliminates the use of toxic organic solvents (Chen et al, 2021). Naringenin was exposed to alkaline conditions, which stimulated deprotonation of its hydroxyl group and improved its water solubility. The alkaline solution containing naringenin was mixed with the nanoliposome solution; thereafter, the pH was adjusted to an acidic condition. The acidic environment enabled the incorporation of naringenin in the hydrophobic part of the nanoliposomes. This produced naringenin-loaded nanoliposomes with 95.34% encapsulation efficiency, which translated to improved bioavailability. Although this method has obvious advantages, it cannot be applied in encapsulation of phytoconstituents (e.g., quercetin) degraded by high alkaline conditions (Peng et al, 2019).

5.6.4 Formulation of Long-Circulating Daidzein-Loaded Liposomes by Ultrasonication and Lipid Film-Hydration Method

Daidzein is a phytoconstituent obtained from soybeans and other legumes. It has multiple therapeutic potential including anti-inflammatory properties with a low bioavailability and rapid degradation (Wang et al, 2019b; Chinta et al, 2013). Daidzein long-circulating liposomes (DLCL) were prepared using the ultrasonication and lipid film-hydration method to enhance bioavailability of daidzein (Wang et al, 2019b). Soybean phosphatidylcholine, cholesterol, DSPE-mPEG2000, and daidzein were dissolved in chloroform-methanol solution. The mixture was evaporated under vacuum at 40°C to obtain a thin film, which was then hydrated by sonication. The liposomal suspension was obtained after filtration. DLCL exhibited the desired slow and prolonged release with encapsulation efficiency of 85.3 ± 3.6% and up to 2.5-fold increase in total absorption when compared with the free drug after administration in mice. The results propose the potential bioavailability enhancement of daidzein by preparation of long-circulating liposomes through ultrasonication and lipid film-hydration method (Wang et al, 2019b).

5.6.5 Formulation of Resveratrol-Loaded Peptide and Sucrose Liposomes

Resveratrol (RES) are phytoconstituents derived from some nuts and fruits, with an anticancer potential. However, like most phytoconstituents they are poorly water-soluble with low bioavailability (Zhao et al, 2020). Addition of sucrose and peptides to liposomes have been reported to improve functional properties and stability of liposomes (Guimarães et al, 2019; Almatroodi, et al, 2020). RES was incorporated in peptide and sucrose liposomes (PSL) to improve its bioavailability in potential treatment of breast cancer (Zhao et al, 2020). The RES-loaded peptide and sucrose liposomes (PSL@RES) were stable, improved bioavailability of resveratrol, and effectively inhibited the growth of cancer cells when compared with the free drug. PSL@RES was evaluated in mice with breast cancer. The therapeutic efficacy of PSL@RES at dosage 5 mg kg^{-1} was more effective than 10 mg kg^{-1} of free RES. Tumor growth was completely suppressed by PSL@RES at a dosage of 10 mg kg^{-1}. The incorporation of RES into peptides reduced the amount of toxic free RES in circulation, thereby improving the safety profile in mice. The study highlighted the potential of bioavailability enhancement of resveratrol through formulation as peptide-sucrose liposomes.

5.6.6 Formulation of EGCG-Loaded Liposomes with Vitamin E

Epigallocatechin gallate (EGCG) is one of the green tea catechins with the most-significant antioxidant activity. However, it is rapidly degraded in the system and possesses poor oral bioavailability and stability (Almatroodi, et al, 2020; Wang et al, 2006). Bioavailability of EGCG can be improved by liposome formulation and preventing degradation of EGCG through oxidation when ingested into the human system. EGCG-loaded liposomes were developed with chloroform-based phosphatidylcholine (PC) or phosphatidylserine (PS) (which are similar components of the cell membrane). The cholesterol solutions were mixed with aqueous solution of EGCG and α-tocopherol (vitamin E-VE) (Wang et al, 2006). The PC-EGCG-VE liposomes were more stable and showed higher encapsulation efficiency (76.8%) when compared with the PS-EGCG-VE liposomes (60.2%). This was attributed to the surface properties of PC, which repelled liposomes to prevent aggregation. The addition of vitamin E to the liposome during formulation protected EGCG from oxidation and increased the amount of EGCG entrapped into the formulation, thereby improving its bioavailability.

5.6.7 Albumin-Coating of Naringenin Liposomes

Albumin-coating of liposomes have been confirmed to improve bioavailability of hydrophobic drugs through formation of an albumin-drug complex which enables encapsulation of the drugs into the aqueous component of the liposome (Cheng et al, 2021; Wiedenhoeft et al, 2020; Liu et al, 2020). Naringenin-loaded albumin-coated liposomes were developed and reported to possess high entrapment efficiency (>85%) (Wang et al, 2020). The albumin-coating reduced liposome aggregation and drug leakage, thereby enhancing the bioavailability of naringenin. The liposome exhibited an overall high efficacy for the treatment of liver fibrosis when compared with other naringenin-based solutions. Albumin-coated liposomes of another phytoconstituent, ellagic acid, have also been successfully developed to its improve bioavailability, uptake, and stability (Taguchi et al, 2021). However, albumin-coating can only be applied for hydrophobic drugs which have affinity for albumin (Wang et al, 2020).

5.7 APPROVAL STATUS OF PHYTO-LIPOSOMES AND PHYTO-NANOLIPOSOMES

Since the FDA approval of the first nanoliposomes (Doxil©) in 1995, many researchers and drug delivery scientists have been in the fore front of developing nanomedicines based on liposomes. At present, the database of the Food and Drugs Administration (FDA) and European Medicines Agency (EMA) list only 14 types of authorized liposomal products with different indications as shown in Table 5.4 (Liu et al, 2022). However, the nanoliposomal products listed in Table 5.4 do not contain phyto-nanoliposomes, indicating that nanoliposomes encapsulating phytoconstituents are yet to gain worldwide approval.

5.8 FUTURE PERSPECTIVES

With the advances in drug delivery systems, it appears that phytochemicals offer opportunities for development of medicines capable of treating different disease conditions. Many phytonanoliposomes have been formulated and tested against different disease conditions (Vaiserman et al, 2021; Jagwani et al, 2020; Oskoueian et al, 2020; Nouri et al, 2020; Aisha et al, 2014). Phytobioactive nanoliposomes, such as *Orthosiphon stamineus* extract nanoliposomes, have higher bioavailability and in vitro antioxidant activity (Takahashi et al, 2009). Curcumin-encapsulated nanoliposomes show higher bioavailability after oral treatment in rats with enhanced antioxidant activity (Mourtas et al, 2014), and in vitro studies of multifunctional curcumin nanoliposomes proved their ability to cross the blood-brain barrier and were effective against Alzheimer's disease (Lushchak et al, 2020).

TABLE 5.4
Nanoliposomal Products Approved by FDA (USA) and EMA (European Union)

Product Name	API	Approved Year/Area	Dosage Form	Adm. Route	Indication
Doxil Caelyx	Doxorubicin hydrochloride (DOX HCl)	1995, US 1996, EU	Suspension	IV	Ovarian cancer, Kaposi's sarcoma, myeloid melanoma
DaunoXome	Daunorubicin	1996, US	Suspension	IV	Kaposi's sarcoma
AmBisome	Amphotericin B (AmpB)	1997, US	Lyo	IV	Systemic fungal infection
DepoCyt DepoCyte	Cytarabine	1999, US 2001, EU	Suspension	IT	Lymphomatous meningitis
Myocet	DOX HCl	2000, EU	3 vials	IV	Breast cancer
Visudyne	Verteporfin	2000, US	Lyo	IV	Wet AMD
DepoDur	Morphine	2000, EU 2004, US	Suspension	Epidural	Postoperative pain
Mepact	MTP-PE	2009, EU	Lyo	IV	Osteosarcoma
Exparel	Bupivacaine	2011, US 2020, EU	Suspension	Local infiltration	Post-surgical analgesia
Marqibo	Vincristine sulfate	2012, US	3 vials	IV	Leukemia
Onivyde	Irinotecan hydrochloride trihydrate	2015, US 2016, EU	Suspension	IV	Pancreatic adenocarcinoma
Vyxeos	Daunorubicin, cytarabine	2017, US 2018, EU	Lyo	IV	Leukemia
Shingrix	Recombinant varicella-zoster virus glycoprotein E	2018, EU	2 vials (powder and suspension)	IM	Against shingles and post-herpetic neuralgia
Arikayce	Amikacin sulfate	2018, US 2020, EU	Suspension	Oral inhalation	Lung disease

Phytobioactives are known to usually activate pathways involving different mechanisms of action leading to treatment of difficult-to-treat diseases. The therapeutic potential of orally administered phytochemicals is, however, significantly limited because of their low gastrointestinal absorption, chemical instability, low hydrophilicity, scarce biodistribution, and poor penetration/accumulation in the body (Patra et al, 2018). These features lead to decrease in the rate and extent of the absorption of oral drugs from solid dosage forms and reducing their bioavailability in the body. Nanoformulations, such as nanoemulsions, nanoliposomes, nanopolymersomes, nanocrystals, and lipid and polymeric nanoparticles, have been shown to provide multiple benefits over conventional formulations (Morea et al, 2021). Among these novel approaches, nanoliposomes offer unique advantages with respect to phytoconstituents because of its ability to accommodate large cargo of the bioactive in the different compartments of the delivery system. Importantly, multiple phytochemicals with synergistic effects can be encapsulated into the same nanoliposome vesicle, and the nanoliposomes can be further functionalized using ligand to enhance targeted delivery to specific cells with improved therapeutic efficacy. It is also very important to factor in the delivery route of phytonanoliposomes when developing such formulations. The oral route is usually the most preferred route outside the pharmaceutical, biochemical, and physiological challenges associated with it, which, expectedly, the nano nature would circumvent, but presenting nanoliposomes for oral delivery would also present its own challenges, such as long-term stability and ability to sustain the long circulation time in vivo. In preclinical models, the phytochemical-based nanoformulations are usually delivered through intravenous or subcutaneous route (Morea et al, 2021). All the same, with the interest and amount of research currently being undertaken, it is expected that products based phytonanoliposomes will be available on the market in the near future.

5.9 CONCLUSIONS

Plant-based drugs have been used for ages in the alleviation of many diseases and a good percentage of medicines on the market that provide the healthcare needs of global population are based on them. However, their presentation usually limits the scope of their application in the treatment of relevant diseases due to their inability to optimally yield the desired concentration in the systemic circulation. Nanotechnology-driven platforms, such as liposomes and nanoliposomes, have advantages of increased bioavailability, prolongation of drug circulation time, and multiple drug loading, all contributing to improved efficacy and decreased toxicity, and have emerged as novel platforms for ensuring better presentation of plant-based drugs, otherwise called phytobioactives. This chapter has addressed the existing gap and research initiatives to synthesize novel phytonanoliposomes and develop them into dosage forms that may solve the problems associated with conventional delivery systems of phytobioactive compounds.

CONFLICT OF INTEREST

The authors declare no conflict of interest.

REFERENCES

Abreu AS, Castanheira EM, Queiroz MJ, Ferreira PM, Vale-Silva LA, and Pinto E. Nanoliposomes for encapsulation and delivery of the potential antitumoral methyl 6-methoxy-3-(4-methoxyphenyl)-1H-indole-2-carboxylate. Nanoscale Research Letters. 2019;6(1): 482. https://doi.org/10.1186/1556-276X-6-482.

Abrishami M, Motamed Shariati M, Malaekeh-Nikouei B, Sadat Tajani A, Mahmoudi A, Abrishami M, et al. Preparation and in vivo evaluation of nanoliposomes containing vancomycin after intravitreal injection in albino rabbits. Iranian Journal of Basic Medical Sciences. 2020;23(4):551–555.

Aguilar-Pérez KM, Avilés-Castrillo JI, Medina DI, Parra-Saldivar R, and Iqbal HMN. Insight into nanoliposomes as smart nanocarriers for greening the twenty-first century biomedical settings. Frontiers in Bioengineering and Biotechnology. 2020;8: 579536. https://doi.org/10.3389/fbioe.2020.579536.

Ahmad N, Ahmad R, Md Alam A, and Ahmad FJ. Enhancement of oral bioavailability of doxorubicin through surface modified biodegradable polymeric nanoparticles. Chemistry Central Journal. 2018;12: 65. https://doi.org/10.1186/s13065-018-0434-1.

Aisha AFA, Majid AMSA, and Ismail Z. Preparation and characterization of nano liposomes of orthosiphon stamineus ethanolic extract in soybean phospholipids. BMC Biotechnology. 2014;14: 23.

Akbarzadeh A, Rezaei-Sadabady R, Davaran S, Joo SW, Zarghami N, Hanifehpour Y, et al. Liposome: Classification, preparation, and applications. Nanoscale Research Letters. 2013;8(1): 102. https://doi.org/10.1186/1556-276X-8-102

Allahou LW, Madani SY, and Seifalian A. Investigating the application of liposomes as drug delivery systems for the diagnosis and treatment of cancer. International Journal of Biomaterials. 2021 (September): e3041969. https://doi.org/10.1155/2021/3041969.

Allen TM, and Cullis PR. Liposomal drug delivery systems: From concept to clinical applications. Advanced Drug Delivery Review. 2013;65(1): 36–48.

Almatroodi, SA, Almatroudi A, Khan AA, et al. Potential therapeutic targets of Epigallocatechin Gallate (EGCG), the most abundant catechin in green tea, and its role in the therapy of various types of cancer. Molecules. 2020;25(14): 3146. https://doi.org/10.3390/molecules25143146.

Altin G, Gültekin-Özgüven M, and Ozcelic B. Chitosan coated liposome dispersions loaded with cacao hull waste extract: Effect of spray drying on physico-chemical stability and in vitro bioaccessibility. Journal of Food Engineering. 2018;223: 91–98.

Alza Corporation. Doxil® (Doxorubicin HCl liposome injection) for intravenous infusion: US prescribing information. Available at: http://www.doxiline.com. [Accessed 2011 Aug 25].

Amiri S, Rezazadeh-Bari M, Alizadeh-Khaledabad M, and Amiri S. New formulation of vitamin C encapsulation by nanoliposomes: Production and evaluation of particle size, stability and control release. Food Science and Biotechnology. 2019;28(2): 423–432.

Arafah A, Rehman MU, Mir TM, et al. Multi-therapeutic potential of naringenin (4′,5,7-Trihydroxyflavonone): Experimental evidence and mechanisms. Plants. 2020;9: 1784. https://doi.org/10.3390/plants9121784.

Ashour A, El Aziz MMA, and Melad ASG. A review on saponins from medicinal plants: Chemistry, isolation, and determination. Journal of Nanomedicine Research. 2019;7(4): 282–288. https://doi.org/10.15406/jnmr.2019.07.00199.

Assadpour S, Shiran M, and Akhtari J. Chitosan coating of anionic liposomes containing sumatriptan succinate: A candidate for nasal administration. Nanomedicine Journal. 2021;8(2): 132–139.

Bajelan E, Vali A, Ostad S, and Dadashzadeh S. Co-delivery of doxorubicin and PSC 833 (Valspodar) by stealth nanoliposomes for efficient overcoming of multidrug resistance. Journal of Pharmacy & Pharmaceutical Sciences: A Publication of the Canadian Society for Pharmaceutical Sciences, Société canadienne des sciences pharmaceutiques. 2012;15: 568–582.

Baradaran B, Mohammadi A, Shamekhi S, Majidazar N, Dilmaghani A, Soofiyani SR, et al. A novel method for the development of plasmid DNA-loaded nanoliposomes for cancer gene therapy. Drug Delivery and Translational Research. 2022;12: 1508–1520.

Barati M, Mirzavi F, Nikpoor AR, Sankian M, Namdar Ahmadabad H, Soleimani A, et al. Enhanced antitumor immune response in melanoma tumor model by anti-PD-1 small interference RNA encapsulated in nanoliposomes. Cancer Gene Therapy. 2021;29(6): 814–824.

Bazrgari F, Khameneh B, Fazly Bazzaz BS, Mahmoudi A, and Malaekeh-Nikouei B. Effect of the nanoliposomal formulations of rifampin and N-acetyl cysteine on staphylococcus epidermidis biofilm. Nanomedicine Journal. 2020;7(2): 131–137.

Beltrán-Gracia E, López-Camacho A, Higuera-Ciapara I, Velázquez-Fernández JB, and Vallejo-Cardona AA. Nanomedicine review: Clinical developments in liposomal applications. Cancer Nanotechnology. 2019;10: 1–40. https://doi.org/10.1186/s12645-019-0055-y.

Bernard AL, Guedeau-Boudeville MA, Marchi-Artzner V, Gulik-Krzywicki T, di Meglio JM, and Jullien L. Shear-induced permeation and fusion of lipid vesicles. Journal of Colloid and Interface Science. 2005;287(1): 298–306.

Bhanu P. 2020. Functional and Preservative Properties of Phytochemicals. London, UK; San Diego, CA, US; Cambridge, MA, US; Kidlington, Oxford: United Kingdom Academic Press, An Imprint of Elsevier.

Bhatnagar S, Bankar NG, Kulkarni MV, and Venuganti VVK. Dissolvable microneedle patch containing doxorubicin and docetaxel is effective in 4T1 xenografted breast cancer mouse model. International Journal of Pharmaceutics. 2019;556: 263–275.

Biosca A, Dirscherl L, Moles E, Imperial S, and Fernàndez-Busquets X. An immuno PEG liposome for targeted antimalarial combination therapy at the nanoscale. Pharmaceutics. 2009;11(7): 341. https://doi.org/10.3390/pharmaceutics11070341.

Bochicchio S, Dalmoro A, Lamberti G, and Barba AA. Advances in nanoliposomes production for ferrous sulfate delivery. Pharmaceutics. 2020;12(5): 445.

Bomana NL, Bally MB, Cullis PR, Mayer LD, and Web MS. Encapsulation of vincristine in liposomes reduces its toxicity and improves its anti-tumor efficacy. Journal of Lipid Research. 1995;5: 523–541. https://doi.org/10.3109/08982109509010240.

Bos JD, and Meinardi MM. The 500 Dalton rule for the skin penetration of chemical compounds and drugs. Experiment Dermatol: Viewpoint. 2000;9: 165–169.

Bulbake U, Doppalapudi S, Kommineni N, and Khan W. Liposomal formulations in clinical use: An updated review. Pharmaceutics. 2017;9(2): 12. https://doi.org/10.3390/pharmaceutics9020012.

Cai W, Geng C, Jiang L, Sun J, Chen B, Zhou Y, et al. Encapsulation of gemcitabine in RGD-modified nanoliposomes improves breast cancer inhibitory activity. Pharmaceutical Development and Technology. 2020;25(5): 640–648.

Carvalho MSD, Hortelão ACL, Calhelha RC, Abreu AS, Coutinho PJG, Queiroz M-JRP, et al. Fluorescence studies on potential antitumor 6-(hetero)arylthieno[3,2-b]pyridine derivatives in solution and in nanoliposomes. Journal of Photochemistry and Photobiology A: Chemistry. 2013;264: 56–66.

Chavda VP, Vihol D, Mehta B, Shah D, Patel M, Vora LK, Pereira-Silva M, and Paiva-Santos AC. Phytochemical-loaded liposomes for anticancer therapy: An updated review. Nanomedicine (Lond.). 2022;17(8): 547–568.

Chaves MA, and Pinho SC. Unpurified soybean lecithins impact on the chemistry of proliposomes and liposome dispersions encapsulating vitamin D3. Food Bioscience. 2020;37: 100700.
Chen C, Han D, Cai C, and Tang X. An overview of liposome lyophilization and its future potential. Journal of Controlled Release. 2010;142: 299–311.
Chen H, Torchilin V, and Langer R. Polymerized liposomes as potential oral vaccine carriers: Stability and bioavailability. Journal of Controlled Release. 1996;42: 263–272.
Chen M, Li R, Gao Y, et al. Encapsulation of hydrophobic and low-soluble polyphenols into nanoliposomes by pH-driven method: Naringenin and naringin as model compounds. Foods. 2021;10: 963. https://doi.org/10.3390/foods10050963.
Chen X, Zou L-Q, Niu J, et al. The stability, sustained release and cellular antioxidant activity of curcumin nanoliposomes. Molecules. 2015;20: 14293.
Cheng C, Peng S, Li Z, et al. Improved bioavailability of curcumin in liposomes prepared using a pH-driven, organic solvent-free, easily scalable process. Royal Society of Chemistry Advances. 2017;7: 25978–25986.
Cheng C-Y, Barro L, Tsai S-T, et al. Epigallocatechin-3-gallate-loaded liposomes favor anti-inflammation of microglia cells and promote neuroprotection. International Journal of Molecular Sciences. 2021;22: 3037. https://doi.org/10.3390/ijms22063037.
Chhikara N, Kushwaha K, Sharma P, Gat Y, and Panghal A. Bioactive compounds of beetroot and utilization in food processing industry: A critical review. Food Chemistry. 2019;272(January): 192–200. https://doi.org/10.1016/j.foodchem.2018.08.022.
Chinta, SJ, Ganesan A, Reis-Rodrigues P, et al. Anti-inflammatory role of the isoflavone diadzein in lipopolysaccharide-stimulated microglia: Implications for Parkinson's disease. Neurotoxicity Research. 2013;23: 145–153.
Cuong DX, Hoan NX, Dong DH, Thuy LTM, Van Thanh N, Ha HT, Tuyen DTT, and Chinh DX. 2019. Tannins: Extraction from Plants. www.intechopen.com. IntechOpen. https://www.intechopen.com/chapters/67043.
Dag D, Guner S, and Oztop MH. Physicochemical mechanisms of different biopolymers' (lysozyme, gum arabic, whey protein, chitosan) adsorption on green tea extract loaded liposomes. International Journal of Biological Macromolecules. 2019;138(October): 473–482. https://doi.org/10.1016/j.ijbiomac.2019.07.106.
Dai Y, Song Y, Xie J, Xiao G, Li X, Li Z, et al. CB1-antibody modified liposomes for targeted modulation of epileptiform activities synchronously detected by microelectrode arrays. ACS Applied Materials & Interfaces. 2020;12(37): 41148–41156.
Dalmoro A, Bochicchio S, Lamberti G, Bertoncin P, Janssens B, and Barba AA. Micronutrients encapsulation in enhanced nanoliposomal carriers by a novel preparative technology. RSC Advances. 2019;9(34): 19800–19812.
Daraee H, Etemadi A, Kouhi M, Alimirzalu S, and Akbarzadeh A. Application of liposomes in medicine and drug delivery. Artificial Cells, Nanomedicine, and Biotechnology. 2016;44(1): 381–391.
Das M, and Huang L. Liposomal nanostructures for drug delivery in gastrointestinal cancers. Journal of Pharmacology and Experimental Therapeutics. 2019;370(3): 647–56.
Davidson EM, Barenholz Y, Cohen R, Haroutiunian S, Kagan, L, and Ginosar Y. High-dose bupivacaine remotely loaded into multivesicular liposomes demonstrates slow drug release without systemic toxic plasma concentrations after subcutaneous administration in humans. Anesthesia & Analgesia. 2010;110: 1018–1023.
De Leo V, Milano F, Mancini E, et al. Encapsulation of curcumin-loaded liposomes for colonic drug delivery in a pH-responsive polymer cluster using a pH-driven and organic solvent-free process. Molecules. 2018;23: 739. https://doi.org/10.3390/molecules23040739.
De Simone V, Dalmoro A, Bochicchio S, Caccavo D, Lamberti G, Bertoncin P, et al. Nanoliposomes in polymeric granules: Novel process strategy to produce stable and versatile delivery systems. Journal of Drug Delivery Science and Technology. 2020;59: 101878.
Dhule S, Penfornis P, Frazier T, Walker R, Feldman J, Tan G, He J, Alb A, John V, and Pochampally R. Curcumin-loaded γ-cyclodextrin liposomal nanoparticles as delivery vehicles for osteosarcoma. Nanomedicine: Nanotechnology, Biollogy and Medicine. 2012;8: 440–451. https://doi.org10.1016/j.nano.2011.07.011.
Dias DA, Urban S, and Roessner U. A historical overview of natural products in drug discovery. Metabolites. 2012;2(2): 303–336. https://doi.org/10.3390/metabo2020303.
Dua JS, Rana PAC, Bhandari DAK, editors. 2012. Liposome: Methods of Preparation and Applications. International Journal of Pharmaceutical Studies and Research. E-ISSN 2229-4619

Egbuna C, Mishra AP, and Goyal MR. 2021. Preparation of Phytopharmaceuticals for the Management of Disorders: The Development of Nutraceuticals and Traditional Medicine. London, United Kingdom; San Diego, CA, US: Elsevier.

Elsayed MMA, Abdallah OY, Naggar VF, and Khalafallah NM. Lipid vesicles for skin delivery of drugs: Reviewing three decades of research. International Journal of Pharmaceutics. 2007;332: 1–16.

European Medicines Agency (EMEA). Summary of product characteristics: Caelyx (Doxorubicin hydrochloride in a pegylated liposomal formulation). Available at: http://www.emea.europa.eu/humandocs/PDFs/EPAR/Caelyx/H-089-PI.en.pdf. [Accessed 2011 Aug 25].

Fan Y, Marioli M, and Zhang K. Analytical characterization of liposomes and other lipid nanoparticles for drug delivery. Journal of Pharmaceutical and Biomedical Analysis. 2021;192: 113642.

Fathi M, Mozafari MR, and Mohebbi M. Nanoencapsulation of food ingredients using lipid based delivery systems. Trends in Food Science & Technology. 2012;23(1): 13–27, ISSN 0924-2244. https://doi.org/10.1016/j.tifs.2011.08.003.

Franzé S, Selmin F, Samaritani E, Minghetti P, and Cilurzo F. Lyophilization of liposomal formulations: Still necessary, still challenging. Pharmaceutics. 2018;10: 139. https://doi.org/10.3390/pharmaceutics10030139.

Gao DW, Tang SN, and Tong Q. Oleanolic acid liposomes with polyethylene glycol modification: Promising antitumor drug delivery. International Journal of Nanomedicine. 2012;7: 3517–3526.

Ghahary A, Tredget EE, Shen Q, Kilani RT, Scott PG, and Takeuchi M. Liposome associated interferon-alpha-2b functions as an anti-fibrogenic factor in dermal wounds in the guinea pig. Molecular and Cellular Biochemistry. 2008;208(1–2): 129–137. https://doi.org/10.1023/a:1007054424400.

Ghassemi S, Haeri A, Shahhosseini S, and Dadashzadeh S. Labrasol-enriched nanoliposomal formulation: Novel approach to improve oral absorption of water-insoluble drug, carvedilol. American Association of Pharmaceutical Scientists PharmSciTech. 2018;19(7): 2961–2970.

Ghosh S, Lalani R, Maiti K, Banerjee S, Patel V, Bhowmick S, et al. Optimization and efficacy study of synergistic vincristine coloaded liposomal doxorubicin against breast and lung cancer. Nanomedicine. 2020;15(26): 2585–2607.

Golmohammadzadeh S, Imani F, Hosseinzadeh H, and Jaafari MR. Preparation, characterization and evaluation of sun protective and moisturizing effects of nanoliposomes containing safranal. Iranian Journal of Basic Medical Sciences. 2011;14(6): 521–533.

Goranthla S, Rapalli VK, Waghule T, et al. Nanocarriers for ocular drug delivery: Current status and translational opportunity. RSC Advances. 2020;20: 27835–27855.

Gorjian H, Raftani Amiri Z, Mohammadzadeh Milani J, and Ghaffari Khaligh N. Preparation and characterization of the encapsulated myrtle extract nanoliposome and nanoniosome without using cholesterol and toxic organic solvents: A comparative study. Food Chemistry. 2021;342: 128342.

Gosangari SL, and Watkin KL. Enhanced release of anticancer agents from nanoliposomes in response to diagnostic ultrasound energy levels. Pharmaceutical Development and Technology. 2012;17(3): 383–388.

Guidelines for Evaluation of Nanopharmaceuticals in India. 2019. Available at: https://www.drishtiias.com/daily-updates/daily-news-analysis/guidelines-for-evaluation-of-nanopharmaceuticals-in-india.

Guimarães D, Noro J, Silva C, et al. Protective effect of saccharides on freeze-dried liposomes encapsulating drugs. Frontiers in Bioengineering and Biotechnology. 2019;7: 424. https://doi.org/10.3389/fbioe.2019.00424.

Haeri A, Sadeghian S, Rabbani S, Anvari MS, Boroumand MA, and Dadashzadeh S. Use of remote film loading methodology to entrap sirolimus into liposomes: Preparation, characterization and in vivo efficacy for treatment of restenosis. International Journal of Pharmaceutics. 2011;414(1): 16–27.

Hamed A, Osman R, Al-Jamal KT, Holayel SM, and Geneidi A-S. Enhanced antitubercular activity, alveolar deposition and macrophages uptake of mannosylated stable nanoliposomes. Journal of Drug Delivery Science and Technology. 2019;51: 513–523.

Han S-B, Won B, Yang S-C, and Kim D-H. Asterias pectinifera derived collagen peptide-encapsulating elastic nanoliposomes for the cosmetic application. Journal of Industrial and Engineering Chemistry. 2021;98: 289–297.

Harmata M. 2021. Strategies and Tactics in Organic Synthesis. Volume 15. Amsterdam: Elsevier.

Hasan M, Elkhoury K, Belhaj N, Kahn C, Tamayol A, Barberi-Heyob M, et al. Growth-inhibitory effect of chitosan-coated liposomes encapsulating curcumin on MCF-7 breast cancer cells. Marine Drugs. 2020;18(4): 217.

Heidarli E, Dadashzadeh S, and Haeri A. State of the art of stimuli-responsive liposomes for cancer therapy. Iranian Journal of Pharmaceutical Research: IJPR. 2017;16(4): 1273–1304. Available at: https://www.ncbi.nlm.nih.gov/pmc/articles/PMC5843293/.

Heravi Shargh V, Luckett J, Bouzinab K, Paisey S, Turyanska L, Singleton WGB, et al. Chemosensitization of temozolomide-resistant pediatric diffuse midline glioma using potent nanoencapsulated forms of a N(3)-propargyl analogue. ACS Applied Materials & Interfaces. 2021;13(30): 35266–35280.

Horwitz SB. Mechanism of action of taxol. Trends in Pharmacological Sciences. 1992;13(4): 134–136.

Hu S, Niu M, Hu F, Lua Y, Qi J, Yin Z, and Wu W. Integrity and stability of oral liposomes containing bile salts studied in simulated and ex vivo gastrointestinal media. International Journal of Pharmaceutics. 2013;441: 693–700.

Huang L, Li R, Wang X, Zhang L, and Zhang B. Preparation and characterization of nanoliposomes loaded with the antimicrobial peptide CGA-N9. International Journal of Peptide Research and Therapeutics. 2021;(4): 2727–2734.

Iwanaga K, Ono S, Narioka K, Morimoto K, Kakemi M, Yamashita S, Nango M, and Oku N. Oral delivery of insulin by using surface coating liposomes: Improvement of stability of insulin in GI tract. International Journal of Pharmaceutics. 1997;157: 73–80.

Jagwani S, Jalalpure S, Dhamecha D, Jadhav K, and Bohara R. Pharmacokinetic and pharmacodynamic evaluation of resveratrol loaded cationic liposomes for targeting hepatocellular carcinoma. ACS Biomaterials Science & Engineering. 2020;6(9): 4969–4984. https://doi.org/10.1021/acsbiomaterials.0c00429.

Jain D, Panda AK, and Majumdar DK. Eudragit S100 entrapped insulin microspheres for oral delivery. American Association of Pharmaceutical Scientists PharmSciTech. 2005;6: E100–E107.

Jakoby J, Beuschlein F, Mentz S, Hantel C, and Süss R. Liposomal doxorubicin for active targeting: Surface modification of the nanocarrier evaluated in vitro and in vivo: Challenges and prospects. Oncotarget. 2015;6(41): 43698–43711. https://doi.org/10.18632/oncotarget.6191.

Jensen GM. The care and feeding of a commercial liposomal product: Liposomal amphotericin B (AmBisome). Journal of Liposome Research. 2017;27: 173–179.

Johnson-Arbor K, and Dubey R. Doxorubicin. National Library of Medicines. 2022. Available at: https://www.ncbi.nlm.nih.gov/books/NBK459232/. [Accessed 2022 Jun 1].

Joshee N, Dhekney SA, and Parajuli P. 2019. Medicinal Plants: From Farm to Pharmacy. Cham, Switzerland: Springer.

Justo OR, and Moraes AM. Kanamycin incorporation in lipid vesicles prepared by ethanol injection designed for tuberculosis treatment. Journal of Pharmacy and Pharmacology. 2005;57(1): 23–30. https://doi.org/10.1211/0022357055092.

Kalepu S, Manthina M, and Padavala V. Oral lipid-based drug delivery systems—An overview. Acta Pharmaceutica Sinica B. 2013;3(6): 361–372.

Kanásová M, and Nesměrák K. Systematic review of liposomes' characterization methods. Monatshefte für Chemie—Chemical Monthly. 2017;148(9): 1581–1593.

Karimi N, Ghanbarzadeh B, Hajibonabi F, Hojabri Z, Ganbarov K, Kafil HS, Hamishehkar H, et al. Turmeric extract loaded nanoliposome as a potential antioxidant and antimicrobial nanocarrier for food applications. Food Bioscience. 2019;29(June): 110–117. https://doi.org/10.1016/j.fbio.2019.04.006.

Karpuz M, Gunay MS, and Ozer AY. 2020a. Liposomes and Phytosomes for Phytoconstituents: In Advances in the Development of Novel Carriers for Bioactives and Biological Agents. Elsevier Inc, Academic Press, pp. 525–553. Available at: https://www.sciencedirect.com/science/article/pii/B9780128196663000183?via%3Dihub. https://doi.org/10.1016/b978-0-12-8-19666-3.00018-3.

Karpuz M, Gunay MS, and Ozer AY. 2020b. "Liposomes and phytosomes for phytoconstituents" in advances and avenues in the development of novel carriers for bioactives and biological agents. Singh MR, Singh D, Kanwar JR et al. (Eds.), Advances and Avenues in the Development of Novel Carriers for Bioactives and Biological Agents. Academic Press, pp. 525–553, ISBN 9780128196663, https://doi.org/10.1016/B978-0-12-819666-3.00018-3.

Khan T, and Gurav P. Phyto nanotechnology: Enhancing delivery of plant based anti-cancer drugs. Frontiers in Pharmacology. 2018;8: 1002. https://doi.org/10.3389/fphar.2017.01002.

Khatib I, Chow MYT, Ruan J., et al. Modeling of a spray drying method to produce ciprofloxacin nanocrystals inside the liposomes utilizing a response surface methodology: Box-behnken experimental design. International Journal of Pharmaceutics. 2021;597: 120277.

Khorasani S, Danaei M, and Mozafari MR. Nanoliposome technology for the food and nutraceutical industries. Trends in Food Science & Technology. 2018;79: 106–115.

Kim AR, Kim SN, Jung IK, Kim HH, Park YH, and Park WS. The inhibitory effect of Scutellaria baicalensis extract and its active compound, baicalin, on the translocation of the androgen receptor with implications for preventing androgenetic alopecia. Planta Medica. 2014;80(2–3): 153–158.

Kim HY, Kang M, Choo YW, et al. Immunomodulatory lipocomplex functionalized with photosensitizer-embedded cancer cell membrane inhibits tumor growth and metastasis. Nano Letters. 2019;19(8): 5185–93.

Kogan M, Feng B, Norden B, Rocha S, and Beke-Somfai T. Shear-induced membrane fusion in viscous solutions. Langmuir. 2014;30(17): 4875–8.

Kokkona M, Kallinteri P, Fatouros D, and Antimisiaris SG. Stability of SUV liposomes in the presence of cholate salts and pancreatic lipases: Effect of lipid composition. European Journal of Pharmaceutical Sciences. 2000;9: 245–252.

Kurek J. 2019. Introductory Chapter: Alkaloids—Their Importance in Nature and for Human Life. www.intechopen.com. IntechOpen. Available at: https://www.intechopen.com/chapters/66742.

Labouta HI, Gomez-Garcia MJ, Sarsons CD, Nguyen T, Kennard J, Ngo W, Terefe K, et al. Surface-grafted polyethylene glycol conformation impacts the transport of PEG-functionalized liposomes through a tumour extracellular matrix model. RSC Advances. 2018;8(14): 7697–7708. https://doi.org/10.1039/c7ra13438j.

Lan L, Braiteh FS, and Kurzrock R. Liposome-encapsulated curcumin. Cancer. 2005;104: 1322–1331.

Large DE, Abdelmessih RG, Fink EA, and Auguste DT. Liposome composition in drug delivery design, synthesis, characterization, and clinical application. Advanced Drug Delivery Reviews. 2021;176: 113851.

Li C, Cui J, Li Y, et al. Copper ion-mediated liposomal encapsulation of mitoxantrone: The role of anions in drug loading, retention and release. European Journal of Pharmaceutical Sciences. 2008 7;34(4–5): 333–344. https://doi.org/10.1016/j.ejps.2008.05.006.

Li C, Zhang XL, Huang XL, Wang XY, Liao GJ, and Chen ZB. Preparation and characterization of flexible nanoliposomes loaded with daptomycin, a novel antibiotic, for topical skin therapy. International Journal of Nnaomedicine. 2013;8: 1292.

Li J, Niu R, Dong L, Gao L, Zhang J, Zheng Y, et al. Nanoencapsulation of curcumin and its protective effects against CCl_4-induced hepatotoxicity in mice. Journal of Nanomaterials. 2019;2019: 7140132.

Li Z, Paulson AT, and Gill TA. Encapsulation of bioactive salmon protein hydrolysates with chitosan-coated liposomes. Journal of Functional Foods. 2015;19: 733–743.

Lin YL, Liu YK, Tsai NM, Hsieh JH, Chen CH, Lin CM, and Liao K-W. A Lipo-PEG-PEI complex for encapsulating curcumin that enhances its antitumor effects on curcumin-sensitive and curcumin-resistant cells. Nanomedicine. 2012, 8: 318–327. https://doi.org/10.1016/j.nano.2011.06.011. [Available 2011 Jun 24].

Liposome Drug Products Chemistry, Manufacturing, and Controls; Human Pharmacokinetics and Bioavailability; and Labeling Documentation. U.S. Department of Health and Human Services Food and Drug Administration Center for Drug Evaluation and Research (CDER) April 2018 Pharmaceutical Quality/CMC. Available at: https://www.fda.gov/media/70837/download.

Liu P, Chen G, and Zhang, J. A review of liposomes as a drug delivery system: Current status of approved products, regulatory environments, and future perspectives. Molecules. 2022;27: 1372. https://doi.org/10.3390/molecules27041372.

Liu W, Liu J, Liu W, Li T, and Liu C. Improved physical and in vitro digestion stability of a polyelectrolyte delivery system based on layer-by-layer self-assembly alginate–chitosan-coated nanoliposomes. Journal of Agricultural and Food Chemistry. 2013;61(17): 4133–4144.

Liu X, Situ A, Kang Y, et al. Irinotecan delivery by lipid coated mesoporous silica nanoparticles shows improved efficacy and safety over liposomes for pancreatic cancer. ACS Nano. 2016;10(2): 2702–15.

Liu YC, Lin MTY, Ng AHC, et al. Nanotechnology for the treatment of allergic conjunctival diseases. Pharmaceuticals. 2020;13: 351.

Lujan H, Griffin WC, Taube JH, and Sayes CM. Synthesis and characterization of nanometer-sized liposomes for encapsulation and microRNA transfer to breast cancer cells. International Journal of Nanomedicine. 2019;14: 5159–5173.

Lushchak O, et al. Geroscience. 2020;42: 117–139. https://doi.org/10.1007/s11357-019-00116-9.

Ma Y, Wang Z, Zhao W, Lu T, Wang R, Mei Q, et al. Enhanced bactericidal potency of nanoliposomes by modification of the fusion activity between liposomes and bacterium. International Journal of Nanomedicine. 2013;8: 2351–2360.

Maherani B, Arab-tehrany E, Kheirolomoom A, Reshetov V, Stebe MJ, and Linder M. Optimization and characterization of liposome formulation by mixture design. Analyst. 2012;137(3): 773–786.

Mahomoodally F, Asif F, Rahmon R, Ishaq A, and Nisar S. A review of the pharmacological potential ad phytochemical profile of weeping Fig. Ficus bejamina. International Journal of Chemical and Biochemical Sciences. 2019;16(2019): 70–75.

Maja L, Željko K, and Mateja P. Sustainable technologies for liposome preparation. The Journal of Supercritical Fluids. 2020;165: 104984.

Manikkath J, Parekh HS, and Mutalik S. Surface-engineered nanoliposomes with lipidated and non-lipidated peptide-dendrimeric scaffold for efficient transdermal delivery of a therapeutic agent: Development, characterization, toxicological and preclinical performance analyses. European Journal of Pharmaceutics and Biopharmaceutics. 2020;156: 97–113.

Mansouri K, Rasoulpoor S, Daneshkhah A, et al. Clinical effects of curcumin in enhancing cancer therapy: A systematic review. BMC Cancer. 2020;20: 791. https://doi.org/10.1186/s12885-020-07256-8.

Mbah CC, and Attama AA. Vesicular carriers as innovative nanodrug delivery formulations. Grumezescu AM (Ed.), Organic Materials as Smart Nanocarriers for Drug Delivery, (Multivolume Set), Chapt. 12, 2018. London: Elselvier Inc., pp. 519–559. https./doi.org/10.1016/B978-0012-B13663-8.00012-9. [Available 2018 Mar 30].

Mbah CC, Builders PF, and Attama AA. Nanovesicular carriers as alternative drug delivery systems: Ethosomes in focus. Expert Opinion on Drug Delivery. 2014;11(1): 45–59.

Memvanga PB, Coco R, and Préat V. An oral malaria therapy: Curcumin-loaded lipid-based drug delivery systems combined with β-arteether. Journal of Controlled Release. 2013;172(3): 904–913. https://doi.org/10.1016/j.jconrel.2013.09.001.

Memvanga PB, and Nkanga CI. Liposomes for malaria management: The evolution from 1980 to 2020. Malaria Journal. 2021;20(1). https://doi.org/10.1186/s12936-021-03858-0.

Merlo-Mas J, Tomsen-Melero J, Corchero J-L, González-Mira E, Font A, Pedersen JN, et al. Application of quality by design to the robust preparation of a liposomal GLA formulation by DELOS-susp method. The Journal of Supercritical Fluids. 2021;173: 105204.

Mezzomo NJ, Becker Borin D, Ianiski F, et al. Creatine nanoliposome reverts the HPA-induced damage in complex II–III activity of the rats' cerebral cortex. Molecular Biology Reports. 2019;46: 5897–5908. https://doi.org/10.1007/s11033-019-05023-y.

Misra A, Jinturkar K, Patel D, Lalani J, and Chrougule M. Recent advances in liposomal dry powder formulations: Preparation and evaluation. Expert Opinion on Drug Delivery. 2009;6: 71–88.

Molaveisi M, Shahidi-Noghabi M, and Naji-Tabasi S. Controlled release and improved stability of vitamin D3 within nanoliposomes stabilized by palmitic acid. Journal of Food Safety. 2021;41(5): e12924.

Morea MP, Pardeshib SR, Pardeshic CV, Sonawanea GA, Shinded MN, Deshmukhe PK, Naikb JB, Kulkarnia AD. Recent advances in phytochemical-based nano-formulation for drug-resistant cancer. Medicine in Drug Discovery. 2021;10: 100082. https://doi.org/10.1016/j.medidd.2021.100082.

Mourtas S, Lazar AN, Markoutsa E, Duyckaerts C, and Antimisiaris SG. Multifunctional nanoliposomes with curcumin-lipid derivative and brain targeting functionality with potential applications for Alzheimer disease. European Journal of Medicinal Chemistry. 2014;80: 175–183.

Mozafari MR. 2010a. Nanoliposomes: Preparation and analysis. Weissig V, (Eds.), Liposomes: Methods in Molecular Biology (Methods and Protocols), Volume 605. Humana Press. https://doi.org/10.1007/978-1-60327-360-2_2.

Mozafari MR. 2010b. Nanoliposomes: Preparation and Analysis. Liposomes: Methods and Protocols, Volume 1: Pharmaceutical Nanocarriers. Totowa, NJ: Humana Press, pp. 29–50.

Mu J, Lin J, Huang P, and Chen X. Development of endogenous enzyme-responsive nanomaterials for theranostics. Chemical Society Review. 2018;47(15): 5554–73.

Naik A, Kalia YN, and Guy RH. Transdermal drug delivery: Overcoming the skin's barrier function. Pharmaceutical Science & Technology Today. 2000;3: 318–326.

Najlah M, Said Suliman A, Tolaymat I, Kurusamy S, Kannappan V, Elhissi AMA, et al. Development of injectable PEGylated liposome encapsulating disulfiram for colorectal cancer treatment. Pharmaceutics. 2019;11(11): 610.

Natarajan JV, Ang M, Darwitan A, Chattopadhyay S, Wong TT, and Venkatraman SS. Nanomedicine for glaucoma: Liposomes provide sustained release of latanoprost in the eye. International Journal of Nanomedicine. 2012;7: 123–131.

Naziris N, Pippa N, and Demetzos C. A novel, nontoxic and scalable process to produce lipidic vehicles. Materials (Basel, Switzerland). 2020;13(21): 5035.

Neupane R, Boddu SHS, Abou-Dahech MS, Bachu RD, Terrero D, Babu RJ, and Tiwari AK. Transdermal delivery of chemotherapeutics: Strategies, requirements, and opportunities. Pharmaceutics. 2021;13, 960. https://doi.org/10.3390/pharmaceutics13070960.

Nguyen HX, Bozorg BD, Kim Y, Wieber A, Birk G, Lubda D, and Banga AK. Poly (vinyl alcohol) microneedles: Fabrication, characterization, and application for transdermal drug delivery of doxorubicin. European Journal of Pharmaceutics and Biopharmaceutics. 2018;129: 88–103.

Ngwa W, Kumar R, Thompson D, et al. Potential of flavonoid-inspired phytomedicines against COVID-19. Molecules. 2020;25: 2707. https://doi.org/10.3390/molecules25112707.

Nikkhah M, Khoshkhoo Z, Hosseini SE, Shotorbani PM, and Basti AA. Physical stability of microliposomes in bene (Pistacia atlantica) oil with different formulations. Quality Assurance and Safety of Crops & Foods. 2020;12: 49.

Niu G, Cogburn B, and Hughes J. Preparation and characterization of doxorubicin liposomes. Methods in Molecular Biology. 2010;624: 211–219. https://doi.org/10.1007/978-1-60761-609-214.

Nouri M. Preparation of nanoliposomes containing Hyssopus officinalis L. and Eryngium caeruleum M.Bieb extracts and investigate their antimicrobial and antimicrobial effects. Journal of Medicinal Plants. 2020;19(75): 118–131. Available at: http://jmp.ir/article-1-2548-en.html.

Olatunde OO, Benjakul S, Vongkamjan K, and Amnuaikit T. Influence of stabilising agents on the properties of liposomal encapsulated ethanolic coconut husk extract. International Journal of Food Science & Technology. 2019;55(2): 702–711. https://doi.org/10.1111/ijfs.14339.

Oskoueian E, Karimi E, Noura R, Ebrahimi M, and Shafaei N. Nanoliposomes encapsulation of enriched phenolic fraction from pistachio hulls and its antioxidant, anti-inflammatory, and anti-melanogenic activities. Journal of Microencapsulation. 2020;37(1): 1–13. https://doi.org/10.1080/02652048.2019.1692941.

Panahi Y, Farshbaf M, Mohammadhosseini M, Mirahadi M, Khalilov R, Saghfi S, et al. Recent advances on liposomal nanoparticles: Synthesis, characterization and biomedical applications. Artificial Cells, Nanomedicine, and Biotechnology. 2017;45(4): 788–799.

Panche AN, Diwan AD, and Chandra SR. Flavonoids: An overview. Journal of Nutritional Science. 2016;5(e47). https://doi.org/10.1017/jns.2016.41.

Park SN. Preparation of quercetin and rutin-loaded ceramide liposomes and drug-releasing effect in the liposome-in-hydrogel complex system. Biochemical and Biophysical Research Communication. 2013;435: 361–366.

Patra JK, Das G, Fraceto LF, et al. Nano based drug delivery systems: Recent developments and future prospects. Journal of Nanobiotechnology. 2018;16: 71. https://doi.org/10.1186/s12951-018-0392-8.

Pattni BS, Chupin VV, and Torchilin VP. New developments in liposomal drug delivery. Chemical Review. 2015;115(19): 10938–66.

Păvăloiu R-D, Sha'at F, Neagu G, Deaconu M, Bubueanu C, Albulescu A, Sha'at M, and Hlevca C. Encapsulation of polyphenols from lycium barbarum leaves into liposomes as a strategy to improve their delivery. Nanomaterials. 2021;11(8): 1938. https://doi.org/10.3390/nano11081938.

Peng S, Zou L, Liu W, Liu C, and McClements DJ. Fabrication and characterization of curcumin-loaded liposomes formed from sunflower lecithin: Impact of composition and environmental stress. Journal of Agricultural and Food Chemistry. 2018;66(46): 12421–12430.

Peng S, Zou L, Zhou W, et al. Encapsulation of lipophilic polyphenols into nanoliposomes using pH-driven method: Advantages and disadvantages. Journal of Agricultural and Food Chemistry. 2019;67: 7506–7511. [CrossRef]

Pinilla CMB, Thys RCS, and Brandelli A. Antifungal properties of phosphatidylcholine-oleic acid liposomes encapsulating garlic against environmental fungal in wheat bread. International Journal of Food Microbiology. 2019;293: 72–78.

Praveen A, Aqil M, Imam SS, Ahad A, Moolakkadath T, and Ahmad FJ. Lamotrigine encapsulated intra-nasal nanoliposome formulation for epilepsy treatment: Formulation design, characterization and nasal toxicity study. Colloids and Surfaces B: Biointerfaces. 2019;174: 553–562.

Rajamanicka M, Yellappa K, Karthikeyan C, and Kumar S. Pharmacokinetics, biodistribution, in vitro cytotoxicity and biocompatibility of Vitamin E TPGS coated trans-resveratrol liposomes. Colloids and Surfaces B: Biointerfaces. 2016;145: 479–491.

Raju TP, Reddy MS, and Reddy VP. Phytosomes: A novel phyto-phospholipid carrier for herbal drug delivery. International Research Journal of Pharmacy. 2011;2: 28–33.

Ramezani M, Malaekeh-nikouei B, Tahreh K, and Mohammad M-N. The effect of linear PEI on characteristics and transfection efficiency of PEI-based cationic nanoliposomes. Iranian Journal of Basic Medical Sciences. 2011;14.

Rastakhiz S, Yazdani M, Shariat S, Arab A, Momtazi-Borojeni AA, Barati N, et al. Preparation of nanoliposomes linked to HER2/neu-derived (P5) peptide containing MPL adjuvant as vaccine against breast cancer. Journal of Cellular Biochemistry. 2019;120(2): 1294–1303.

Rasti B, Jinap S, Mozafari MR, and Yazid AM. Comparative study of the oxidative and physical stability of liposomal and nanoliposomal polyunsaturated fatty acids prepared with conventional and Mozafari methods. Food Chemistry. 2012;135(4): 2761–2770.

Refaat H, Naguib YW, Elsayed MMA, et al. Modified spraying technique and response surface methodology for the preparation and optimization of propolis liposomes of enhanced anti-proliferative activity against human melanoma cell line A375. Pharmaceutics. 2019;28(11):558. https://doi.org/10.3390/pharmaceutics11110558. PMID: 31661945; PMCID: PMC6921042.

Renukuntla J, Vadlapudi AD, Patel A, Sai HS, Boddu Sai HS, and Mitra AK. Approaches for enhancing oral bioavailability of peptides and proteins. International Journal of Pharmaceutics. 2013, April 15;447(0): 75–93. https://doi.org/10.1016/j.ijpharm.2013.02.030.

Risaliti L, Pini G, Ascrizzi R, Donato R, Sacco C, Bergonzi MC, et al. Artemisia annua essential oil extraction, characterization, and incorporation in nanoliposomes, smart drug delivery systems against Candida species. Journal of Drug Delivery Science and Technology. 2020;59: 101849.

Rodriguez EB, Almeda RA, Vidallon MLP, and Reyes CT. Enhanced bioactivity and efficient delivery of quercetin through nanoliposomal encapsulation using rice bran phospholipids. Journal of the Science of Food and Agriculture. 2019;99(4): 1980–1989. https://doi.org/10.1002/jsfa.9396.

Saengkrit N, Saesoo S, Srinuanchai W, Phunpee S, and Ruktanonchai U. Influence of curcumin-loaded cationic liposome on anticancer activity for cervical cancer therapy. Colloids and Surfaces B: Biointerfaces. 2014;114: 349–356.

Salem M, Rohani S, and Gillies ER. Curcumin, a promising anti-cancer therapeutic: A review of its chemical properties, bioactivity and approaches to cancer cell delivery. Royal Society of Chemistry Advances. 2014;4: 10815–10829.

Sarabandi K, Mahoonak AS, Hamishehkar H, Ghorbani M, and Jafari SM. Protection of casein hydrolysates within nanoliposomes: Antioxidant and stability characterization. Journal of Food Engineering. 2019;251: 19–28.

Savaghebi D, Barzegar M, and Mozafari MR. Manufacturing of nanoliposomal extract from sargassum boveanum algae and investigating its release behavior and antioxidant activity. Food Science & Nutrition. 2019;8(1): 299–310. https://doi.org/10.1002/fsn3.1306.

Seyed Mohammad Gheibi H, Mahmoud RJ, Mahdi H, Peter EP, and Amirhossein S. Liposome circulation time is prolonged by CD47 coating. Protein & Peptide Letters. 2020;27(10): 1029–1037.

Shakeel K, Raisuddin S, Ali S, Imam SS, Rahman MA, Jain GK, et al. Development and in vitro/in vivo evaluation of artemether and lumefantrine co-loaded nanoliposomes for parenteral delivery. Liposome Research. 2019;29(1): 35–43.

Shalaby TI, El-Refaie WM, Shams El-Din RS, and Hassanein SA. Smart ultrasound-triggered doxorubicin-loaded nanoliposomes with improved therapeutic response: A comparative study. Journal of Pharmaceutical Sciences. 2020;109(8): 2567–2576.

Shariare MH, and Kazi M. Phospholipid based nano drug delivery systems of phytoconstituents. Smart Drug Delivery. IntechOpen. 2021a: 1–20. http://doi.org/10.5772/intechopen.101040.

Shariare MH, Noor HB, Khan JH, Uddin J, Ahamad SR, Altamimi MA, Alanazi FK, and Kazi M. Liposomal drug delivery of Corchorus olitorius leaf extract containing phytol using Design of Experiment (DoE): In-vitro anticancer and in-vivo anti-inflammatory studies. Colloid and Surfaces B: Biointerfaces. 2021b, 199: 111543.

Shariare MH, Rahman M, Lubna SR, Roy RS, Abedin J, Marzan AL, Altamimi MA, Ahamad SR, Ahmad A, Alanazi FK, and Kazi M. Liposomal drug delivery of Aphanamixis polystachya leaf extracts and its neurobehavioral activity in mice model. Scientific Reports. 2020;10: 6938. https://doi.org/10.1038/s41598-020-63894-9.

Sheoran R, Khokra SL, and Dureja H. Preparation, optimization and characterization of letrozole nanoliposomes. International Journal Pharmaceutical Sciences and Research. 2020;11(3): 1373–1379.

Shreya AB, Pandey A, Nikam AN, Patil PO, Sonawane R, Deshmukh PK, et al. One-pot development of spray dried cationic proliposomal dry powder insufflation: Optimization, characterization and bio-interactions. Journal of Drug Delivery Science and Technology. 2021;61: 102298.

Silindir M, Erdogan S, Ozer AY, and Maia S. Liposomes and their applications in molecular imaging. Journal of Drug Targeting. 2012;20: 401–415.

Singh AP, and Kumar S. 2019a. Applications of Tannins in Industry. www.intechopen.com. IntechOpen. pages 1–19. Available at: https://www.intechopen.com/chapters/69199.

Singh M, Devi S, Rana VS, Mishra BB, Kumar J, and Ahluwalia V. Delivery of phytochemicals by liposome cargos: Recent progress, challenges and opportunities. Journal of Microencapsulation. 2019b;36(3): 215–235. https://doi.org/10.1080/02652048.2019.1617361.

Singh MR. 2020. Advances and Avenues in the Development of Novel Carriers for Bioactives and Biological Agents. London; San Diego, CA; Cambridge, MA: Oxford Academic Press, An Imprint of Elsevier.

Singla AK, Garg A, and Aggarwal D. Paclitaxel and its formulations. International Journal of Pharmaceutics. 2002;235(1–2): 179–192.

Socaciu MA, Diaconeasa Z, and Socaciu C. Simple and fast procedure to incorporate doxorubicine in small unilamellar liposomes: Effects on liposome size and zeta potential. Studia UBB Chemia. 2019;64(3): 181–192.

Sogut O, Sezer UA, and Sezer S. Liposomal delivery systems for herbal extracts. Journal of Drug Delivery Science and Technology. 2021;61: 102147. https://doi.org/10.1016/j.jddst.2020.102147.

Srinivas NR. Baicalin, an emerging multi-therapeutic agent: Pharmacodynamics, pharmacokinetics, and considerations from drug development perspectives. Xenobiotica. 2010;40(5): 357–367.

Stone NR, Bicanic T, Salim R, and Hope W. Liposomal amphotericin B (AmBisome®): A review of the pharmacokinetics, pharmacodynamics, clinical experience and future directions. Drugs. 2016;76(4): 485–500. https://doi.org/10.1007/s40265-016-0538-7.

Suri R, Neupane YR, Kohli K, and Jain GK. Polyoliposomes: Novel polyol-modified lipidic nanovesicles for dermal and transdermal delivery of drugs. Nanotechnology. 2020;31(35): 355103.

Tacara O, Sriamornsak P, and Dass CR. Doxorubicin: An update on anticancer molecular action, toxicity and novel drug delivery systems. J. Pharm. Pharmacol. 2012;65: 157–170. https://doi.org/10.1111/j.2042-7158.2012.01567.x.

Taguchi K, Okamoto Y, Matsumoto K, et al. When albumin meets liposomes: A feasible drug carrier for biomedical applications. Pharmaceuticals. 2021;14: 296. https://doi.org/10.3390/ph14040296.

Takahashi M, Uechi S, Takara K, Asikin Y, and Wada K. Evaluation of an oral carrier system in rats: Bioavailability and antioxidant properties of liposome-encapsulated curcumin. Journal of Agriculture and Food Chemistry. 2009;57(19): 9141–9146.

Takeuchi H, Matsui Y, Sugihara H, Yamamoto H, and Kawashima Y. Effectiveness of submicron-sized, chitosan-coated liposomes in oral administration of peptide drugs. International Journal of Pharrmaceutics. 2005;303: 160–170.

Tamaddon L, Mohamadi N, and Bavarsad N. Preparation and characterization of mucoadhesive loratadine nanoliposomes for intranasal administration. Turkish Journal of Pharmaceutical Sciences. 2021;18(4): 492–497.

Tang WL, Tang WH, Chen WC, et al. Development of a rapidly dissolvable oral pediatric formulation for mefloquine using liposomes. Molecular Pharmaceutics. 5;14(6): 1969–1979. https://doi.org/10.1021/acs.molpharmaceut.7b00077.

Tekade RK 2019. Basic Fundamentals of Drug Delivery. London; San Diego, CA; Cambridge, MA: Oxford Academic Press, An Imprint Of Elsevier.

Thomford N, Senthebane D, Rowe A, Munro D, Seele P, Maroyi A, and Dzobo K. Natural products for drug discovery in the 21st century: Innovations for novel drug discovery. International Journal of Molecular Sciences. 2018;19(6): 1578. https://doi.org/10.3390/ijms19061578.

Triana-Martínez F, Picallos-Rabina P, Da Silva-Álvarez S, Pietrocola F, Llanos S, Rodilla V, Soprano E, et al. Identification and characterization of cardiac glycosides as senolytic compounds. Nature Communications. 2019;10(1). https://doi.org/10.1038/s41467 019 12888 x.

Tundis R, Loizzo MR, and Menichini F. An overview on chemical aspects and potential health benefits of limonoids and their derivatives. Critical Reviews in Food Science and Nutrition. 2013;54(2): 225–250. https://doi.org/10.1080/10408398.2011.581400.

Utreja P, Jain S, and Tiwary AK. Localized delivery of paclitaxel using elastic liposomes: Formulation development and evaluation. Drug Delivery. 2011;18(5): 367–376. https://doi.org/10.3109/10717544.2011.558527.

Vaiserman A, Koliada A, and Lushchak O. Phyto-nanotechnology in anti-aging medicine. Aging (Albany NY). 2021 Apr 30;13(8): 10818–10820. https://doi.org/10.18632/aging.203026.

Voronin GL, Hettiarachchi CA, and Harte FM. High pressure jet spray drying of condensed skim milk results in powders with enhanced interfacial properties. Journal of Food Engineering. 2021;292: 110249.

Wang G, Wang JJ, Yang GY, Du SM, Zeng N, Li DS, et al. Effects of quercetin nanoliposomes on C6 glioma cells through induction of type III programmed cell death. International Journal of Nanomedicine. 2012;7: 271–280.

Wang, Y., & Grainger, D. W. Lyophilized liposome-based parenteral drug development: Reviewing complex product design strategies and current regulatory environments. Advanced Drug Delivery Reviews. 2019. https://doi.org/10.1016/j.addr.2019.03.003.

Wang J, Ding Y, and Zhou W. Albumin self-modified liposomes for hepatic fibrosis therapy via SPARC-dependent pathways. International Journal of Pharmaceutics. 2020;574: 118940.

Wang L, Cao J, Li C, et al. Efficacy and safety of mitoxantrone hydrochloride liposome injection in Chinese patients with advanced breast cancer: A randomized, open-label, active-controlled, single-center, phase II clinical trial. Investigational New Drugs. 2021. 9: 11. https://doi.org/10.1007/s10637-021-01182-7.

Wang Q, Liu W, Wang J, et al. Preparation and pharmacokinetic study of daidzein long-circulating liposomes. Nanoscale Research Letters. 2019a;14: 321. https://doi.org/10.1186/s11671-019-3164-y.

Wang R, Zhou W, and Wen RA. Kinetic study of the thermal stability of tea catechins in aqueous systems using a microwave reactor. Journal of Agriculture and Food Chemistry. 2006;54: 5924–5932.

Wang S, Su R, Nie S, Sun M, Zhang J, Wu D, and Moustaid-Moussa N. Application of nanotechnology in improving bioavailability and bioactivity of diet-derived phytochemicals. The Journal of Nutritional Biochemistry. 2014;25(4): 363–376. https://doi.org/10.1016/j.jnutbio.2013.10.002.

Wang Y, and Grainger DW. Lyophilized liposome-based parenteral drug development: Reviewing complex product design strategies and current regulatory environments. Advanced Drug Delivery Reviews. 2019b. https://doi.org/10.1016/j.addr.2019.03.003.

Wang Z-Y, Zhang H, Yang Y, Xie X-Y, Yang Y-F, Li Z, et al. Preparation, characterization, and efficacy of thermosensitive liposomes containing paclitaxel. Drug Delivery. 2016;23(4): 1222–1231.

Wani MC, Taylor HL, Wall ME, Coggon P, and McPhail AT. Plant antitumor agents. VI: The isolation and structure of taxol, a novel antileukemic and antitumor agent from Taxus brevifolia. Journal of American Chemical Society. 1971;93(9): 2325–2327.

Wei X, Wu H, Lu Q, Xu J, and Xu Y. Nebulized liposomal gadobenate dimeglumine contrast formulation for magnetic resonance imaging of larynx and trachea. International Journal of Nanomedicine. 2011;6: 3383–3391.

Wei Y, Guo J, Zheng X, Wu J, Zhou Y, Yu Y, Ye Y, Zhang L, and Zhao L. Preparation, pharmacokinetics and biodistribution of baicalin-loaded liposomes. International Journal of Nanomedicine. 2014;9: 3623–3630. https://doi.org/10.2147/IJN.S66312.

Wen H, Jung H, and Li X. Drug delivery approaches in addressing clinical pharmacology-related issues: Opportunities and challenges. The AAPS Journal. 2015;17(6): 1327–1340. https://doi.org/10.1208/s12248-015-9814-9.

Werle M, and Takeuchi H. Chitosan-aprotinin coated liposomes for oral peptide delivery: Development, characterisation and in vivo evaluation. International Journal of Pharmaceutics. 2009;370: 26–32.

Wiedenhoeft T, Braun T, Springer R, et al. The basement membrane in a 3D breast acini model modulates delivery and anti-proliferative effects of liposomal anthracyclines. Pharmaceuticals. 2020;13: 256.

Wiedersberg S, and Guy RH. Transdermal drug delivery: 30+ years of war and still fighting! Journal of Controlled Release. 2014;190: 150–156.

Wilkhu JS, McNeil SE, Anderson DE, Kirchmeier M, and Perrie Y. Development of a solid dosage platform for the oral delivery of bilayer vesicles. European Journal Pharmaceutical Sciences. 2017;108: 71–77.

Wilson WW, Wade MM, Holman SC, and Champlin FR. Status of methods for assessing bacterial cell surface charge properties based on zeta potential measurements. Journal of Microbiological Methods. 2001;43(3): 153–164.

Worsham RD, Thomas V, and Farid SS. Potential of continuous manufacturing for liposomal drug products. Biotechnology Journal. 2018;14(2): 1700740. https://doi.org/10.1002/biot.201700740.

Xiao J, Capanoglu E, Jassbi AR, and Miron A. Advance on the flavonoid C-glycosides and health benefits. Critical Reviews in Food Science and Nutrition. 2016;56(Suppl 1): S29–45. https://doi.org/10.1080/10408398.2015.1067595.

Yang H, Wu X, Zhou Z, Chen X, and Kong M. Enhanced transdermal lymphatic delivery of doxorubicin via hyaluronic acid based transfersomes/microneedle complex for tumor metastasis therapy. International Journal of Biological Macromolecules. 2019;125: 9–16.

Yang S, Liu W, Liu C, Liu W, Tong G, Zheng H, et al. Characterization and bioavailability of Vitamin C nanoliposomes prepared by film evaporation-dynamic high pressure microfluidization. Journal of Dispersion Science and Technology. 2012;33(11): 1608–1614.

Yata VK, Ranjan S, Dasgupta N, and Lichtfouse E. 2021. Nanopharmaceuticals, Volume 1: Environmental Chemistry for a Sustainable World. Volume 48. Cham, Switzerland: Springer.

Yoshimoto M, Tamura R, and Natsume, T. Liposome clusters with shear stress-induced membrane permeability. Chemistry and Physics of Lipids. 2013;174: 8–16.

Yu JY, Chuesiang P, Shin GH, et al. Post-processing techniques for the improvement of liposome stability. Pharmaceutics. 2021;13: 1023 https://doi.org/10.3390/pharmaceutics13071023.

Yuan H, Ma Q, Ye L, and Piao G. The traditional medicine and modern medicine from natural products. Molecules. 2016;21(5): 559. https://doi.org/10.3390/molecules21050559.

Yuan Z. Liposomal quercetin efficiently suppresses the growth of solid tumors in murine models. Clinical Cancer Research. 2006;12: 3193–3199.

Zagana P, Mourtas S, Basta A, and Antimisiaris SG. Preparation, physicochemical properties, and in vitro toxicity towards cancer cells of novel types of arsonoliposomes. Pharmaceutics. 2020;12(4): 327.
Zhao YN, Cao YN, Sun J, et al. Anti-breast cancer activity of resveratrol encapsulated in liposomes. Journal of Materials Chemistry B. 2020;8: 27–37.
Zhong-Jian C, Si-Cong Y, Xue-Liang L, et al. Nanobowl-supported liposomes improve drug loading and delivery. Nano Letters. 2020;20: 4177–87.
Zhou Y, Ning Q, Yu D-N, Li W-G, and Deng J. Improved oral bioavailability of breviscapine via a Pluronic P85-modified liposomal delivery system. Journal of Pharmacy and Pharmacology. 2014;66(7): 903–911.

6 Liposomes and Exosome-Like Nanoparticles for Curcumin Delivery to the Gut and to Treat Inflammatory Bowel Disease

Vincenzo De Leo, Lucia Catucci, Anna Maria Maurelli, Valeria Daniello, Massimo Conese, and Sante Di Gioia

6.1 CURCUMIN AND CURCUMINOIDS

Curcumin, also called diferuloylmethane or (1E,6E)-1,7-bis(4-hydroxy-3-methoxyphenyl)-1,6-heptadiene-3,5-dione, is the main o-methoxyphenol derived from the rhizome of plants of the genus *Curcuma*, mainly *Curcuma longa* L., from which the spice turmeric is produced (Barclay et al. 2000). It belongs to a class of plant secondary metabolites known as diarylheptanoid. This type of compounds consist of two aromatic rings connected by 1,6-heptadiene-3,5-dione moiety (Pouliquen, 2014). Curcumin corresponds to a molecular formula $C_{21}H_{20}O_6$, with a molecular weight of 368.37 g/mol and melting point of 183°C (Feng, Wei, et al. 2017). Together with two structurally similar molecules, demethoxy-curcumin and bis-demethoxy-curcumin, it forms a group of molecules collectively known as curcuminoids, which make up 2 to 8% of the oleoresin extracted from plant rhizomes (Braga et al. 2003) (Figure 6.1). Curcumin comes as a yellow, hydrophobic powder, mostly insoluble in water and not very soluble in hydrocarbon solvents (Zielinska et al. 2020). Curcumin and curcuminoids exert a marked antioxidant activity, even if curcumin is the most (bio) active among the curcuminoids. Initially this antioxidant activity was attributed to the central active $-CH_2-$ group (Jovanovic et al. 1999), but subsequent studies have clarified that curcumin behaves like a classic chain-breaking phenolic antioxidant molecule, as non-phenolic synthetic curcuminoids have shown no antioxidant activity (Barclay et al. 2000).

Due to its structure, curcumin is subject to keto-enol tautomerism. The greater antioxidant activity is associated with the keto form, while the enol form is more exposed to degradation phenomena (Rege et al. 2019).

Curcumin's spectroscopic dependence on solvent has been demonstrated, since it exhibits various absorption λ_{max} in different solvents. Curcumin occurs in keto form in polar and acidic media, but it rapidly degrades in non-polar, neutral, or basic environments. Therefore, under physiological conditions it rapidly degrades into vanillin, feruloyl methane, ferulic acid, and ferulic aldehyde, among others (Zielinska et al. 2020; Lopez-Malo et al. 2020). In vivo curcumin is rapidly metabolized. The main metabolic pathways are conjugation and reduction. Conjugation occurs for orally administered curcumin, with the formation of glucuronide and curcumin sulfates in the intestine and liver. Reduction processes transform curcumin into hexahydrocurcumin and octahidrocurcumin when administered by systemic routes. In turn these are converted into monoglucuronic conjugates, or curcumin glucuronide, dihydrocurcumin glucuronide, and tetrahydrocurcumin glucuronide (Joe et al. 2004; Zielinska et al. 2020). Additionally, curcumin is rather photosensitive and

DOI: 10.1201/9781003300588-6

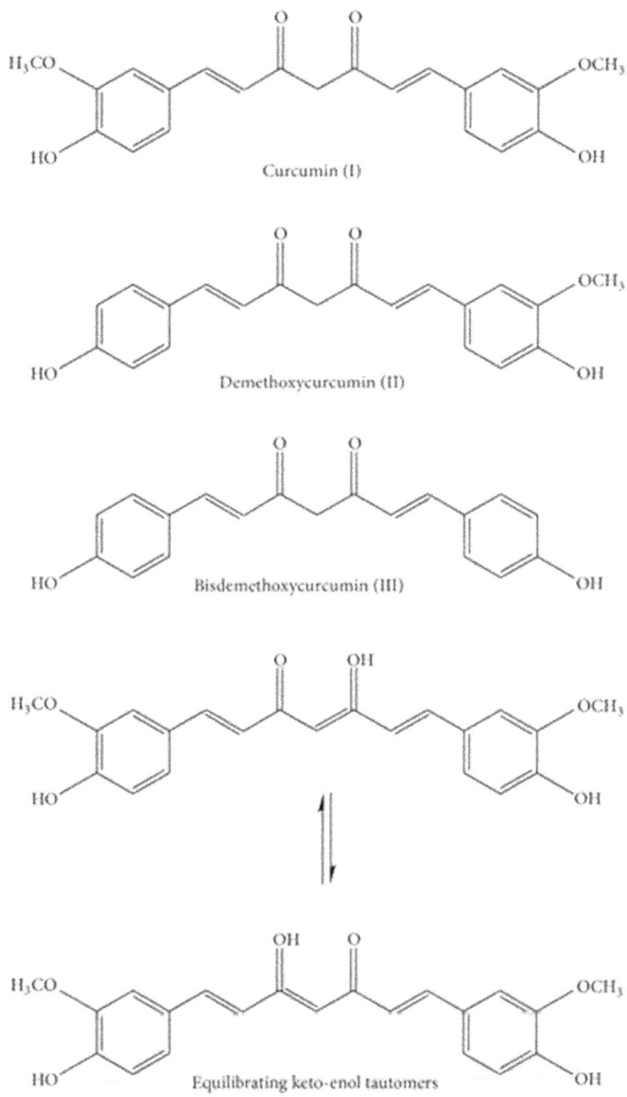

FIGURE 6.1 Curcumin I, II, and III (curcumin, demethoxycurcumin, and bisdemethyoxycurcumin) and curcumin keto-enol tautomers (reproduced under Creative Common license from Kaminaga et al. 2003).

should therefore be protected from direct light (Saraf et al. 2014). On the contrary, the temperature has minimal effect on its stability even at 250°C.

Curcumin of varying degrees of purity is commercially available. Except for analytical applications, commercial curcumin always contains minor fractions of the other two curcuminoids due to the difficulty of purification from these very similar molecules. For example, NCB-02 is a standardized *Curcuma longa* extract that contains 78% curcuminoids, 72% of which is curcumin, 18.08% demethoxy curcumin and 9.42% bis-demethoxy curcumin (Venkataranganna et al. 2007).

6.2 CURCUMIN'S BIOLOGICAL EFFECTS

Inflammation is a biological response that occurs when our body is invaded by pathogens (Artis et al. 2015; Isailovic et al. 2015) or our cells are involved in processes causing cell damage (Artis et al. 2015; Lucas et al. 2006; Rock et al. 2011; Fernandes et al. 2015; Heppner et al. 2015; Loane et al. 2016; Waisman et al. 2015). Many studies have demonstrated that curcumin can modulate several molecular targets, which are involved in the pathogenesis of inflammatory processes and neoplastic diseases. Curcumin has strong anti-inflammatory and antioxidant properties via multiple actions (Deogade et al. 2015). Mainly, this molecule downregulates genes coding for several pro-inflammatory cytokines (*e.g.*, IL-1 and TNF-α by inhibiting transcription factors, such as NF-κB [NF nuclear factor kappa-light-chain-enhancer of activated B cells]). Oxidative stress is another important aspect to take in account since reactive oxygen species (ROS) can trigger the inflammatory process and, in turn, inflammation causes an over production of ROS. Interestingly, there is a strong link between antioxidant molecules and their anti-inflammatory properties. The presence of particular chemical structures, such as double bonds, β-diketo group, phenyl rings with hydroxyl, and o-methoxy groups, make curcumin a strong antioxidant (Wright 2002; Priyadarsini 2009; Menon et al. 2007). This molecule can scavenge ROS (e.g., superoxide radicals, hydrogen peroxide, and nitric oxide) and inhibit lipid peroxidation (Ak et al. 2008). This latter activity results from the increase of many intracellular ROS-scavenging enzymes activity, such as superoxide dismutase, catalase, glutathione peroxidase, and heme oxygenase. Curcumin can also upregulate glutathione transferases and their mRNAs (Xu et al. 2018), which results in an increase of reduced glutathione. On the contrary, curcumin inhibits enzymes involved in the generation of ROS, such as lipoxygenase (LOX), cyclooxygenase-2 (COX-2), and xanthine oxidase (Rahman et al. 2009). Moreover, on the basis of its lipophilic structure, curcumin is able to abbreviate the oxidation chain length as it can scavenge peroxyl-radicals (Priyadarsini 2009).

The anticancer effects of curcumin are mainly directed against processes, such as angiogenesis, cancer cell growth, and cell apoptosis. Angiogenesis is one of the main processes in cancer development. This process is orchestrated by a range of proangiogenic factors and signaling pathways and can involve the participation of cancer cells as well as cancer stromal cells. Several research groups have demonstrated that curcumin decreases the generation of angiogenic factors, such as VEGF and bFGF. The reduction of VEGF is a consequence of the negative regulation of transcription factors such as NF-κB and activator protein (AP)-1 (a heterodimer of c-Fos and c-Jun), as well as of VEGFR and the PI3K/Akt signaling pathway modulation (Yance et al. 2006; Mishra et al. 2015; Man et al. 2018). Angiogenesis is a complex process characterized by the "sprouting" of new vessels from existing ones. During the first step of this process, the basement membrane of blood vessels is degraded, and the extracellular matrix (ECM) undergoes a remodeling process. Thereafter, endothelial cells migrate and proliferate, and new matrix components are generated. Matrix metalloproteinases (MMPs) constitute an important class of proteases, which participate in several processes, such as ECM degradation, tumor angiogenesis, and subsequent metastatization, while tissue inhibitors of metalloproteinases (TIMPs) downregulate their activity. Therefore, angiogenesis is a process that can be modulated, via MMPs, by the balance between pro- and anti-angiogenic factors. Interestingly, curcumin inhibits MMPs and increase the TIMP-1 activity, causing the stabilization of ECM (Yance et al. 2006).

Another important characteristic of curcumin is its ability to induce the apoptotic response through p53-dependent pathway. p53 is one of the most frequently altered genes in human tumors (Kandoth et al. 2013) and represents an obstacle to tumor development. p53 is able to act as a "sensor" of stress, in response to which it evokes diverse responses so as to protect both cell and its genome. It is interesting to observe that p53 activity is correlated to some miRNAs (Hermeking 2007). It has been demonstrated that curcumin can induce apoptosis in non-small-cell lung cancer by miR-192–5p and miR-215, which activate the p53 (Ye et al. 2015). On the other hand, some studies have shown that in HT-29 colon cancer, the induction of apoptosis by curcumin is p53-independent (Watson et al. 2010).

The cell cycle deregulation is another important aspect in cancer development. The accumulation of mutations into tumor cells causes a dysfunctional activation of the cell cycle with a cell proliferation, which is completely autonomous respect to the absence or presence of appropriate mitogenic or anti-mitogenic signaling. Each phase of the cell cycle is regulated by coordinate interactions between serine/threonine kinases (CDKs, or cyclin-dependent kinases) and their regulatory subunits called cyclins. These kinases are regulated in a timely manner since each cyclin undergoes a synthesis/degradation step at each phase of cell cycle. It has been demonstrated that curcumin can interfere with the CDK/cyclin interactions. It is interesting to observe how the treatment of MDA-MB-231 cell line (breast cancer cells) with curcumin causes the degradation of CDKs/cyclin complexes and induces the downregulation of cyclin D1, which is necessary for passage through the G1/S phase, thus leading to cell cycle arrest at G1 (Chiu et al. 2009).

Curcumin has been shown to act against cancers characterized by an overexpression of the proto-oncogene *Ras* (rat sarcoma). It is out of the question that somatic, gain-of-function mutations of genes encoding the Ras proteins are essential in altering the signaling networks controlling cellular proliferation, differentiation, or survival. Curcumin downregulates the *Ras* proteins and upregulate ERK, resulting in the inhibition of proliferation of the human gastric carcinoma AGS cells (Cao et al. 2015). Moreover, it has been demonstrated that curcumin treatment can interfere with the intracellular signaling pathways, for example, causing a shift from a *Ras*-induced MEK/ERK pro-proliferative pathway toward a p38MAPK/JNK1 pro-death signaling (Silvestre et al. 2023). Curcumin also revealed to be able to modulate Wnt/β-catenin signaling pathway, which is overexpressed in several human cancers. In a colon-carcinoma cell line, curcumin treatment was able to suppress Wnt/β-catenin pathways via miR-130a (Dou et al. 2017).

Besides these mentioned activities, curcumin is characterized by other important properties, such as its antimicrobial activity, its effect on immune system, thrombocytes aggregation, detoxification enzymes, cell proliferation, and cell death (Joe et al. 2004; Burge et al. 2019). Concerning curcumin antibacterial activity (Moghadamtousi et al. 2014; Praditya et al. 2019; Trigo-Gutierrez et al. 2021), it has been shown that the by-products obtained after isolation of curcumin were shown to be endowed with properties against several bacteria belonging to different types: *Bacillus* spp., *Staphylococcus aureus*, *Escherichia coli*, and *Pseudomonas aeruginosa* (Negi et al. 1999). Curcumin exhibited lower antibacterial properties against *S. aureus* and *E. coli* but showed higher activity against *B. cereus* and *Yersinia enterocolitica* than did curcumin-β-diglucoside, a water-soluble sugar derivative of curcumin (Parvathy et al. 2009). Curcumin derivatives, such as indium curcumin and indium diacetyl curcumin, displayed also promising antibacterial activity, against *S. aureus, S. epidermis, E. coli*, and *P. aeruginosa* (Tajbakhsh et al. 2008). Others have shown a significant antibacterial activity against *Helicobacter pylori* in in vitro and in vivo studies in mice (De et al. 2009). Curcumin inhibition of cell division of *E. coli* and *B. subtilis* was demonstrated to be due to the inhibitory effect on the polymerization of the FtsZ cytoskeletal protein (Kaur et al. 2010). Curcumin had significant antibacterial effects against clinical multidrug-resistant strains of *Shigella dysenteriae* and *Campylobacter jejuni* pathogens (Kareem et al. 2020). Regarding the microbiota, oral administration of curcumin to C57BL/6 mice resulted in a decrease of the microbial richness and diversity (Shen et al. 2017). Interestingly, the oral administration of curcumin, in rats fed with high-fat diet, induces a beneficial change in gut microbiota, for example, with the growth of bacteria producing molecules able to protect intestinal mucosa (Feng, Wang, et al. 2017).

Curcumin has beneficial effects onto various stages of wound healing, *i.e.*, hemostasis, inflammation, proliferation, and remodeling (Choudhary et al. 2018; Akbik et al. 2014; Tejada et al. 2016). Curcumin enhances cutaneous wound healing by granulation tissue deposition, production of matrix proteins, re-epithelialization, and deposition of collagen-fibers. In vivo studies have shown that topical application or systemic administration of curcumin modulate both inflammatory response and oxidative stress to avoid delay in the wound healing process (Gopinath et al. 2004; Jagetia et al. 2005; Kulac et al. 2013). Curcumin promotes fibroblast transformation in myofibroblasts (Mohanty et al. 2012), an essential step in the formation of granulation tissue, and enhances wound healing by increasing the formation of granulation tissue, angiogenesis, collagen deposition, and TGF-β1 expression in wounds in rats and guinea pigs (Sidhu et al. 1998). In a full-thickness excision wound model in rats, topically administered curcumin determined a faster healing rate, as shown by increased rates of epithelialization, wound contraction, and improved tensile strength, likely due to its antioxidant activity (Panchatcharam et al. 2006). These studies

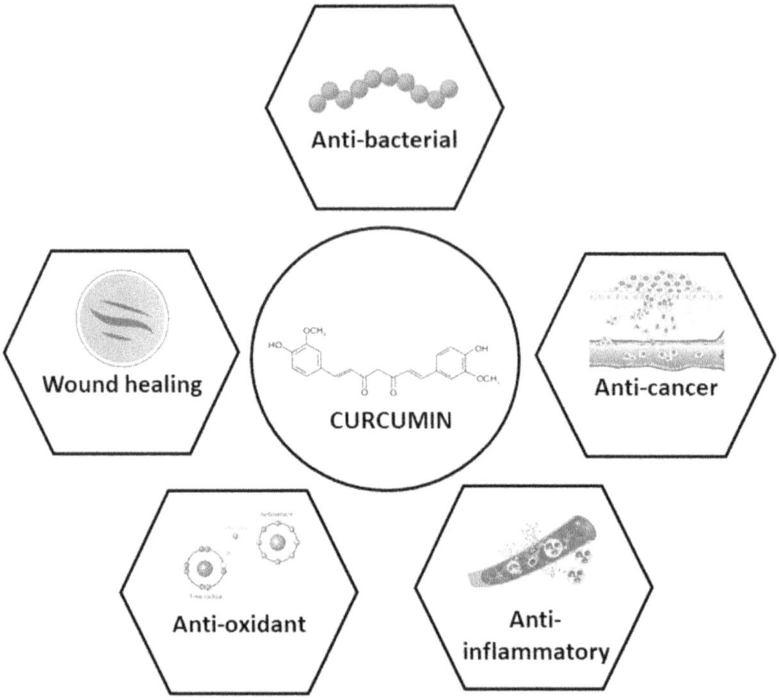

FIGURE 6.2 Different functional and biological features of curcumin.

has demonstrated that topical treatment with curcumin are more effective than oral administration owing to the bioavailability of the drug at the wound site (Merrell et al. 2009; Sidhu et al. 1999; Sidhu et al. 1998; Mani et al. 2002).

Overall, the antioxidant, antibacterial, and pro-healing effects of curcumin are important in the resolution of IBD (Figure 6.2) and, therefore, have been intensively explored in this disease.

6.3 INFLAMMATORY BOWEL DISEASE

Inflammatory bowel disease (IBD) comprise Crohn's disease (CD) and ulcerative colitis (UC), chronic relapsing conditions of unknown origin affecting more than 3.5 million people and with increasing incidence worldwide (Ng et al. 2017). In IBD, the dysregulation of host immune responses leads to chronic intestinal inflammation in genetically susceptible individuals. Indeed, both diseases are associated with genetic factors, alterations in epithelial cell barrier functions, immunopathology related to abnormal T-cell reactions to commensal microflora (microbiota), and other luminal antigens (Vermeire et al. 2012). However, CD and UC differ for some pathological features, including location of lesions, inflammatory involvement of mucosa and submucosa, presence of granulomas, and formation of fistulae and abscesses (Table 6.1). For instance, the so-called skip lesions, determined by the involvement of some haustral segments but not others, are a hallmark of Crohn's disease.

To date, although more than 240 IBD risk genes have been found, 30% of the susceptibility to disease development is explained by these risk alleles, suggesting a major contribution by environmental factors, mainly affecting the gut microbiota (Zhao et al. 2019). The exposure to antibiotics, hygiene, medication, and diet affects the IBD onset and incidence (Ponder et al. 2013).

TABLE 6.1
Characteristics of Crohn's Disease and Ulcerative Colitis

Incidence	**Crohn's Disease**	**Ulcerative Colitis**
Age of onset	Any age; 10–30 years most common	Any age; 10–40 years most common
Family history	More common	Less common
Gender (prevalence)	About equal in women and men	Equal in women and men
Pathology		
Location of lesions	All of GI tract: mouth to anus, "skip" lesions common	Colon and rectum, no "skip" lesions
Inflammation and ulceration	Entire intestinal wall involved	Mucosal layer involved
Granulomas	Common	Rare
Fistulae and abscesses	Common	Rare
Strictures and possible obstruction	Common	Rare

Modified from (McCance et al. 2013)

Genetic factors that are involved in CD and UC include those regulating the intestinal barrier, inflammatory mediators, autophagy, apoptosis, and the innate and adaptive immunity (Packey et al. 2008). Table 6.2 lists some of the most significant, bearing in mind that for CD, gene polymorphisms have been related to both diagnosis and prognosis of the disease (Tsianos et al. 2012). The *CARD15/NOD2* (caspase recruitment domain family member 15/nucleotide-binding-oligomerization-domains) gene mutations have the strongest association with CD (35% to 45% of cases). CARD15/NOD2 is a pattern recognition receptor (PRR) recognizing muramyl dipeptide, which is present in the peptidoglycans of gram-positive and gram-negative bacteria. Moreover, NOD2 plays important roles in the expression of α-defensins and promotes the antibacterial IL-23-IL-1-IL-17 axis (see the following). Genome-wide association studies (GWAS) have found that many other gene polymorphisms are associated with CD, including those involved in epithelial barrier integrity, autophagy of bacteria (i.e., *ATG16L1* and *IRGM*), and adaptive immune response (e.g., *IL23R*), all relevant for the immunopathogenesis of CD (Burge et al. 2019; Tsianos et al. 2012). Besides the same genes involved in CD, ulcerative colitis has been associated with polymorphisms in the multidrug resistance-1 (*MDR-1*) gene, that encodes p-glycoprotein, which is a transmembrane efflux pump involved in clearing bacterial xenotoxins (Ho et al. 2005). Polymorphisms of the *PTPN22* gene, a major risk factor for human autoimmune diseases, are differentially associated with CD and UC; a PTPN22 gain-of-function allele, R620W, is associated with reduced risk of CD, whereas the loss-of-function allele, R263Q, is only associated with UC (Diaz-Gallo et al. 2011).

The pathogenesis of IBD may be associated with an overly aggressive response to normal flora bacteria in genetically predisposed individuals. Bacterial translocation across a defective mucosal barrier and the imbalanced regulation of the intestinal immune response are the two principal hallmarks. In general, multiple factors, such as genetic susceptibility, lifestyle factors, including altered microbiome, intestinal barrier dysfunction, and diet and antibiotic use (Uhlig et al. 2018), result in an increased mucosal inflammatory response to luminal microbiota and disruption of tight junctions leading to breakdown of the intestinal barrier (Burge et al. 2019; Uhlig et al. 2018). In IBD, Paneth cells play a prominent role. The loss-of-function polymorphisms in NOD2 determine that Paneth cells release fewer antimicrobial peptides (AMPs), determining more bacteria to pass through the intestinal epithelium and interact with adaptive immune cells (Salzman et al. 2007). Moreover, alterations in autophagy of pathogens (xenophagy) lead to less microbial clearance and

TABLE 6.2
List of Main IBD-Associated Functional Genes

Function	IBD type
Bacterial Sensing	
CARD15/NOD2	CD
CARD9	CD, UC
Autophagy	
ATG16L1	CD
IRGM	CD
Epithelial Permeability	
CDH1	CD, UC
MDR1	UC
Antigen Presentation	
HLA—DRB1	UC
IL17-IL23R Pathway	
IL23R:	
R381Q (protection against disease)	CD, UC
Other polymorphisms (extensive disease)	UC
IL-12B	CD, UC
JAK2	CD, UC
STAT3	UC
IgA Production	
PTPN22:	
R620W (risk of disease)	UC
R263Q (protection against disease)	UC
Inflammatory Mediators	
IL-10 (anti-inflammatory)	UC
TNFSF15 (pro-inflammatory)	CD, UC
Signaling Pathways	
SMAD3 (recurrent surgery)	CD
Apoptosis	
Fas-ligand	CD
Caspase 9	CD

Modified from (Zhao et al. 2019)

reduced disposal of injured tissues (Iida et al. 2017). T cells contribute significantly to the onset and progression of IBD, with some differences of T cell polarization by disease (Mazieiro et al. 2018). In CD, Th1 and Th17 are the predominant phenotypes (Fuss et al. 1996; Parronchi et al. 1997; Romagnani 1994). Th1 cells are important for macrophage activation directed to elimination of intracellular pathogens and are characterized by large amounts of interferon (IFN)-γ, TNF-α, and IL-2; Th17 cells are involved in neutrophil and monocyte recruitment aimed to the elimination of extracellular bacteria and fungi, through the production of IL-17. In UC, the adaptive immune response is skewed toward Th2 and Th9. IL-13, produced by Th2 cells, induces apoptosis in intestinal epithelial cell, contributing to tissue destruction (Bamias et al. 2015). IL-9, a pleiotropic signature cytokine of Th9 cells, impairs the intestinal barrier function and exacerbates UC-associated tissue damage (Shohan et al. 2018). Abnormal differentiation of macrophages and dendritic cells (DCs) toward a pro-inflammatory phenotype, which depends on PRR ligation, governs the harmful activation of T cells. Thus, exploiting the differentiation toward a M2 macrophage phenotype may represent a new therapeutic approach for colitis (Castro-Dopico et al. 2020).

The breach in the intestinal barrier is further worsened by the over-release of pro-inflammatory cytokines by intestinal epithelial cells (IECs) and immune cells, especially TNF-α, IL-1β, IL-13, and IFN-γ (Bischoff et al. 2014). Cytokines determine the tight junction disruption and loss of IECs by apoptosis (Zeissig et al. 2007). Signal transduction pathways that are mainly involved in cytokine synthesis and release are NF-κB and AP1/mitogen-activated protein kinase (AP-1/MAPK). NF-κB and AP-1/MAPK pathways can be triggered by different stimuli, among which toll-like receptors (TLRs), immune sensors of microbiota in the gut, are thought to be critical in IBD (Lu et al. 2018). Of these, TLR4 is an important member that recognizes bacterial lipopolysaccharide (LPS) and instigates signaling to activate NF-κB by recruiting myeloid differentiation factor 88 (MyD88), eventually leading to the synthesis of various pro-inflammatory mediators. In particular, inactive cytosolic NF-κB is associated with its inhibitor IκB. Following different stimuli, phosphorylated IκB is targeted to rapid degradation. This releases NF-κB into the nucleus, driving the transcription of many genes, including cytokines. The nuclear translocation of NF-κB is negatively regulated by peroxisome proliferator-activated receptor gamma (PPARγ), a nuclear receptor with anti-inflammatory properties. Thus, PPARγ-deficient mice were more susceptible to colitis induced by dextran sulfate sodium (DSS) (Shah et al. 2007). Among other intracellular signal transduction molecules involved in the expression of numerous inflammatory cytokine genes in IECs and leukocytes, signal transducers and activators of transcription (STAT), and particularly STAT3, have been described to have a role in IBD pathogenesis (Mudter et al. 2005; Musso et al. 2005; Fu 2006).

6.4 CURCUMIN IN IBD

Drugs already used in the clinic to treat IBD, such as sulfasalazine (SASP), mesalamine, and corticosteroids, all inhibit NF-κB activation, but they present many undesirable side effects (Wahl et al. 1998; Weber et al. 2000; Bantel et al. 2000; Scheinman et al. 1995; Auphan et al. 1995). More recently, other medicines have been introduced, namely, anti-TNF-α drugs (Cohen et al. 2017), antibiotics (Nitzan et al. 2016), probiotics (Derwa et al. 2017), and immunosuppressants (Zabana et al. 2019), but they are costly and often involve significant side effects, and their use is limited by low effectiveness and specificity. Importantly, colorectal cancer is a sequela of longstanding IBD involving at least 1/3 of the colon (Clarke et al. 2019). Therefore, drugs that cause minimal side effects are largely available, do not induce refractoriness, maintain remission, are urgently desired for the IBD treatment.

Curcumin exerts pleiotropic therapeutic effects through acting on a vast array of molecular targets that are assumed to play a role in IBD pathogenesis, including signal transduction pathways and transcription factors, pro-inflammatory cytokines, and enzymes (Burge et al. 2019; Vecchi Brumatti et al. 2014; Karthikeyan et al. 2021). These actions underpin the beneficial effects of curcumin on TLR/NF-κB/AP-1/MAPK signal pathways, cytokine profiles, and immune cell maturation and differentiation. Various experimental models of IBD have shown possible anti-inflammatory effects of curcumin, including free radical scavenging, antioxidant increase, myeloperoxidase, COX-1, COX-2, LOX, TNF-α, IFN-γ, and iNOS inhibition, as well as regulation of multiple signaling pathways, particularly kinases (MAPK and ERK) and transcription factors NF-κB. Curcumin has been also shown to regulate dendritic cell and macrophage activation profiles and consequently modify the differentiation status of T cells toward less harmful phenotypes, e.g., Tregs (Table 6.3). Overall, these works demonstrated the beneficial effects of curcumin on disease activity, as judged by his to pathological scores, neutrophil infiltration (measured as increase in myeloperoxidase activity in the mucosa), and levels of malondialdehyde (an indicator of lipid peroxidation), as well as inhibitory actions on NF-κB, MAPK, and STAT pathways and downstream pro-inflammatory cytokine expression. The alterations in the inflammatory milieu exerted by curcumin determine an inhibition of pro-inflammatory CD8 + CD11c + DCs and M1 macrophages (Zhao et al. 2016; Kang et al. 2021).

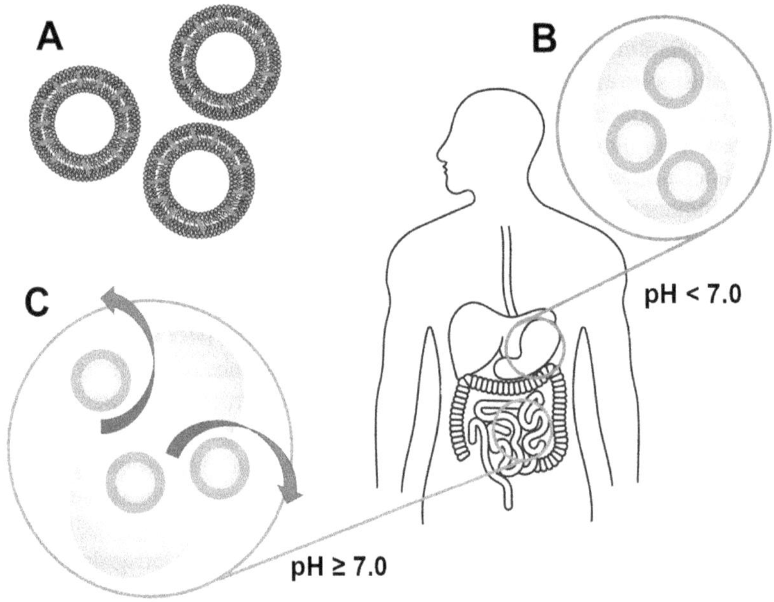

FIGURE 6.3 pH-sensitive liposomes and curcumin delivery to the gut. (A) Curcumin molecules embedded in the lipid bilayer of liposomes. (B) Curcumin liposomes are protected in the gastric environment (pH < 7.0) by the polymeric coating of Eudragit S100. (C) At neutral-alkaline pH, the polymer coating dissolves, releasing the curcumin liposomes, which can thus interact with the intestinal mucosa. The sketches provide a pictorial and qualitative description and the dimensions of the different components are not to scale.

in just over 3 hours (De Leo, Milano, et al. 2018). MTT assay confirmed the safeness of these samples for the enteric Caco-2 cell line. Curcumin-loaded liposomes were also assessed for their uptake by Caco-2 cells, generally used as in vitro model for drug uptake and transportation through the intestinal mucosa. Cytofluorimetric analysis demonstrated that liposome uptake by Caco-2 cells occurred by 2 hours at 37°C. Extending the incubation period, the lengthened interplay carried to a greater rate of assimilated vesicles, which can rise the curcumin load of intestinal cells. Curcumin-loaded liposomes have been shown to be able to significantly reduce the degree of ROS generated by H_2O_2 in Caco-2 cells, confirming the in vitro antioxidant capabilities of the prepared system (De Leo et al. 2020). Multicompartment liposomes, made of Eudragit® S100 and hyaluronan with diameter ranging from 220 to 287 nm, were shown to safeguard vesicles from the hard conditions of the gastrointestinal apparatus (i.e., pH and ionic strength changes), as in vivo supported by the larger curcumin accumulation in the bowel area when curcumin was transported by liposomes respect to the free drug administration. Moreover, these vesicles were capable to preserve Caco-2 cells from oxidative stress (Catalan-Latorre et al. 2016).

Gel core liposomes (GCLs) loaded with curcumin have also been proposed for IBD. GCLs are polymer-based liposomal formulation of active payload that will gel following administration and thus can protect curcumin as well as prolong its release. GCLs were prepared by thin-film hydration method, in which the dried lipid film containing curcumin was hydrated with aqueous solution of gel forming polymer (Carbopol 974P). The entrapment efficiency of optimized formulation was found to be 81.21%. Also, for these systems, colon targeting ability was achieved by coating GCL with Eudragit S100. Vesicle sizes of uncoated and coated GCLs were found to be about 500 and 850 nm, respectively. Different release media representing each stage of the GI tract were used to evaluate the release of curcumin from coated and uncoated GCLs, observing a sustained release in both cases. A drug release mechanism by diffusion, as proposed by Korsmeyer-Peppas model (Dash et al. 2010), was also observed. Uncoated and Eudragit-coated GCL were then subjected to animal studies carried out on colitis caused by acetic acid in rats. The liposomal formulation has been shown to significantly reduce the symptoms of colitis and inflammation of the colon. Rat colon histopathology reports, in fact, showed that rats treated with acetic acid suffered severe ulceration with necrosis while rats treated with the Eudragit-coated GLC formulation showed mild colitis compared to those treated with simple curcumin suspension or uncoated GCL (Munot et al. 2020).

6.6 EXOSOME DELIVERY OF CURCUMIN

Exosomes belong to a large family of cell-derived vesicles defined as extracellular vesicles (EVs) (Lugli et al. 2015; Abels et al. 2016), which are secreted by all types of prokaryotic and eukaryotic cells (Silverman et al. 2011; Ellis et al. 2010). The dimensions of exosomes range from 50 nm up to 120 nm. The first step in their formation is characterized by inward budding of the plasma membrane, which causes the formation of intermediate endosome vesicles, the multivesicular bodies (MVBs). The fate of MVBs is regulated by their contents: they can fuse with lysosomes to be degraded or combine with the plasma membrane to form exosomes that are released in the extracellular space (Colombo et al. 2014). The composition of exosomes is rather complex, depending on both the donor cells and their state (physiological or pathological) (Zhang, Liu, et al. 2019). Exosomes can contain lipid molecules, RNA, DNA, and various more complex structures, such as receptor proteins, enzymatic activities, transcription factors, and extracellular matrix proteins, that are inside or on the surface of exosomes (D'Asti et al. 2012). These EVs represent a good system to deliver drugs into cells since they have good bioavailability and low toxicity. Their composition, which is similar to the cell membrane, makes them well tolerated (Bang et al. 2012) and with a low immunogenicity (Hood 2016). It has been demonstrated how exosomes are able to permeate tumor masses more efficiently as compared to the permeation rate of liposomes (Kohlhapp et al. 2015). Since exosomes have reduced dimensions, they can overcome every biological barrier. The first study concerning the possible role of exosomes as a drug delivery system was published in 2011. In this paper, exosomes derived from dendritic cells were used to deliver siRNA across the blood-brain barrier (BBB) (Alvarez-Erviti et al. 2011).

Previous studies showed that exosomes could represent an advantageous system to deliver curcumin since they can enhance its anti-inflammatory activity as well as its in vitro stability and its in vivo bioavailability (Sun et al. 2010). The treatment of IBD represents another important application of exosomes for therapeutics purposes. During colonic inflammation the epithelium is progressively disrupted, mucus becomes thicker than normal, and the transit through colon is prolonged. Although this inflamed environment is clearly challenging, it has been demonstrated that nanoparticles can overcome such barriers, inducing an augmentation within the inflamed tissue. This causes an increase in the quantity of the drug in the colon with a consequent improvement in efficacy and a reduced likelihood of systemic side effects (Beloqui et al. 2016; Ocansey et al. 2020). Recently, it has been demonstrated that cow-milk-derived exosomes can ameliorate the delivery of curcumin in terms of cellular uptake and intestinal permeability (Vashisht et al. 2017). More precisely, in this study, for the first time, the role of each step of digestion on the stability of exosomal curcumin was determined. On the basis of these experiments, it has been demonstrated that when curcumin is delivered by milk exosomes, its stability, for example, in presence of degradative enzymes, is improved as compared to free curcumin. The human intestinal tumor cell line Caco-2 was used to study the uptake of these milk exosomes encapsulating curcumin. The Caco-2 cell line is an optimal in vitro model to study the process of absorption and the transport across the human intestinal epithelium for pharmacological research (Hidalgo et al. 1989; Vashisht et al. 2017). By fluorescence microscopy, it has been shown that bovine milk exosome uptake occurs via endocytosis, and exosomal curcumin localizes in the cytosol of treated Caco-2 cells. Moreover, this study demonstrated that the curcumin-loaded bovine milk exosomes can pass through the differentiated

Camacho-Barquero, L., I. Villegas, J. M. Sanchez-Calvo, E. Talero, S. Sanchez-Fidalgo, V. Motilva, and C. Alarcon de la Lastra. 2007. "Curcumin, a Curcuma longa constituent, acts on MAPK p38 pathway modulating COX-2 and iNOS expression in chronic experimental colitis." *Int Immunopharmacol* 7 (3): 333–342.

Cao, A. L., Q. F. Tang, W. C. Zhou, Y. Y. Qiu, S. J. Hu, and P. H. Yin. 2015. "Ras/ERK signaling pathway is involved in curcumin-induced cell cycle arrest and apoptosis in human gastric carcinoma AGS cells." *J Asian Nat Prod Res* 17 (1): 56–63.

Carvalho Henriques, M., M. Amparo, F. Faustino, and S. Santos Braga. 2020. "Curcumin innovative delivery forms: Paving the 'yellow brick road' of antitumoral phytotherapy." *Appl Sci* 10: 8990.

Castro-Dopico, T., A. Fleming, T. W. Dennison, J. R. Ferdinand, K. Harcourt, B. J. Stewart, Z. Cader, Z. K. Tuong, C. Jing, L. S. C. Lok, R. J. Mathews, A. Portet, A. Kaser, S. Clare, and M. R. Clatworthy. 2020. "GM-CSF calibrates macrophage defense and wound healing programs during intestinal infection and inflammation." *Cell Rep* 32 (1): 107857.

Catalan-Latorre, A., M. Ravaghi, M. L. Manca, C. Caddeo, F. Marongiu, G. Ennas, E. Escribano-Ferrer, J. E. Peris, O. Diez-Sales, A. M. Fadda, and M. Manconi. 2016. "Freeze-dried eudragit-hyaluronan multicompartment liposomes to improve the intestinal bioavailability of curcumin." *Eur J Pharm Biopharm* 107: 49–55.

Chen, X., L. Q. Zou, J. Niu, W. Liu, S. F. Peng, and C. M. Liu. 2015. "The stability, sustained release and cellular antioxidant activity of curcumin nanoliposomes." *Molecules* 20 (8): 14293–14311.

Cheng, A. L., C. H. Hsu, J. K. Lin, M. M. Hsu, Y. F. Ho, T. S. Shen, J. Y. Ko, J. T. Lin, B. R. Lin, W. Ming-Shiang, H. S. Yu, S. H. Jee, G. S. Chen, T. M. Chen, C. A. Chen, M. K. Lai, Y. S. Pu, M. H. Pan, Y. J. Wang, C. C. Tsai, and C. Y. Hsieh. 2001. "Phase I clinical trial of curcumin, a chemopreventive agent, in patients with high-risk or pre-malignant lesions." *Anticancer Res* 21 (4B): 2895–2900.

Chiu, T. L., and C. C. Su. 2009. "Curcumin inhibits proliferation and migration by increasing the Bax to Bcl-2 ratio and decreasing NF-kappaBp65 expression in breast cancer MDA-MB-231 cells." *Int J Mol Med* 23 (4): 469–475.

Choudhary, V., and H. G. Shivakumar. 2018. "A review on curcumin: Wound healing properties and biomarkers of wound healing." *Int Res J Pharm* 9 (9): 1–5.

Clarke, W. T., and J. D. Feuerstein. 2019. "Colorectal cancer surveillance in inflammatory bowel disease: Practice guidelines and recent developments." *World J Gastroenterol* 25 (30): 4148–4157.

Cohen, B. L., and D. B. Sachar. 2017. "Update on anti-tumor necrosis factor agents and other new drugs for inflammatory bowel disease." *BMJ* 357: j2505.

Colgan, S. P., C. A. Parkos, J. B. Matthews, L. D'Andrea, C. S. Awtrey, A. H. Lichtman, C. Delp-Archer, and J. L. Madara. 1994. "Interferon-gamma induces a cell surface phenotype switch on T84 intestinal epithelial cells." *Am J Physiol* 267 (2 Pt 1): C402–410.

Colombo, M., G. Raposo, and C. Thery. 2014. "Biogenesis, secretion, and intercellular interactions of exosomes and other extracellular vesicles." *Annu Rev Cell Dev Biol* 30: 255–289.

Cooney, J. M., M. P. Barnett, Y. E. Dommels, D. Brewster, C. A. Butts, W. C. McNabb, W. A. Laing, and N. C. Roy. 2016. "A combined omics approach to evaluate the effects of dietary curcumin on colon inflammation in the Mdr1a(-/-) mouse model of inflammatory bowel disease." *J Nutr Biochem* 27: 181–192.

Pouliquen, D.L. (Ed.). 2014. *Curcumin: Synthesis, Emerging Role in Pain Management and Health Implications*. Hauppauge, NY, USA: Nova Science Publishers, Inc.

D'Asti, E., D. Garnier, T. H. Lee, L. Montermini, B. Meehan, and J. Rak. 2012. "Oncogenic extracellular vesicles in brain tumor progression." *Front Physiol* 3: 294.

Dash, S., P. N. Murthy, L. Nath, and P. Chowdhury. 2010. "Kinetic modeling on drug release from controlled drug delivery systems." *Acta Pol Pharm* 67 (3): 217–223.

De Leo, V., F. Milano, A. Agostiano, and L. Catucci. 2021. "Recent advancements in polymer/liposome assembly for drug delivery: From surface modifications to hybrid vesicles." *Polymers(Basel)* 13 (7).

De Leo, V., F. Milano, A. Paiano, R. Bramato, L. Giotta, R. Comparelli, S. Ruscigno, A. Agostiano, C. Bucci, and L. Catucci. 2017. "Luminescent CdSe@ZnS nanocrystals embedded in liposomes: A cytotoxicity study in HeLa cells." *Toxicol Res(Camb)* 6 (6): 947–957.

De Leo, V., F. Milano, E. Mancini, R. Comparelli, L. Giotta, A. Nacci, F. Longobardi, A. Garbetta, A. Agostiano, and L. Catucci. 2018. "Encapsulation of curcumin-loaded liposomes for colonic drug delivery in a pH-responsive polymer cluster using a pH-driven and organic solvent-free process." *Molecules* 23 (4).

De Leo, V., S. Di Gioia, F. Milano, P. Fini, R. Comparelli, E. Mancini, A. Agostiano, M. Conese, and L. Catucci. 2020. "Eudragit S100 entrapped liposome for curcumin delivery: Anti-oxidative effect in Caco-2 cells." *Coatings* 10 (2): 114.

De Leo, V., S. Ruscigno, A. Trapani, S. Di Gioia, F. Milano, D. Mandracchia, R. Comparelli, S. Castellani, A. Agostiano, G. Trapani, L. Catucci, and M. Conese. 2018. "Preparation of drug-loaded small unilamellar liposomes and evaluation of their potential for the treatment of chronic respiratory diseases." *Int J Pharm* 545 (1–2): 378–388.

De, R., P. Kundu, S. Swarnakar, T. Ramamurthy, A. Chowdhury, G. B. Nair, and A. K. Mukhopadhyay. 2009. "Antimicrobial activity of curcumin against Helicobacter pylori isolates from India and during infections in mice." *Antimicrob Agents Chemother* 53 (4): 1592–1597.

Deguchi, Y., A. Andoh, O. Inatomi, Y. Yagi, S. Bamba, Y. Araki, K. Hata, T. Tsujikawa, and Y. Fujiyama. 2007. "Curcumin prevents the development of dextran sulfate Sodium (DSS)-induced experimental colitis." *Dig Dis Sci* 52 (11): 2993–2998.

Deogade, S., and S. Ghate. 2015. "Curcumin: Therapeutic applications in systemic and oral health." *Int J Biol Pharm Res* 6 (4): 281–290.

Derwa, Y., D. J. Gracie, P. J. Hamlin, and A. C. Ford. 2017. "Systematic review with meta-analysis: The efficacy of probiotics in inflammatory bowel disease." *Aliment Pharmacol Ther* 46 (4): 389–400.

Di Gioia, S., M. N. Hossain, and M. Conese. 2020. "Biological properties and therapeutic effects of plant-derived nanovesicles." *Open Med(Wars)* 15 (1): 1096–1122.

Diaz-Gallo, L. M., L. Espino-Paisán, K. Fransen, M. Gómez-García, S. van Sommeren, C. Cardeña, L. Rodrigo, J. L. Mendoza, C. Taxonera, A. Nieto, G. Alcain, I. Cueto, M. A. López-Nevot, N. Bottini, M. L. Barclay, J. B. Crusius, A. A. van Bodegraven, C. Wijmenga, C. Y. Ponsioen, R. B. Gearry, R. L. Roberts, R. K. Weersma, E. Urcelay, T. R. Merriman, B. Z. Alizadeh, and J. Martin. 2011. "Differential association of two PTPN22 coding variants with Crohn's disease and ulcerative colitis." *Inflamm Bowel Dis* 17 (11): 2287–2294.

Diomede, F., L. Fonticoli, S. Guarnieri, Y. Della Rocca, T. S. Rajan, A. Fontana, O. Trubiani, G. D. Marconi, and J. Pizzicannella. 2021. "The effect of liposomal curcumin as an anti-inflammatory strategy on lipopolysaccharide e from porphyromonas gingivalis treated endothelial committed neural crest derived stem cells: Morphological and molecular mechanisms." *Int J Mol Sci* 22 (14).

Dou, H., R. Shen, J. Tao, L. Huang, H. Shi, H. Chen, Y. Wang, and T. Wang. 2017. "Curcumin suppresses the colon cancer proliferation by inhibiting Wnt/beta-catenin pathways via miR-130a." *Front Pharmacol* 8: 877.

Duda, M., K. Cygan, and A. Wisniewska-Becker. 2020. "Effects of curcumin on lipid membranes: An EPR spin-label study." *Cell Biochem Biophys* 78 (2): 139–147.

Ellis, T. N., and M. J. Kuehn. 2010. "Virulence and immunomodulatory roles of bacterial outer membrane vesicles." *Microbiol Mol Biol Rev* 74 (1): 81–94.

Feng, T., Y. Wei, R. J. Lee, and L. Zhao. 2017. "Liposomal curcumin and its application in cancer." *Int J Nanomedicine* 12: 6027–6044.

Feng, W., H. Wang, P. Zhang, C. Gao, J. Tao, Z. Ge, D. Zhu, and Y. Bi. 2017. "Modulation of gut microbiota contributes to curcumin-mediated attenuation of hepatic steatosis in rats." *Biochim Biophys Acta Gen Subj* 1861 (7): 1801–1812.

Fernandes, J. V., R. N. Cobucci, C. A. Jatoba, T. A. Fernandes, J. W. de Azevedo, and J. M. de Araujo. 2015. "The role of the mediators of inflammation in cancer development." *Pathol Oncol Res* 21 (3): 527–534.

Fu, X. Y. 2006. "STAT3 in immune responses and inflammatory bowel diseases." *Cell Res* 16 (2): 214–219.

Fuss, I. J., M. Neurath, M. Boirivant, J. S. Klein, C. de la Motte, S. A. Strong, C. Fiocchi, and W. Strober. 1996. "Disparate CD4+ lamina propria (LP) lymphokine secretion profiles in inflammatory bowel disease. Crohn's disease LP cells manifest increased secretion of IFN-gamma, whereas ulcerative colitis LP cells manifest increased secretion of IL-5." *J Immunol* 157 (3): 1261–1270.

Gopinath, D., M. R. Ahmed, K. Gomathi, K. Chitra, P. K. Sehgal, and R. Jayakumar. 2004. "Dermal wound healing processes with curcumin incorporated collagen films." *Biomaterials* 25 (10): 1911–1917.

Gupta, A. S., S. J. Kshirsagar, M. R. Bhalekar, and T. Saldanha. 2013. "Design and development of liposomes for colon targeted drug delivery." *J Drug Target* 21 (2): 146–160.

Heppner, F. L., R. M. Ransohoff, and B. Becher. 2015. "Immune attack: The role of inflammation in Alzheimer disease." *Nat Rev Neurosci* 16 (6): 358–372.

Hermeking, H. 2007. "p53 enters the microRNA world." *Cancer Cell* 12 (5): 414–418.

Hidalgo, I. J., T. J. Raub, and R. T. Borchardt. 1989. "Characterization of the human colon carcinoma cell line (Caco-2) as a model system for intestinal epithelial permeability." *Gastroenterology* 96 (3): 736–749.

Ho, G. T., E. R. Nimmo, A. Tenesa, J. Fennell, H. Drummond, C. Mowat, I. D. Arnott, and J. Satsangi. 2005. "Allelic variations of the multidrug resistance gene determine susceptibility and disease behavior in ulcerative colitis." *Gastroenterology* 128 (2): 288–296.

Hood, J. L. 2016. "Post isolation modification of exosomes for nanomedicine applications." *Nanomedicine(Lond)* 11 (13): 1745–1756.

Hsu, C. H., and A. L. Cheng. 2007. "Clinical studies with curcumin." *Adv Exp Med Biol* 595: 471–480.

Huang, M., C. Liang, C. Tan, S. Huang, R. Ying, Y. Wang, Z. Wang, and Y. Zhang. 2019. "Liposome co-encapsulation as a strategy for the delivery of curcumin and resveratrol." *Food Funct* 10 (10): 6447–6458.

of inflammatory bowel disease in the 21st century: A systematic review of population-based studies." *Lancet* 390 (10114): 2769–2778.

Ng, Z. Y., J. Y. Wong, J. Panneerselvam, T. Madheswaran, P. Kumar, V. Pillay, A. Hsu, N. Hansbro, M. Bebawy, P. Wark, P. Hansbro, K. Dua, and D. K. Chellappan. 2018. "Assessing the potential of liposomes loaded with curcumin as a therapeutic intervention in asthma." *Colloids Surf B Biointerfaces* 172: 51–59.

Nitzan, O., M. Elias, A. Peretz, and W. Saliba. 2016. "Role of antibiotics for treatment of inflammatory bowel disease." *World J Gastroenterol* 22 (3): 1078–1087.

Nones, K., Y. E. Dommels, S. Martell, C. Butts, W. C. McNabb, Z. A. Park, S. Zhu, D. Hedderley, M. P. Barnett, and N. C. Roy. 2009. "The effects of dietary curcumin and rutin on colonic inflammation and gene expression in multidrug resistance gene-deficient (mdr1a-/-) mice, a model of inflammatory bowel diseases." *Br J Nutr* 101 (2): 169–181.

Ocansey, D. K. W., L. Zhang, Y. Wang, Y. Yan, H. Qian, X. Zhang, W. Xu, and F. Mao. 2020. "Exosome-mediated effects and applications in inflammatory bowel disease." *Biol Rev Camb Philos Soc* 95 (5): 1287–1307.

Packey, C. D., and R. B. Sartor. 2008. "Interplay of commensal and pathogenic bacteria, genetic mutations, and immunoregulatory defects in the pathogenesis of inflammatory bowel diseases." *J Intern Med* 263 (6): 597–606.

Panchatcharam, M., S. Miriyala, V. S. Gayathri, and L. Suguna. 2006. "Curcumin improves wound healing by modulating collagen and decreasing reactive oxygen species." *Mol Cell Biochem* 290 (1–2): 87–96.

Parronchi, P., P. Romagnani, F. Annunziato, S. Sampognaro, A. Becchio, L. Giannarini, E. Maggi, C. Pupilli, F. Tonelli, and S. Romagnani. 1997. "Type 1 T-helper cell predominance and interleukin-12 expression in the gut of patients with Crohn's disease." *Am J Pathol* 150 (3): 823–832.

Parvathy, K. S., P. S. Negi, and P. Srinivas. 2009. "Antioxidant, antimutagenic and antibacterial activities of curcumin-beta-diglucoside." *Food Chem* 115: 265–271.

Ponder, A., and M. D. Long. 2013. "A clinical review of recent findings in the epidemiology of inflammatory bowel disease." *Clin Epidemiol* 5: 237–247.

Praditya, D., L. Kirchhoff, J. Bruning, H. Rachmawati, J. Steinmann, and E. Steinmann. 2019. "Anti-infective properties of the golden spice curcumin." *Front Microbiol* 10: 912.

Priyadarsini, K. I. 2009. "Photophysics, photochemistry and photobiology of curcumin: Studies from organic solutions, bio-mimetics and living cells." *J Photochem Photobiol C: Photochem Rev* 10: 81–95.

Rahman, I., and S. K. Biswas. 2009. In *Regulation of Inflammation,Redox, and Glucocorticoid Signaling by Dietary Polyphenols*, edited by Y. J. Surh, Z. Dong, E. Cadenas, and L. Packer. Boca Raton: CRC Press.

Rege, S. A., M. Arya, and S. A. Momin. 2019. "Structure activity relationship of tautomers of curcumin: A review." *Ukrainian Food Journal* 8 (1): 45–59.

Rock, K. L., J. J. Lai, and H. Kono. 2011. "Innate and adaptive immune responses to cell death." *Immunol Rev* 243 (1): 191–205.

Romagnani, S. 1994. "Lymphokine production by human T cells in disease states." *Annu Rev Immunol* 12: 227–257.

Ruemmele, F. M., C. Gurbindo, A. M. Mansour, R. Marchand, E. Levy, and E. G. Seidman. 1998. "Effects of interferon gamma on growth, apoptosis, and MHC class II expression of immature rat intestinal crypt (IEC-6) cells." *J Cell Physiol* 176 (1): 120–126.

Salh, B., K. Assi, V. Templeman, K. Parhar, D. Owen, A. Gomez-Munoz, and K. Jacobson. 2003. "Curcumin attenuates DNB-induced murine colitis." *Am J Physiol Gastrointest Liver Physiol* 285 (1): G235–243.

Salzman, N. H., M. A. Underwood, and C. L. Bevins. 2007. "Paneth cells, defensins, and the commensal microbiota: A hypothesis on intimate interplay at the intestinal mucosa." *Semin Immunol* 19 (2): 70–83.

Saraf, S., A. Gupta, C. D. Kaur, M. Jangde, and S. Saraf. 2014. "Dermatological consequences of photosensitization with an approach to treat them naturally." *Pak J Biol Sci* 17 (2): 167–172.

Sasaki, S., H. Yoneyama, K. Suzuki, H. Suriki, T. Aiba, S. Watanabe, Y. Kawauchi, H. Kawachi, F. Shimizu, K. Matsushima, H. Asakura, and S. Narumi. 2002. "Blockade of CXCL10 protects mice from acute colitis and enhances crypt cell survival." *Eur J Immunol* 32 (11): 3197–3205.

Scheinman, R. I., A. Gualberto, C. M. Jewell, J. A. Cidlowski, and A. S. Baldwin, Jr. 1995. "Characterization of mechanisms involved in transrepression of NF-kappa B by activated glucocorticoid receptors." *Mol Cell Biol* 15 (2): 943–953.

Shah, Y. M., K. Morimura, and F. J. Gonzalez. 2007. "Expression of peroxisome proliferator-activated receptor-gamma in macrophage suppresses experimentally induced colitis." *Am J Physiol Gastrointest Liver Physiol* 292 (2): G657–666.

Shen, L., L. Liu, and H. F. Ji. 2017. "Regulative effects of curcumin spice administration on gut microbiota and its pharmacological implications." *Food Nutr Res* 61 (1): 1361780.

Shohan, M., S. Elahi, H. Shirzad, M. Rafieian-Kopaei, N. Bagheri, and E. Soltani. 2018. "Th9 cells: Probable players in ulcerative colitis pathogenesis." *Int Rev Immunol* 37 (4): 192–205.

Sidhu, G. S., A. K. Singh, D. Thaloor, K. K. Banaudha, G. K. Patnaik, R. C. Srimal, and R. K. Maheshwari. 1998. "Enhancement of wound healing by curcumin in animals." *Wound Repair Regen* 6 (2): 167–177.

Sidhu, G. S., H. Mani, J. P. Gaddipati, A. K. Singh, P. Seth, K. K. Banaudha, G. K. Patnaik, and R. K. Maheshwari. 1999. "Curcumin enhances wound healing in streptozotocin induced diabetic rats and genetically diabetic mice." *Wound Repair Regen* 7 (5): 362–374.

Silverman, J. M., and N. E. Reiner. 2011. "Exosomes and other microvesicles in infection biology: Organelles with unanticipated phenotypes." *Cell Microbiol* 13 (1): 1–9.

Silvestre, F., C. Santos, V. Silva, A. Ombredane, W. Pinheiro, L. Andrade, M. Garcia, T. Pacheco, G. Joanitti, G. Luz, and M. Carneiro. 2023. "Pharmacokinetics of curcumin delivered by nanoparticles and relationship with antitumor efficacy: A systematic review." *Pharmaceutics* 16 (7): 943.

Sugimoto, K., H. Hanai, K. Tozawa, T. Aoshi, M. Uchijima, T. Nagata, and Y. Koide. 2002. "Curcumin prevents and ameliorates trinitrobenzene sulfonic acid-induced colitis in mice." *Gastroenterology* 123 (6): 1912–1922.

Sun, D., X. Zhuang, X. Xiang, Y. Liu, S. Zhang, C. Liu, S. Barnes, W. Grizzle, D. Miller, and H. G. Zhang. 2010. "A novel nanoparticle drug delivery system: The anti-inflammatory activity of curcumin is enhanced when encapsulated in exosomes." *Mol Ther* 18 (9): 1606–1614.

Tajbakhsh, S., K. Mohammad, I. Deilami, K. Zandi, M. Fouladvand, E. Ramedani, and G. Asayesh. 2008. "Antibacterial activity of indium curcumin and indium diacetylcurcumin." *Afr J Biotechnol7* (21): 3832–3835.

Tejada, S., A. Manayi, M. Daglia, S. F. Nabavi, A. Sureda, Z. Hajheydari, O. Gortzi, H. Pazoki-Toroudi, and S. M. Nabavi. 2016. "Wound healing effects of curcumin: A short review." *Curr Pharm Biotechnol* 17 (11): 1002–1007.

Thangapazham, R. L., A. Puri, S. Tele, R. Blumenthal, and R. K. Maheshwari. 2008. "Evaluation of a nanotechnology-based carrier for delivery of curcumin in prostate cancer cells." *Int J Oncol* 32 (5): 1119–1123.

Tong, Q., E. V. Vassilieva, A. I. Ivanov, Z. Wang, G. T. Brown, C. A. Parkos, and A. Nusrat. 2005. "Interferon-gamma inhibits T84 epithelial cell migration by redirecting transcytosis of beta1 integrin from the migrating leading edge." *J Immunol* 175 (6): 4030–4038.

Tonnesen, H. H., and J. Karlsen. 1985. "Studies on curcumin and curcuminoids. VI. Kinetics of curcumin degradation in aqueous solution." *Z Lebensm Unters Forsch* 180 (5): 402–404.

Trigo-Gutierrez, J. K., Y. Vega-Chacon, A. B. Soares, and E. G. O. Mima. 2021. "Antimicrobial activity of curcumin in nanoformulations: A comprehensive review." *Int J Mol Sci* 22 (13).

Tsianos, E. V., K. H. Katsanos, and V. E. Tsianos. 2012. "Role of genetics in the diagnosis and prognosis of Crohn's disease." *World J Gastroenterol* 18 (2): 105–118.

Uhlig, H. H., and F. Powrie. 2018. "Translating immunology into therapeutic concepts for inflammatory bowel disease." *Annu Rev Immunol* 36: 755–781.

Ukil, A., S. Maity, S. Karmakar, N. Datta, J. R. Vedasiromoni, and P. K. Das. 2003. "Curcumin, the major component of food flavour turmeric, reduces mucosal injury in trinitrobenzene sulphonic acid-induced colitis." *Br J Pharmacol* 139 (2): 209–218.

Varshney, G. K., S. R. Kintali, P. K. Gupta, and K. Das. 2016. "Effect of bilayer partitioning of curcumin on the adsorption and transport of a cationic dye across POPG liposomes probed by second-harmonic spectroscopy." *Langmuir* 32 (40): 10415–10421.

Vashisht, M., P. Rani, S. K. Onteru, and D. Singh. 2017. "Curcumin encapsulated in milk exosomes resists human digestion and possesses enhanced intestinal permeability in vitro." *Appl Biochem Biotechnol* 183 (3): 993–1007.

Vecchi Brumatti, L., A. Marcuzzi, P. M. Tricarico, V. Zanin, M. Girardelli, and A. M. Bianco. 2014. "Curcumin and inflammatory bowel disease: Potential and limits of innovative treatments." *Molecules* 19 (12): 21127–21153.

Venkataranganna, M. V., M. Rafiq, S. Gopumadhavan, G. Peer, U. V. Babu, and S. K. Mitra. 2007. "NCB-02 (standardized Curcumin preparation) protects dinitrochlorobenzene-induced colitis through down-regulation of NFkappa-B and iNOS." *World J Gastroenterol* 13 (7): 1103–1107.

Vermeire, S., G. Van Assche, and P. Rutgeerts. 2012. "Classification of inflammatory bowel disease: The old and the new." *Curr Opin Gastroenterol* 28 (4): 321–326.

Wahl, C., S. Liptay, G. Adler, and R. M. Schmid. 1998. "Sulfasalazine: A potent and specific inhibitor of nuclear factor kappa B." *J Clin Invest* 101 (5): 1163–1174.

Waisman, A., R. S. Liblau, and B. Becher. 2015. "Innate and adaptive immune responses in the CNS." *Lancet Neurol* 14 (9): 945–955.

Wang, C., Z. Han, Y. Wu, X. Lu, X. Tang, J. Xiao, and N. Li. 2021. "Enhancing stability and anti-inflammatory properties of curcumin in ulcerative colitis therapy using liposomes mediated colon-specific drug delivery system." *Food Chem Toxicol* 151: 112123.

Wang, D., M. S. Veena, K. Stevenson, C. Tang, B. Ho, J. D. Suh, V. M. Duarte, K. F. Faull, K. Mehta, E. S. Srivatsan, and M. B. Wang. 2008. "Liposome-encapsulated curcumin suppresses growth of head and neck squamous cell carcinoma in vitro and in xenografts through the inhibition of nuclear factor kappaB by an AKT-independent pathway." *Clin Cancer Res* 14 (19): 6228–6236.

Wang, Q., X. Zhuang, J. Mu, Z. B. Deng, H. Jiang, L. Zhang, X. Xiang, B. Wang, J. Yan, D. Miller, and H. G. Zhang. 2013. "Delivery of therapeutic agents by nanoparticles made of grapefruit-derived lipids." *Nat Commun* 4: 1867.

Watson, J. L., R. Hill, P. B. Yaffe, A. Greenshields, M. Walsh, P. W. Lee, C. A. Giacomantonio, and D. W. Hoskin. 2010. "Curcumin causes superoxide anion production and p53-independent apoptosis in human colon cancer cells." *Cancer Lett* 297 (1): 1–8.

Weber, C. K., S. Liptay, T. Wirth, G. Adler, and R. M. Schmid. 2000. "Suppression of NF-kappaB activity by sulfasalazine is mediated by direct inhibition of IkappaB kinases alpha and beta." *Gastroenterology* 119 (5): 1209–1218.

Wei, C., J. Y. Wang, F. Xiong, B. H. Wu, M. H. Luo, Z. C. Yu, T. T. Liu, D. F. Li, Q. Tang, Y. X. Li, D. G. Zhang, Z. L. Xu, H. T. Jin, L. S. Wang, and J. Yao. 2021. "Curcumin ameliorates DSS-induced colitis in mice by regulating the Treg/Th17 signaling pathway." *Mol Med Rep* 23 (1).

Wright, J. S. 2002. "Predicting the antioxidant activity of curcumin and curcuminoids." *J Mol Structure: THEOCHEM* 591: 207–217.

Wu, K., F. Xing, S. Y. Wu, and K. Watabe. 2017. "Extracellular vesicles as emerging targets in cancer: Recent development from bench to bedside." *Biochim Biophys Acta Rev Cancer* 1868 (2): 538–563.

Xiao, H. T., J. Peng, D. D. Hu, C. Y. Lin, B. Du, S. W. Tsang, Z. S. Lin, X. J. Zhang, F. P. Lueng, Q. B. Han, and Z. X. Bian. 2015. "Qing-dai powder promotes recovery of colitis by inhibiting inflammatory responses of colonic macrophages in dextran sulfate sodium-treated mice." *Chin Med* 10: 29.

Xu, X. Y., X. Meng, S. Li, R. Y. Gan, Y. Li, and H. B. Li. 2018. "Bioactivity, health benefits, and related molecular mechanisms of curcumin: Current progress, challenges, and perspectives." *Nutrients* 10 (10).

Yance, D. R., Jr., and S. M. Sagar. 2006. "Targeting angiogenesis with integrative cancer therapies." *Integr Cancer Ther* 5 (1): 9–29.

Yang, J. Y., X. Zhong, S. J. Kim, D. H. Kim, H. S. Kim, J. S. Lee, H. W. Yum, J. Lee, H. K. Na, and Y. J. Surh. 2018. "Comparative effects of curcumin and tetrahydrocurcumin on dextran sulfate sodium-induced colitis and inflammatory signaling in mice." *J Cancer Prev* 23 (1): 18–24.

Ye, M., J. Zhang, J. Zhang, Q. Miao, L. Yao, and J. Zhang. 2015. "Curcumin promotes apoptosis by activating the p53-miR-192–5p/215-XIAP pathway in non-small cell lung cancer." *Cancer Lett* 357 (1): 196–205.

Yue, W., Y. Liu, X. Li, L. Lv, J. Huang, and J. Liu. 2019. "Curcumin ameliorates dextran sulfate sodium-induced colitis in mice via regulation of autophagy and intestinal immunity." *Turk J Gastroenterol* 30 (3): 290–298.

Zabana, Y., L. Rodriguez, T. Lobaton, J. Gordillo, A. Montserrat, R. Mena, B. Beltran, M. Dotti, O. Benitez, J. Guardiola, E. Domenech, E. Garcia-Planella, X. Calvet, M. Piqueras, M. Aceituno, F. Fernandez-Banares, and M. Esteve. 2019. "Relevant infections in inflammatory bowel disease, and their relationship with immunosuppressive therapy and their effects on disease mortality." *J Crohns Colitis* 13 (7): 828–837.

Zeissig, S., N. Burgel, D. Gunzel, J. Richter, J. Mankertz, U. Wahnschaffe, A. J. Kroesen, M. Zeitz, M. Fromm, and J. D. Schulzke. 2007. "Changes in expression and distribution of claudin 2, 5 and 8 lead to discontinuous tight junctions and barrier dysfunction in active Crohn's disease." *Gut* 56 (1): 61–72.

Zeng, Z., L. Zhan, H. Liao, L. Chen, and X. Lv. 2013. "Curcumin improves TNBS-induced colitis in rats by inhibiting IL-27 expression via the TLR4/NF-kappaB signaling pathway." *Planta Med* 79 (2): 102–109.

Zhang, L., H. Xue, G. Zhao, C. Qiao, X. Sun, C. Pang, and D. Zhang. 2019. "Curcumin and resveratrol suppress dextran sulfate sodium-induced colitis in mice." *Mol Med Rep* 19 (4): 3053–3060.

Zhang, M., C. Deng, J. Zheng, J. Xia, and D. Sheng. 2006. "Curcumin inhibits trinitrobenzene sulphonic acid-induced colitis in rats by activation of peroxisome proliferator-activated receptor gamma." *Int Immunopharmacol* 6 (8): 1233–1242.

Zhang, M., C. S. Deng, J. J. Zheng, and J. Xia. 2006. "Curcumin regulated shift from Th1 to Th2 in trinitrobenzene sulphonic acid-induced chronic colitis." *Acta Pharmacol Sin* 27 (8): 1071–1077.

Zhang, M., and D. Merlin. 2017. "Su1836—curcuma longa-derived nanoparticles reduce colitis and promote intestinal wound repair by inactivating the NF-kB pathway." *Gastroenterology* 152 (5, Supplement 1): S567.

Zhang, X., J. Wu, B. Ye, Q. Wang, X. Xie, and H. Shen. 2016. "Protective effect of curcumin on TNBS-induced intestinal inflammation is mediated through the JAK/STAT pathway." *BMC Complement Altern Med* 16 (1): 299.

Zhang, Y., Y. Liu, H. Liu, and W. H. Tang. 2019. "Exosomes: Biogenesis, biologic function and clinical potential." *Cell Biosci* 9: 19.

Zhao, H. M., F. Han, R. Xu, X. Y. Huang, S. M. Cheng, M. F. Huang, H. Y. Yue, X. Wang, Y. Zou, H. L. Xu, and D. Y. Liu. 2017. "Therapeutic effect of curcumin on experimental colitis mediated by inhibiting CD8(+) CD11c(+) cells." *World J Gastroenterol* 23 (10): 1804–1815.

Zhao, H. M., R. Xu, X. Y. Huang, S. M. Cheng, M. F. Huang, H. Y. Yue, X. Wang, Y. Zou, A. P. Lu, and D. Y. Liu. 2016. "Curcumin suppressed activation of dendritic cells via JAK/STAT/SOCS signal in mice with experimental colitis." *Front Pharmacol* 7: 455.

Zhao, M., and J. Burisch. 2019. "Impact of genes and the environment on the pathogenesis and disease course of inflammatory bowel disease." *Dig Dis Sci* 64 (7): 1759–1769.

Zheng, B., and D. J. McClements. 2020. "Formulation of more efficacious curcumin delivery systems using colloid science: Enhanced solubility, stability, and bioavailability." *Molecules* 25 (12).

Zielinska, A., H. Alves, V. Marques, A. Durazzo, M. Lucarini, T. F. Alves, M. Morsink, N. Willemen, P. Eder, M. V. Chaud, P. Severino, A. Santini, and E. B. Souto. 2020. "Properties, extraction methods, and delivery systems for curcumin as a natural source of beneficial health effects." *Medicina(Kaunas)* 56 (7).

7 Lipid-Based Nanoparticles
SLN and NLC—a Versatile Delivery Vehicle for Phytodrugs

S. Mohana Lakshmi and C.K. Ashok Kumar

7.1 INTRODUCTION

Nanotechnology has made significant advances, particularly in the disciplines of material science and photonics, medication delivery, and the food industry. Nanomedicine (application of nanotechnology in medicine) has offered a major push for the creation of numerous types of drug-carrying nanoparticles (Germain et al., 2020).

The physical, chemical, and biological processes that affect bioavailability of bio-components, such as drug release and ADME (absorption, distribution, metabolism, and excretion), have been highlighted in research to help boost the bioavailability of drug substances. Till recently, nanocarriers used as alternative drug delivery techniques have grabbed attention due to their enhanced drug penetration and distribution into particular areas of the skin. These nanocarriers might lead to improved half-lives, accurate short- or long-period release kinetics, and similar site-specific medicinal ingredient delivery (Ragelle et al., 2017). Nanocarriers are particulate colloids with dimensions ranging from 1 to 1,000 nm. Medical nanocarriers typically have a size range of 1–100 nm. The interaction between the polar-charged phospholipid body and the solvent or between the non-polar uncharged lipid groups in the solvent will determine the type of nanocarrier.

7.1.1 Lipid-Based Nanocarriers

Lipid-based nanocarriers (LNs) are suitable as drug carriers because of their unique physicochemical features, which include nanoparticles with sold lipid core matrix with better compatibility to the biological environment (Jahangir et al., 2020). These LNs, which are made up of homogeneous lipid bilayers or solid cores, have the ability to entrap a variety of cytotoxic medicines. The lipophilic medication will be trapped in the lipid leaflets, whereas the hydrophilic drug will be encapsulated in the watery area. Furthermore, LNs transport medications safely to the tumor site, release them gradually, and subsequently break them down (Wacker, 2013) Liposomes, niosomes, self-nano- and microemulsifying drug delivery systems (SNEEDS and SMEEDS), micelles, nanoemulsions, polymeric nanoparticles, and albumin-based formulations are all examples of lipid-based nanomaterials that have recently exploded in popularity (Bordat et al., 2019)

7.1.2 Liposomes

Liposomes are phospholipid vesicle systems created in the late 20th century, are made up of a phospholipid bilayer comparable to that found in human cells' plasma membrane. As a result, liposomes are biocompatible and can aid medication diffusion across plasma membranes. Liposomes are self-assembled vesicles made up of one or more concentric lipid bilayers around an aqueous core (Guimaraes et al., 2021).

Liposomes have a conventional structure enclose with a non-polar lipid bilayer and a polar central core, with sizes ranging from 20 nm to more than 1 μm. As a result, liposomes can

DOI: 10.1201/9781003300588-7

encapsulate lipophilic medications in a lipid bilayer while holding and stabilizing hydrophilic pharmaceuticals in the hydrophilic core. As a result, liposomes are capable of transporting both non-polar and polar drugs through the water-soluble lumen and outer lipid in bilayers, respectively, adding to their flexibility. However, large drug loading of hydrophobic medicines is problematic due to the restricted bilayer area of liposomes. A precise compromise must be struck between expected higher drug-loading capacity and liposome size distribution of the particles and their relative stability.

Furthermore, liposomes have the benefit of simple customization and targeting ability since the surface may be changed with suitable moieties (or ligands) to directly engage a target moiety of a specific cell, system, or organ. Other chemicals, such as cholesterol, can be added to their formulations in addition to phospholipids, boosting the stability of these nanoparticles in blood by decreasing their fluidity and increasing the permeability of hydrophobic medicines across the bilayer membrane (Yingchoncharoen et al., 2016)

The key phospholipid (PL) components are phosphatidyl ethanolamine (PE), phosphatidylcholine (PC), and phosphatidylinositol. Natural herbal PLs, like soybean and sunflower PLs, were commonly employed for liposomal production (PI). However, because to the high degree of unsaturation, herbal PLs are more vulnerably oxidized and have worse stability physically and upon storage (Wu et al., 2020). Animal-based phospholipids, derived from milk and eggs, and marine source, are also gaining popularity for liposome manufacture. Marine sources have been shown to have greater PUFA levels, making them a viable functional component (Burri et al., 2012).

7.1.3 Solid-Lipid Nanoparticles

In the 1990s, solid-lipid nanoparticles (SLNs) were designed to tandem the benefits of nanocarriers with polymers, such as high drug loading, controlled drug administration, lipid emulsion biocompatibility, and improved bioavailability of the drugs (Yuan et al., 2007)

SLNs are a type of colloidal drug delivery system that is made up of physiological lipids that are solid at normal temperature. SLN systems are frequently made with biocompatible chemicals such as fatty acids, triglycerides, bio waxes, and steroids. A blend of surfactants or polymers stabilizes this matrix. These particles range in size from 50 to 1,000 nm. SLNs are ideal for covering with functionalized ligand moieties, antibodies, and other functional groups due to their tiny size and wide surface area (Rostami et al., 2014).

Easy manufacture, pharmaceutical stability, higher drug percentage, effective drug release and higher stability, regulated release of both lipophilic and hydrophilic medications, protection of labile drugs, cheap cost, and nontoxicity are just a few of the benefits of SLNs. Furthermore, because to its high lipid content, the SLN system may effectively encapsulate anticancer medicines and other compounds with low water solubility (Wong et al., 2007). SLNs, on the other hand, have several drawbacks, such as a limited drug-loading capacity and drug ejection owing to crystallization during storage.

7.1.4 Nanostructured Lipid Carriers

Nanostructured lipid carriers (NLCs) have been extensively researched as an effective drug delivery system. Because of their greater stability, efficient drug loading, and better biocompatibility, they are regarded effective than previous classic lipid-based nanocarriers like the liposomes and SLNs (Haider et al., 2020).

Researchers modified SLNs by substituting a percentage of liquid lipids in place of solid lipids to create NLCs that have improved biocompatibility and formulation qualities. NLCs are biocompatible drug delivery systems characterized by a rigid shape that adds to unique capabilities as compared to other formulations (Müller et al., 2011). The mean particle sizes are quite comparable to SLNs, ranging from 10 to 1,000 nm.

Fatty acids, steroids, waxes, and triglycerides, such as ethyl oleate, glyceryl tricaprylate, isopropyl myristate, and glyceryl dioleate, are among the many lipids employed in topical lipid nanoparticulate compositions. Pharmaceuticals have a higher solubility in oil than solid lipids; hence, conventional NLCs have a better encapsulation capacity for drugs than SLNs (Yuan et al., 2007). Due to the presence of liquid lipids in the NLC formulation, NLCs offer a larger drug loading-capacity and may help prevent drug ejection during storage by preventing lipid crystallization. They're good for site-specific targeting, have good drug release control, and have low in vivo toxicity. NLCs, on the other hand, have the drawback of being difficult to surface functionalize (Puri et al., 2009). However, there are significant drawbacks, such as drug expulsion during storage due to polymorphic transition of the lipid from the nanocarrier matrix and poor loading capacity.

7.1.5 Nanoemulsions

In the healthcare and cosmetics sectors, nanoemulsions, also known as dispersed systems, with a droplet size ranging from 100 nm and lower, have becoming more appealing due to the unique properties of nanosized droplets, such as their enormous surface area (Ashaolu et al., 2021). Nanoemulsions may be classified into three categories based on their components: water in oil (W/O), oil in water (O/W), and multiple emulsions with inter-dispersed micro dispersions sandwiched oil and aqueous phases (W/O/W and O/W/O) (Zhang et al., 2018)

Nanoemulsions are also classified as anionic, neutral, or cationic nanoemulsions based on the charge on the surface of the nanosized droplets. Lipophiles are commonly dissolved in vegetable oils, such as sesame oil, soybean oil, and cottonseed oil (Constantinides et al., 2004)

Nanoemulsions, in comparison to other nanocarriers, are simple to make and do not necessitate the use of non-polar solvents/co-solvents, especially when using vegetable oils with lower toxicity. Nanoemulsions are utilized to encapsulate anticancer drugs like paclitaxel, retinoic acid, curcumin, and other antitumor medicines because of the benefits listed earlier (Chinsriwongkul et al., 2010). Nanoemulsions are thermodynamically unstable, which can result in destabilization processes, including coalescence and flocculation.

7.1.6 Microemulsions

Microemulsions are dispersions of 2 immiscible phases in the form of colloids that are stabilized by surfactants supplied in larger quantities (>20%) than nanoemulsions (3–10%). Thermodynamic stability is conferred on microemulsions as a result of this. Microemulsions often have smaller mean diameters than nanoemulsions. Though the microemulsions often have size ranges less than 100 nm, there is no defined size threshold that accurately differentiates microemulsions from nanoemulsions (McCements, 2012)

Another distinguishing feature of microemulsions is their ability to take on non-spherical forms, such as hexagonal or worm-like, which might affect the flow properties of the systems. Furthermore, while not well investigated, it is suggested that the form of nanocarriers influences their interaction with membranes of cells participating in the endocytosis (Decuzzi and Ferrari, 2008). Microemulsions have been created in recent years to improve oral bioavailability and medication delivery to the brain (Sang et al., 2005; Guo et al., 2019)

7.1.7 Nanocapsules

Nanocapsules are used to increase the BA of a wide range of medications when taken orally (Roger et al., 2011). Lipid nanocapsules are nanocarriers with vesicles having lipid and water phase separated by a polymeric shell and stabilized by surfactants. The nanocapsules were around 100 nm in diameter and are positively charged. Many physicochemical features of nanocarriers, and hence their biological identity, are influenced by their surface properties. Thus, the lipid polymer outer shell of the nanocapsules influences their characteristics, such as surface charge or hydrophilicity, as well as their gastrointestinal behavior, such as interactions with mucus and/or enzymes. Derivatives of polyethylene glycol (PEG), chitosan, caprolactone, hyaluronic acid, PLA, and Eudragit are some of the polymers commonly employed in the manufacture of lipid nanocapsules.

7.1.8 Niosomes

Niosomes are bilayer structures that entrap bioactive chemicals, comparable to liposomes. The bilayer of niosomes, on the other hand, is composed up of non-ionic surfactants apart from charged phospholipids (Devaraj et al., 2002). Niosomes are utilized to carry both hydrophilic drug substances (in the aqueous layer) and hydrophobic molecules (in the surfactant lipid bilayer) like liposomes. Efficiency of biodegrading and biological compatibility are two benefits of niosomes, as are their changeable vesicle features. Niosomes are a cost-effective alternative to liposomes due to their low cost, highly stable, simplicity, and storage of non-ionic surfactants. Niosome-forming surfactants are also biodegradable, biocompatible, nonimmunogenic, and nontoxic.

7.1.9 Self-Nano- and Microemulsifying Drug Delivery Systems

Anhydrous nanoemulsions are self-nanoemulsifying drug delivery systems (SNEDDS). SNEDDS are medication, oil, and surfactant mixes. Co-surfactants or solubilizers might be added if needed. The fundamental feature of SNEDDS is that they generate nanoemulsions deliberately when poured into polar or aqueous phase with gentle mixing, such as during GI motility (Guo et al., 2019).

The high ratio of surfactants in SNEDDS facilitates the system to emulsify by lowering the energy limits that are required for oil droplet dispersion to zero or negative values. Similarly, microemulsions in aqueous media are used by self-microemulsifying drug delivery systems (SMEDDS) (Khan et al., 2012).

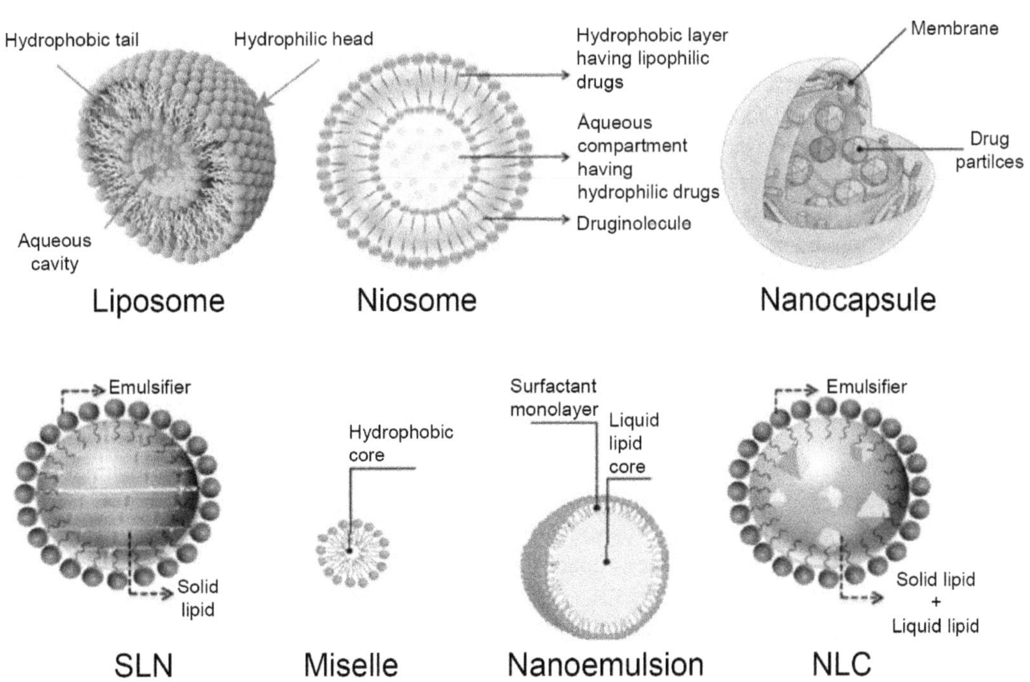

FIGURE 7.1 Different lipid-based nanocarriers for the delivery of phytodrugs.

Because of their composition, they have various benefits over ordinary nanoemulsions. First, because they don't contain water, they have superior long-term stability profiles and may be administered in less amounts. They can also be included in dose carriers like gelatin or HPMC (hydroxypropyl methylcellulose) capsules, which increases their palatability and, as a result, patient compliance (Date et al., 2010). Furthermore, traditional SNEDDS and SMEDDS may have significant drawbacks, such as long-term component incompatibility with capsule shells or notable precipitation of the drug during storage under low-temperature.

7.2 NANOCARRIER-BASED DELIVERY OF PHYTOCHEMICALS

The relevance of phytochemicals for the prevention of chronic illnesses has piqued the interest of the general public and health experts, resulting in a surge in demand for phytochemical products. Phytochemical product translation from lab to market is arduous, including tedious and winding path requiring fundamental research, formulation of the drug molecule, complex preclinical studies, and regulatory clearances. Numerous aspects contribute to the effectiveness of phytochemical substances, including therapeutic efficacy, physicochemical stability, and BA. Due to lack of solubility in the GI system, restricted permeability of the intestine, and first-pass metabolism in the liver, many bioactive substances have low bioavailability. The bioactive ingredient must primarily dissolve in the GI fluid and pass stable before reaching the circulation, yet poorly soluble chemicals (lipophilic compounds) have meager absorption and resulting bioavailability due to their improper dissolving properties. The second component that limits bioavailability to a large amount is intestinal as well as the first pass metabolism in liver (Arzani et al., 2015).

Many novel drug delivery systems have all been investigated as ways to increase phytochemical component bioaccessibility and bioavailability. The goal of these technical and chemical alterations to phytochemical compounds is to improve their solubility, absorption site, colloidal system design (micelles and vesicles), and nanosystem utilization (Rein et al., 2013). Hydrophobic properties are found in a wide spectrum of vital bioactives/phytochemicals found in nature, like polyphenols and flavonoid compounds, volatile oils and fatty acids, and fat-soluble vitamins, and these hydrophobic molecules are used as drugs to treat various diseases. Bioactive chemicals' poorly soluble properties eventually diminish bioavailability and sustainability (Rezaei et al., 2019). A lipid-based carrier may be an effective way to preserve bioactive substances in the GI tract while also increasing their bioavailability.

Bioactive ingredients, such as antioxidants, phenolic compounds, natural food colorants, essential oils, antimicrobial agents, minerals, flavors, and vitamins could be tagged using nanocarriers to deliver the targeted drugs and phytochemicals into site of action and protect the molecules against digestible metabolic processes (Assadpour and Mahdi Jafari, 2019). Food lipids, which are commonly employed to entrap the nutrients and functional components, have been shown to preserve, regulate, and enhance the activity of bioactive chemicals (Caddeo et al., 2018). Wrapping polyunsaturated fatty acids in liposomes improves their stability and absorption, as they are easily impacted by heat and light. Nanoencapsulation of essential oils has a number of advantages, including ease of handling, stability, oxidation resistance, improved dispersion, solubility, and controlled release, as well as minimal or no detrimental influence on the organoleptic qualities of related food products (Prakash et al., 2018).

Bai et al. created coix seed oil liposomes in 2011, which enhanced the stability of the oil, increased its solubility in water, and dramatically increased its absorption in the gut. When antioxidant components are wrapped in liposomes, the antioxidant resistance to the environment is improved, and the antioxidant function is enhanced to some extent. Resveratrol liposomes, according to earlier studies, have a greater ROS-quenching capacity than resveratrol and can exhibit stronger antioxidant activity in cells (Vanaja et al., 2013).

It was also suggested that the delivery of lutein via the lipid-based nanocarriers that are stabilized by modified starch or stearate of sugars mixed in nutraceutical drinks might be suitable for hydrophobic bioactive chemicals (Sedaghat Doost et al., 2020). In 2021, Lee et al. suggested the effects of lipids carriers on the stability of lipid nanocarriers that are exposed to the fluctuations in temperature and salinity and discovered that lipid nanocarriers are very stable and there was no oxidation or degradation in the active ingredient caryophylline was noticed.

Lactoferrin is a peptide with antioxidant, anti-inflammatory, antibacterial, and antiviral properties that has sparked a lot of attention because of its critical function in the immune system. It is, however, vulnerable to gastrointestinal enzyme hydrolysis, which reduces its oral bioavailability. Lactoferrin encapsulation in liposomes has been studied as a way to get around this constraint. Indeed, a phase 2 and 3 clinical investigation investigating the effectiveness and safety of lactoferrin-loaded liposomes taken orally in COVID-19 patients with mild to moderate illness and asymptomatic people is under underway (clinical trial id. NCT04475120).

Quercetin is a flavonoid that has anti-obesity, anticancer, anti-inflammatory, and antioxidant properties, among other things. The primary disadvantages of using it in the food industry are its poor water solubility and chemical stability. Quercetin has a limited bioavailability and a short biological half-life due to its weak water solubility. Cyclodextrin-coated liposomes encapsulating quercetin had a greater encapsulation capacity and enhanced physicochemical qualities, according to Azzi et al. [46]. After storage period of 1 year (at 4°C), the scientists discovered that the quercetin liposomes were normal (as measured by particle size distribution and drug concentration). Surprisingly, a larger percentage of quercetin (98%) was preserved in cyclodextrin complexes in liposomes after 12 months (Cai et al., 2013).

Curcumin is a polyphenolic molecule with antioxidant, anti-inflammatory, antibacterial, and cancer-fighting effects. Curcumin's widespread usage is limited, however, due to its poor water solubility, rapid metabolism, and very low permeability as a result of susceptibility of P-glycoprotein efflux. The enhancement of the stability of curcumin stability in liposomes can be achieved by converting the formulation liquid to a dry liposomal powder form (Ipar et al., 2019; Pu C et al., 2019).

The influence of milk and krill PLs on curcumin-loaded liposomes was investigated, and the authors discovered that the liposomes of bovine milk PLs were more stable, had smaller particle sizes, and had a higher zeta potential. The retention of curcumin is also greater in milk PL formulation than in krill PL formulation, which is due to the large quantity of phosphatidylcholine in krill PLs, which is vulnerable to breaking down or leaking, which results in loss of curcumin from the phospholipids (Wu et al., 2020).

Researchers used a unique nanofiber weaving process to manufacture liposomal curcuminoids in powder form. The burst release (first rupture) of drug from the liposomal curcumin powder was detected within 12 hours, followed by a steady and continuous release. The release of curcumin from the formulation was increased by a somewhat acidic pH, and the strong contact between the drug and phospholipids resulted in reduced desorption (Gopi et al., 2018). Curcumin was preserved against degradation by Pluronic-modified liposomes, and over half of the drug was maintained post exposure to 80°C. Curcumin's bio-accessibility was also raised to 43.3%, whereas the basic curcumin liposome was found to be just 26.9% bioaccessible (Li et al., 2018).

Green tea contains around eight polyphenol compounds (catechins) and is popular among healthcare researchers because of its possible health advantages, which include antioxidant, anti-diabetic, and anti-inflammatory characteristics. Catechins, on the other hand, are poorly absorbed by the body. Catechin in a liposomal form enhances absorption and stability in the GI tract of the body. The pharmacokinetic features of chitosan-coated catechin liposomes were studied in Wistar rats by Ezzat et al. The rapid hike in absorption of chitosan-coated catechin liposomes in the intestine was discovered in an in situ intestinal perfusion research (Ezzat et al., 2019). Similarly, chitosan-coated catechin liposomes had 1.37-fold higher bioavailability than classic liposomes and 2.12-fold more bioavailability than free catechin.

Fornasier has recently produced catechin hexosomes (a tube-like hexagon arrangement of lipids) for topical distribution and observed an enhanced catechin content in pig skin. Bile salts were incorporated into the design of the hexasomes in order to increase their permeability features, and it was demonstrated that hexasomes containing bile salts had relatively deeper transdermal penetration of catechin than those that did not contain bile salts (Fornasier et al., 2021).

The most significant flavonoid in green tea is epigallocatechin gallate (EGCG), but it loses its stability quickly in the GI tract and is very poorly absorbed by epithelial cells. Liang reported on the stability of EGCG niosomes in GI using their z-average diameter. The stability of the niosomes was maintained at (60 nm) after 60 minutes of exposure in a simulated gastric environment, while particle size grew from 60 to 178 nm in 2 hours in the intestinal condition (Liang et al., 2016).

Resveratrol is a non-flavonoid polyphenol drug found in fruits like grapes and berries. It is well-known for its anticarcinogenic, anti-inflammatory, anti-obesity, antioxidant, and anti-aging activities. On the negative side, resveratrol is sensitive to light and weakly soluble in the lower GI tract, with little absorption in the intestine, while artemisinin decreased the survival of HT-29 cells to 16% and 37%, respectively, after 24 hours of exposure to 10 and 20 g mL1. The increased tumor cell mortality was linked to resveratrol's various biological targets, which all impact cellular proliferation and growth (Jagwani et al., 2020).

A combination of antitumor bioactives, resveratrol, and artemisinin in the form of liposomes were tested, and spikes in the absorption of the drugs were noticed when compared to the free drugs (Caddeo et al., 2021). Hzr-Kad et al. recently enhanced the solubility and bioaccessibility of pollen phenolic extract (PPE) by 2-fold when compared to ordinary liposomes. However, during the proliposome synthesis stage, PPE was dissolved in a few drops of alcohol, which may limit the general implementation of this technology on an industrial scale (HizirKadi et al., 2020).

According to Tavano et al., the technique of co-encapsulation can have an effect on the physicochemical properties as well as the entrapment efficiencies of niosomal formulations that contain co-encapsulated compounds (such as gallic acid and curcumin or ascorbic acid and quercetin). Curcumin and gallic acid co-encapsulated in niosomes demonstrated a lower profile of release than the corresponding free compounds when the conditions of a simulated gastrointestinal environment were replicated. After being subjected to the pH of the digestive tract for around 2 hours, both the drugs were found to be released at the same rate, indicating the complex interaction between the drugs in the liposomes (Tavano et al., 2014).

Formulation	Phytochemical	Lipid Type	Effect Exhibited	Reference
Liposome	Curcumin	Phosphatidylcholine, cholesterol	Curcumin release was monitored for 4 days, along with lower proportion (37%) of curcumin release.	Reddy et al., 2019
	Curcumin	Bovine milk phospholipids, krill phospholipids	Liposomes made from krill phosphate are more digested and had better BA than liposomes made from bovine milk phosphate.	Wu et al., 2020
	Quercetin	Phosphatidylcholine	Liposomes coated with Eudragit were shown to be harmless in intestinal cells, causing no cytotoxicity.	Caddeo et al., 2019
	Quercetin and linseed oil	Phosphatidylcholine	In the simulated gastrointestinal digestion, liposomal formulations showed low stability (flocculated).	Huang et al., 2020
	Resveratrol	Tripeptide lipid CDO	Resveratrol was shown to be more effective, with 10 mg kg1 entirely stopping tumor development.	Zhao et al., 2020
	Vitamin C, β-carotene	Cholesterol	In the digestive phase, about 70% of bioactive chemicals were slowly released.	Liu et al., 2020
SLN	Curcumin	Tristearin	SLNs stabilized with PEG10SE had a 6-fold increase in oral bioavailability.	Ban et al., 2020
	Curcuminoids	Precirol® ATO5 and Compritol® 888 ATO	SLNs has a 4.48-fold increase in CMax.	Dolatabadi et al., 2021

Formulation	Phytochemical	Lipid Type	Effect Exhibited	Reference
NLC	Curcumin	Beeswax	The cytotoxicity effects of MCT oil digestion products were ascribed to a reduction in NLC cell viability.	Gonçalves et al., 2021
	Curcumin	Stearic acid and capric triglycerides	Release of curcumin from NLCs was about 41% in sim gastric media for 2 hours.	Behbahani et al., 2019
	Lutein	Octenyl modified starch, sucrose monostearate	Lutein demonstrated heightened susceptibility to applied stressors.	SedaghatDoost et al., 2020
SEDDS	Rutin	Glycerol monostearate	When compared to rutin suspensions, the SEDDS formulation had an approximately 1.76-times rise in BA.	Wang et al., 2017
	Ferulic acid	Labrasol	The SEDDS increased ferulic acid oral bioavailability by 1.74 times.	Liu et al., 2020a
Niosome	Epigallocatechin gallate	Niosomal carriers	Improved ()-epigallocatechin gallate's chemical stability and antioxidant activity in intestinal tract conditions.	Liang et al., 2016
Lipid nanocarrier	β-caryophellene	Medium-chain triglycerides (MCT)	The lipid nanocarriers showed considerable coalescence during storage.	Lee et al., 2021

7.2.1 NLCs

The development of an advanced lipid-based carrier system has a number of pros over conventional lipid formulations, such as nanoemulsions and liposomes, including preventing non polar solvents, enabling bulk production processes, and facilitating the protection of loaded thermolabile therapeutic agents. This is all achieved by using lipids in solid forms (SL) instead of an oily phase used in manufacture of emulsions (Abdelbary et al., 2013). SLNs acting as drug carriers, however, were hampered by low efficiency of drug loading and an increased drug ejection from the lipid system because of their transitions into various polymorphs during storage (Müller et al., 2002).

Müller et al. produced NLCs in 2002 utilizing a blend of solid and liquid lipids (SL and LL), which form an amorphous solid matrix at room temperature and in vivo body temperature to address these restrictions. The integration of the latter into the system is the most important step since it considerably improved the formulation's characteristics as compared to SLNs. Though the crystalline matrix is formed in SLNs, it limits the drug-loading capacity of the formulations and results in the expulsion of drug due to the low spatial space. Liquid lipids contribute to the arrangement of lattice characterized with prominent imperfections inside of the solid matrix of the crystals. This allows a greater drug-loading ability (Mehnert et al., 2001). Investigations were heavily done on the NLCs considering them as alternatives for SLNs, nano emulsions, microparticulate liposomes, and polymeric lipid based particles in recent years (Jaiswal et al., 2016).

In addition, NLCs had to widen the range and to overcome most of the drawbacks of traditional lipid-based carriers. NLCs in the solid forms at ambient climate, for example, improved their stability, which is a key hurdle in lipid-based formulations (Müller et al., 2000). The emergence of viable technologies for large-scale NLC manufacturing alleviated the costly technological necessity for mass liposome manufacture. Furthermore, unlike other lipid-based formulations, NLCs are biocompatible systems with a hard shape that adds to their unique features (Müller et al., 2011).

Because of their well-established biocompatibility and safety characteristics, NLCs have been extensively explored as delivery methods for a number of medicinal and aesthetic applications. NLCs have shown very huge spike in the trend for drug carriers, particularly for labile hydrophilic and hydrophobic medicines, by dramatically boosting encapsulation efficiency, shielding the formulations from degrading in the human body, increasing the net bioavailability by managing drug release. The application of NLCs to carry chemotherapeutic drugs received special attention because of their ability to improve the stability of the carried drugs, as well as the therapeutic efficacy of those drugs is also revoked due to low pKa characteristics (Tikekar et al., 2012).

There are various important ingredients in the assembly of the NLCs, which are discussed as follows.

7.2.2 Lipids

Lipid is the major ingredient of NLCs that controls the loading capacity of the active ingredient, duration of action, formulation stability, etc. NLCs are made up of solid lipids, such as fatty acids, triglycerides, diglycerides, monoglycerides, and steroids (Noor et al., 2017). Prior to using lipids in the creation of nanoparticulate carriers, it is critical to choose the right ones. Various features of nanocarriers are influenced by the kind and structure of lipid. Solubility/apparent partition coefficient of drugs in the lipid phase has been proposed as the most fitting parameter for selecting a suitable lipid in practice.

7.2.3 Surfactants

Surfactant type and concentration have an impact on the quality and efficacy of NLCs. The choice of surfactant has been found to have a significant impact on the toxicity, stability, and crystalline character of NLCs.

The amount of dissolution and permeability of the drug are also affected by surfactant systems. An ideal surfactant is selected based on their mode of administering, HLB value, particle size impact, and lipid modification. Because of their amphipathic character, emulsifiers are adsorbed on the junction of the two phases, where emulsifiers lower the surface tension between the two different phases (Jaiswal et al., 2016).

7.2.4 Other Excipients

Another kind of excipient employed in the formulation of NLC is surface-modifiers, which are intended to reduce phagocytic absorption by macrophages in the reticuloendothelial system (RES). To enhance the residence period of therapeutic molecules in systemic circulation, coating of the lipids was done using hydrophilic polymers, such as poloxamines, PEG, and poloxamers. Other benefits of surface modification include improved stability and compatibility in the biological environment, targeting of medication, and greater permeability through epithelia (Uner et al., 2007).

7.2.5 Preparation of NLCs

7.2.5.1 High-Pressure Homogenization

The use of HPH (high-pressure homogenization) technology has demonstrated that it is a dependable and efficient way for the production of lipid nanoparticles. In contrast to the methods that came before it, this method may also be used for the production of LNs on a massive scale. Hot homogenization methods and cold homogenization procedures were both developed throughout time. Prior to the high-pressure heating step, the active medicinal component is either dissolved or distributed in the melted lipid. High pressure (between 100 and 2,000 bar) is used to move the fluid around in the homogenizer. The typical particle size is measured in microns or less. The process of homogenization has a number of benefits, including better drug loading, greater product stability, large-scale production, and elimination of the need for organic solvents, yet it is challenging to implement because of the high pressure and temperature conditions required (Ekambaram et al., 2012).

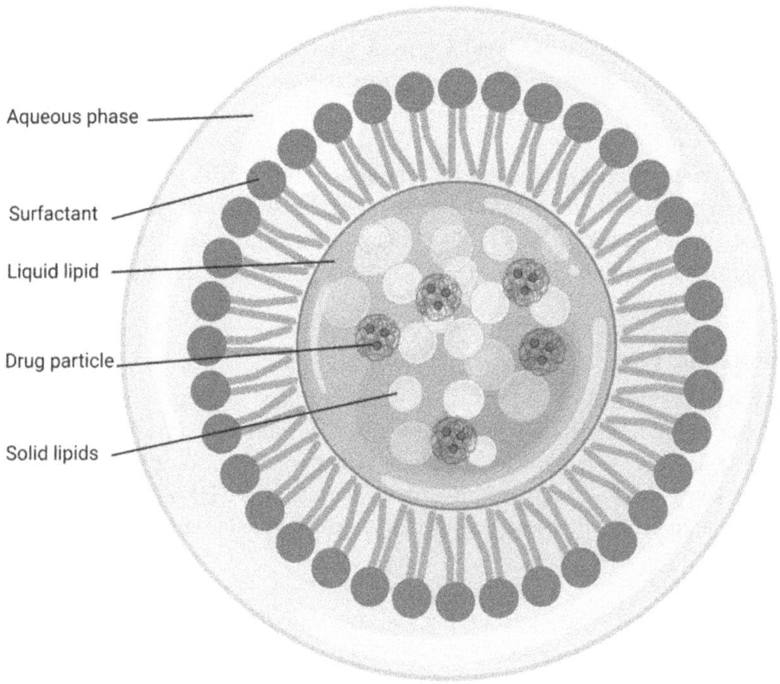

FIGURE 7.2 Composition of nanostructured lipid carriers (NLCs).

7.2.5.2 Ultrasonication

Ultrasonication is one of the procedures for producing LNPs. This phase disperses the lipid phase in the solvent phase, containing high amounts of surfactants. The surfactant content is viewed as a drawback. Another disadvantage of this approach is that it does not generate a small size distribution of particles, which leads to instability in the storage. Unlike hot and cold homogenization, this approach makes use of basic tools that may be obtained in any lab (Patwekar et al., 2014).

7.2.5.3 Solvent Evaporation/Emulsification

During this process, the lipid is dissolved in an organic solvent that does not dissolve in water. Following this step, an emulsion consisting of surfactants and having an aqueous phase is produced. The solvent is extracted from the emulsion by a process of evaporation that takes place at a lower pressure. Nanoparticles become more dispersed in the aqueous phase as a result of evaporation (using lipid precipitation process in the aqueous phase). This method, as opposed to cold homogenization, will not be put under heat stress; nonetheless, the organic solvent that is utilized in this process is a disadvantage of this method. The solid lipid and surfactant that are utilized both play a role in determining the size of the particles (Jagdevappa et al., 2013).

7.2.5.4 Spray Drying

This technology is an alternative to the process of lyophilization for producing pharmaceuticals from SLN aqueous dispersion. This method, rather than lyophilization, is a cheaper process for producing lipids, although it is not widely employed. Because of the application of force and high temperature, aggregation of the particles takes place. Lipids that has melting points more than 70°C have been shown to be appropriate for spray drying in prior research (Mukherjee et al., 2009).

7.2.5.6 Types of the Developed NLCs

According to the lipidic structure of the NLCs, there are three different types of NLCs that may be distinguished by the composition of the lipid mixture, the manufacturing process, or both.

7.2.5.7 Imperfect Type

The lipid matrix is imperfected in this kind by employing lipids with diverse chemical characteristics, like the length of the carbon chains and the degree of saturation. When these lipids are mixed, disruption of the crystal lattice takes place, and crystallization is disrupted. As a result, the lipid matrix may hold more medicine and is less likely to be evacuated during storage than if a single lipid is used.

7.2.5.8 Amorphous Type

Due to the fact that crystallization of lipid matrix is takes place in amorphous forms that are poorly ordered, greater drug loading can be achieved due to the formation of amorphous matrix that resembles a structure less solid. This pattern can be created by mixing solid lipid with triglycerides of medium chains like hydroxyl octacosanyl hydroxyl stearate (HOCH) or isopropyl myristate (IPM). The lipid solid state and transition temperature are confirmed using NMR and DSC analysis.

7.2.5.9 Multiple Type

This particular type of solid lipid matrix possesses a great number of nanosized liquid oils that are capable of dissolving the drug completely. As a direct consequence of this, a greater quantity of the drug is encapsulated. Additionally, because the tiny oil droplets are bound by the solid lipid matrix, the medicine is delivered in a controlled release pattern, and there is minimal leakage of the drug (Battaglia and Gallarate, 2012).

Specific crystallization circumstances result in a very disordered product. Because of imperfections in the structure of the lipid matrix, crystals have gaps between the triglyceride fatty acid chains, allowing medicines to enter the matrix more easily (Sarathchandiran, 2012). Crystals occur while cooling, and particular lipids mixtures must be utilized to avoid them (Jagdevappa et al., 2013).

NLCs cause a rise in the amount of liquid lipid present in the nanoparticles, which causes the nanoparticles to stack on top of one another and generate a shape that is not conventional. This results in an imperfect lattice and an amorphous structure. The lipid material that is present in NLCs has a greater application potential than the lipid material that is present in SLNs. This is due to the fact that increasing the liquid lipid can enhance the solubility of active compounds, and active chemicals can help with better encapsulation. NLCs improve medication stability and loading capacity while also reducing drug leakage.

7.2.5.10 PEG-Conjugated NLCs

Because of its appealing potential to enhance the quantity of medication entering into mucosal layers and delay the circulation period of incorporated pharmaceuticals, NLCs are an excellent lipid-based delivery technology for phytochemicals, which is supported by the literature (Yang et al., 2012).

However, reports prove that the majority of delivery systems that are administered in IV route, including NLCs, were quickly recognized as foreign bodies by the polymorphic leukocytes (PLs) and mononuclear phagocytic system (MPS) and were brought down and finally cleared from the blood (Gref et al., 1995). Often bacterial plasmid-based drugs and other plasma proteins interact with the MPS and PLs effectively and degrade into metabolic products. This process is called as opsonization. The number of proteins interacting and their specificity change based on the properties of nanoparticles that carry drugs. There are various kinds of proteins that interact with specific receptors on the MPS and PLs cell surfaces thus supporting opsonization as an important step in drug and particle clearance (Yan et al., 2005). Positively charged nanoparticles have shown the greatest amount of affinity by organs reticuloendothelial system (RES). They are thus preferably interactive with plasma proteins carrying negative charge as a result it shows increased absorption of drugs in RES (Behzadi et al, 2017). For short-half-life medications, designing a drug carrier that prevents detection by the MPS or the RES is ideal. It has been shown that PEG-modified nanoparticles can inhibit RES absorption and enhance circulation half-life (Stolnik et al., 1995).

PEGylation, or the usage of PEG, is a typical method for enhancing the characteristics of nanocarriers, such as nanovesicles (NVs)/nanoparticles (NPs), for in vivo injection. PEG can reduce NP sequestration by RES organs since it is a hydrophilic biocompatible stealth polymer (liver, spleen, kidneys). It provides steric hindrance to opsonins (proteins that are responsible for opsonization) that are tagged on NPs/NVs to be consumed by MPS by blocking their contact with macrophages, which is required for phagocytosis (Lázaro et al., 2017). PEG may enhance time of circulation of drug in the blood stream and also binding to the RES organs, allowing NPs to be more readily absorbed by their target organs or tissues, such as tumor tissues (Chen et al., 2016).

PEG might preferentially attach to other types of proteins that operate as dysopsonins (proteins that are capable of blocking/resisting the opsonization of the drugs) or other proteins that undergo opsonization in the place of the drug by which PEGs extend the half-life of the drugs in the systemic circulation. PEGylated lipoplexes were shown to have a greater level of proteins in the serum binding than nonPEGylated lipid complexes (Zhnag et al., 2014). Various research works were established to link the degree of absorption of proteins, which is an indication of opsonization, along with the half-life of drugs in the blood, to detect the decline in the binding of proteins to the nanoparticles that are encapsulated in the PEGylated contents (Semple et al., 1998).

Another intriguing feature of nanocarriers that differ PEGylated from other non-PEGylated nanocarriers is greater capacity to penetrate the blood-brain barrier (BBB) and potency to deposit in brain tissue, presumably through endocytosis mediated by receptors or transcytosis pathways that need to be further investigated. The density of the PEG coating also plays a role in the process of diffusion across the brain parenchyma; specifically, the higher the PEG coating density, the higher the brain parenchymal diffusion (Suk et al., 2016). In this regard, nanocarriers that are PEGylated have showed higher deposition in pathological conditions defined by hyperpermeability of the BBB, such as the inflammations induced by multiple sclerosis and gliosarcoma (Jenkins et al., 2016).

In spite of these benefits, recent research has brought to light the drawbacks of PEGylation, also known as PEG dilemmas. These drawbacks include the inability of nanocarriers to interact with tissues in the pathological region, inability of the nanocarriers to escape from the endosomes, and also elevation in the blood drug clearance after administration of PEGylated nanocarriers. Despite these drawbacks, PEGylation is still widely used in the pharmaceutical industry. (Fang et al., 2017).

It has been suggested that the utilization of cleavable PEG that are sensitive to the stimulus (cPEG), from which the PEG may be eliminated from the surfaces of the nanocarriers in specified situations had been identified as a solution to the problems that have been caused by PEG. A few examples of processes that are responsive to the stimulus that trigger the process of cleavage within the micro-environmental settings of a tumor include the utilization of an acidic pH, specific enzymes, and reductive potential.

As previously stated, emphasizing the benefits of PEGylation, the polar based polymer ensures the stearic hindrance and masking of surface charge, which prevents tagging protein adsorption (opsonization), which is required for RES macrophage uptake; however, PEGylation's capacity for preventing opsonization had been identified as very limited, and sequential RES clearance and opsonization was reported various times. PEGylation was shown to be ineffective in RES avoidance because equivalent numbers of both nanoparticles that are deposited in the spleen and the distribution in the tissues of RES altered only marginally between nanoparticles that are PEGylated and non-PEGylated (Voon et al., 2016). In other words, phagocytes that are in systemic circulation and macrophages inside of the tissues like the Kupffer cells in the liver and macrophages in the spleen remove the bulk of NPs from the blood, and they are finally deposited in the spleen and liver (Xu et al., 2013).

The molecular weight, PEG polymer identity, composition and hydrophobicity, and density and stability of the PEG in the coating layer, NP core characteristics, and repeated administration are all factors that influence NP destiny. If these requirements aren't carefully monitored, they might have a negative impact on PEG coating (Perche and Torchilin, 2013). However, the PEGylated

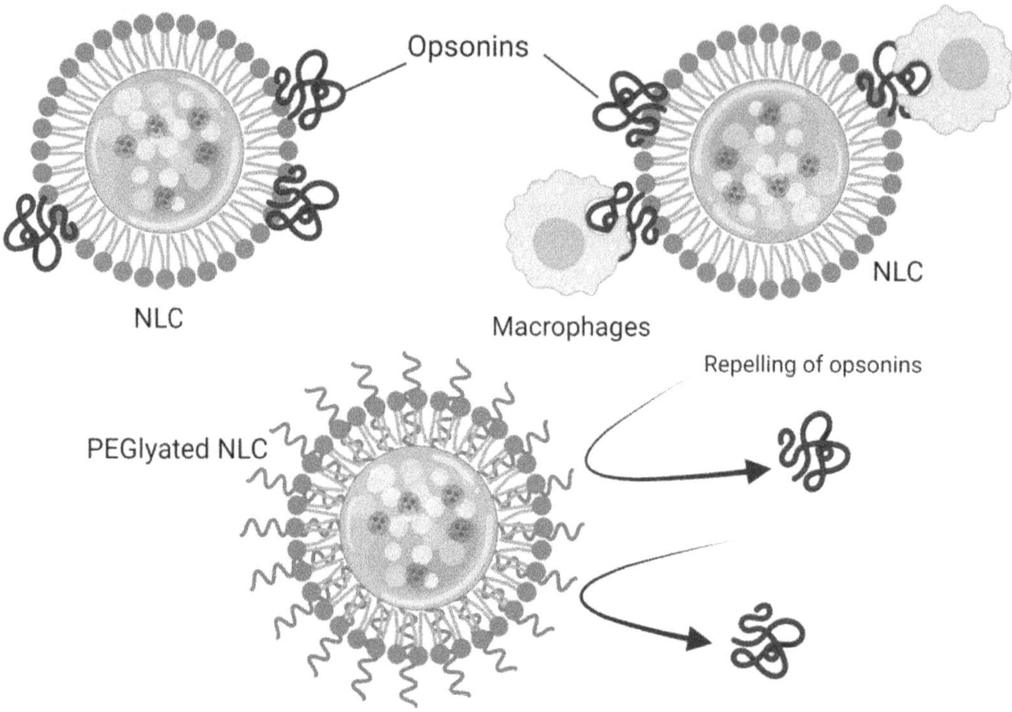

FIGURE 7.3 PEGylated NLCs for the delivery of phytodrugs.

nanocarriers' limited capacity for RES evasion is tagged to several factors, including nanocarrier size enlargement, PEG desorption from nanocarrier surfaces, antiPEG antibody production, and faster drug clearance from systemic circulation due to the reaction of the immune system.

The use of nanocarriers that are PEGylated is expected to limit opsonization and binding to the proteins in the serum, resulting in longer circulation durations for nanoparticles and therapeutic proteins. However, several investigations have found that PEGylation has only minor, if any, effects on half-life of the drug in systemic circulation and binding to the serum proteins. Other studies have found that PEGylation lowers absorption and delivery to target tissues, particularly for therapeutic compounds that must be delivered intracellularly (e.g., siRNA, genes) (Verhoef et al., 2013). Furthermore, it is already well recognized that proteins and therapeutic drugs incorporating nanocarriers that are tagged with PEGylated segments promote the complement activation and the development of antibodies against PEG, which speeds up systemic clearance and decreases pharmacological effectiveness, despite the fact that a recent assessment indicated that the absence of effective antibody detection methods prevents clear findings about specific antibodies to PEG (Schellekens et al., 2013).

7.2.5.11 Second-Generation NLCs

NLCs are one form of modified SLNs that contain a lipid phase containing solid and liquid lipids, which are usually termed as second-generation lipid nanoparticles. These two phases form a solid matrix that holds a drug by enhancing its loading capacity and overall stability (Beloqui et al., 2014). NLCs differ from SLNs in the following ways: higher loading capacity for particular medications, low water content in the dispersion, and the ability to avoid or limit drug ejection during storage. However, no substantial difference in biotoxicity has been documented between SLNs and NLCs. NLCs' disordered crystal structure aids in preventing drug load leakage and provides a greater drug payload (Pardeike et al., 2009).

7.2.5.12 NLCs for Cancer Drugs

Raloxifene HCl (RLX) is an estrogen receptor binder that helps postmenopausal women reduce their risk of breast cancer. Due to limited water solubility and significant hepatic metabolism, RLX has a low bioavailability following oral administration (Patil et al., 2013). Glyceryl caprylate as a liquid lipid and glycerylmonostearate as a solid lipid in a weight ratio of 15:85 and 1% w/v PVA as stabilizer were used to make RLX-loaded NLCs. In comparison to the free drug, the Cmax of the NLCs of RLX increased significantly. Furthermore, the AUC, which represents the absorption range, was almost four times greater in the RLX-loaded-NLCs, possibly because of the minute particle size and the avoidance of the first-pass hepatic metabolism via the transport channel through lymphatic system (Shah et al., 2016).

Dacarbazine (DCZ) is a drug that is used to treat Hodgkin's disease, metastatic melanoma, and sarcoma of the soft tissues. However, its high lipophilicity and shorter half-life make it difficult to use. The integration of DCZ in NLCs yielded a drug release in two phases, with 50% of the medication expelled in the 2 hours and the balance released over the next 30 hours (Almousallam et al., 2015).

Many studies have demonstrated that curcumin (Cur) has a strong cytotoxic impact on a range of cancers when used alone or in combination with chemotherapy. However, due to its weak oral bioavailability and fast elimination, its therapeutic impact is limited. NLCs consisting of oleic acid and tripalmitin (weight ratios of 50:50) along with 3% (w/v) of polysorbate 80 were synthesized and the results suggested a significant improvement in the bioavailability and deposition of the drug in the brain enabling better treatment. In contrast to the control group, NLCs (Cur-NLC) demonstrated improved pharmacokinetic characteristics, with the medication's half-life extended from 3 to 6 hours and AUC assessments showing a 6-fold rise in blood levels of the drug. Furthermore, drug biodistribution experiments revealed that Cur-NLC caused considerable drug accumulation in the brain (Chen et al., 2016).

Researchers developed NLCs comprising a mixture of vincristine (Vcr) and doxorubicin (Dox) to combat resistance to drugs and the risk of lymph cancer relapse following treatment. The integration of these medicines in NLCs lead to the drug release being sustained for up to 48 hours and 16 hours, respectively, for Vcr and Dox. During tumor growth, the extended release was benefited with a persistent anticancer activity. In a lymph cancer animal model, all three combinations of NLCs and drug solutions along with free drugs were evaluated, and the findings revealed that NLCs containing both the drugs displayed regulated release and effective anticancer action (Dong et al., 2016).

Wang created NLCs that were incorporating paclitaxel and doxorubicin to see if they may have a synergistic impact in lung cancer treatment. NLCs containing the combination of drugs in 1:1 ratio showed three times lower IC50 than that of single-drug-loaded NLCs and is over nine times less than the free drug, according to in vitro cytotoxicity experiments on NCL-H460 large-cell lung cancer cells. Co-delivery of both chemotherapeutic medicines boosted the tumor-targeting ability and increased the inhibitory impact of both treatments in an NSCL cancer mouse model, according to in vivo research (Wang et al., 2016).

Tamoxifen was found to be more effective in the therapy and prevention of breast cancer when NLCs were loaded with melatonin. Melatonin-loaded NLCs were optimized in this work utilizing various kinds and quantities of liquid and solid lipids and surfactants. The optimal formula was combined with a low dosage of TAM, and the results revealed that this therapy suppressed MCF-7 cell growth more efficiently than TAM alone, resulting in a two times more incidence of apoptosis and a 10% reduction in cell proliferation (Sabzichi et al., 2016).

It has been demonstrated that NLCs incorporating chemotherapeutic agents are capable of displaying stronger anticancer activity than the free medication. After oral treatment in 4T1 mammary carcinoma in mice, optimized thymoquinone-loaded NLCs had poor anticancer efficacy than free thymoquinone, despite the mice treated with NLCs having less mortality rate than those mice administered with DOX (Ong et al., 2018).

Many studies have even shown that lipid-based nanocarriers that are positively charged adjunction with cationic surfactants like CTAB (cetyltrimethylammonium bromide) have appropriate cellular tolerability (CTAB). At concentrations more than 1 mg/mL, scientists found that SLNs produced with CTAB are less toxic (Almeida et al., 2017).

NLC Formulation	Type of Cancer	IC50 Compared to Free Drug	% Inhibition Compared to Free Drug	Reference
Docetaxel and LidocaineNLC	Murine melanoma (B16)	Lowered by half	Equivalent activity	(de Moura et al., 2021)
Paclitaxel-NLC	Non-small-cell lung carcinoma (H460)	Lowered by 3 times	2.5 times better activity	(Wang et al., 2016)
Tamoxifen-NLC	Breast cancer (4T1)	No change in IC50	—	(How et al., 2013)
Etoposide-NLC	Gastric cancer (SGC7901)	Lowered by 9 times	2 times better activity	(Jiang et al., 2016)
Cisplatin-NLC	Head cancer (FaDu)	Lowered by 9 times	4 times better activity	(Yang et al., 2017)
Curcumin-NLC	Brain cancer (A172)	Lowered by 4 times	4 times better activity	(Chen et al., 2016)

7.3 CONCLUSION

Among all the lipid nanoparticles, SLNs and NLCs are most prominent type of carriers that are efficient in carrying phytopharmaceuticals with a better payload, promising pharmacological activity and modulated release kinetics of the drug and also preventing the physical and chemical degradation of the incorporated phytopharmaceutical agent. Their advantage of carrying both hydrophilic and hydrophobic drugs makes them versatile to deliver wide range of phytochemical drugs. The better biocompatibility and less toxicity have produced a good medium to delivery drugs, and researchers supported their enhanced anticancer activity. Further, the span of applications of SLNs and NLCs to deliver all the phytodrugs can be spread through thorough research and investigations. Thus, this would bring a new vision on the NLCs and SLNs as potential phytodrug carriers, enabling better treatment.

REFERENCES

Abdelbary G, Haider M. In vitro characterization and growth inhibition effect of nanostructured lipid carriers for controlled delivery of methotrexate. *Pharm Dev Technol.* 2013;18:1159–1168. doi: 10.3109/10837450.2011.614251.

Almeida H, Lobão P, Frigerio C, Fonseca J, Silva R, Sousa Lobo JM, Amaral MH. Preparation, characterization and biocompatibility studies of thermoresponsive eyedrops based on the combination of nanostructured lipid carriers (NLC) and the polymer pluronic F-127 for controlled delivery of ibuprofen. *Pharm Dev Technol.* 2017;22:336–349. doi: 10.3109/10837450.2015.1125922.

Almousallam M, Moia C, Zhu H. Development of nanostructured lipid carrier for dacarbazine delivery. *Int Nano Lett.* 2015;5:241–248. doi: 10.1007/s40089-015-0161-8.

Arzani G, Haeri A, Daeihamed M, Bakhtiari-Kaboutaraki H, Dadashzadeh S. Niosomal carriers enhance oral bioavailability of carvedilol: Effects of bile salt-enriched vesicles and carrier surface charge. *Int J Nanomed.* 2015;10:4797.

Ashaolu TJ. Nanoemulsions for health, food, and cosmetics: A review. *Environ Chem Lett.* 2021;19:3381–3395. doi: 10.1007/s10311-021-01216-9.

Assadpour E, Mahdi Jafari S. A systematic review on nanoencapsulation of food bioactive ingredients and nutraceuticals by various nanocarriers. *Crit Rev Food Sci Nutr.* 2019;59:3129–3151. doi: 10.1080/10408398.2018.1484687.

Bai C, Peng H, Xiong H, Liu Y, Zhao L, Xiao X. Carboxymethylchitosan-coated proliposomes containing coix seed oil: Characterisation, stability and in vitro release evaluation. *Food Chem.* 2011;129:1695–1702. doi: 10.1016/j.foodchem.2011.06.033.

Ban C, Jo M, Park YH, Kim JH, Han JY, Lee KW, Kweon DH, Choi YJ. Enhancing the oral bioavailability of curcumin using solid lipid nanoparticles. *Food Chem.* 2020;302:125328.

Battaglia L, Gallarate M. Lipid nanoparticles: State of the art, new preparation methods and challenges in drug delivery. *Expert Opin Drug Deliv.* 2012;9(5):497–508.

Behbahani ES, Ghaedi M, Abbaspour M, Rostamizadeh K, Dashtian K. Curcumin loaded nanostructured lipid carriers: In vitro digestion and release studies. *Polyhedron.* 2019;164:113–122.

Behzadi S, Serpooshan V, Tao W, Hamaly MA, Alkawareek MY, Dreaden EC, et al. Cellular uptake of nanoparticles: A journey inside the cell. *Chem Soc Rev.* 2017;46:4218–4244.

Beloqui A, Solinis MA, des Rieux A, Preat V, Rodriguez-Gascon A. Dextran-protamine coated nanostructured lipid carriers as mucus-penetrating nanoparticles for lipophilic drugs. *Int J Pharm.* 2014;468(1–2):105–111. doi: 10.1016/j.ijpharm.2014.04.027.

Bordat A, Boissenot T, Nicolas J, Tsapis N. Thermoresponsive polymer nanocarriers for biomedical applications. *Adv Drug Deliv Rev.* 2019;138:167–192. doi: 10.1016/j.addr.2018.10.005.

Burri L, Hoem N, Banni S, Berge K. Marine omega-3 phospholipids: Metabolism and biological activities. *Int J Mol Sci.* 2012;13:15401–15419.

Caddeo C, Gabriele M, Fernàndez-Busquets X, Valenti D, Fadda AM, Pucci L, Manconi M. Antioxidant activity of quercetin in Eudragit-coated liposomes for intestinal delivery. *Int J Pharm.* 2019;565:64–69.

Caddeo C, Gabriele M, Nácher A, Fernàndez-Busquets X, Valenti D, Fadda AM, Pucci L, Manconi M. Resveratrol and artemisinin eudragit-coated liposomes: A strategy to tackle intestinal tumors. *Int J Pharm.* 2021;592:120083.

Caddeo C, Pucci L, Gabriele M, Carbone C, Fernàndez-Busquets X, Valenti D, et al. Stability, biocompatibility and antioxidant activity of PEG-modified liposomes containing resveratrol. *Int J Pharm.* 2018;538:40–47. doi: 10.1016/j.ijpharm.2017.12.047.

Cai X, Fang Z, Dou J, Yu A, Zhai G. Bioavailability of quercetin: Problems and promises. *Curr Med Chem.* 2013;20:2572–2582.

Chen S, Yang K, Tuguntaev RG, Mozhi A, Zhang J, Wang PC, et al. Targeting tumor microenvironment with PEG-based amphiphilic nanoparticles to overcome chemoresistance. *Nanomedicine.* 2016;12:269–286.

Chen Y, Pan L, Jiang M, Li D, Jin L. Nanostructured lipid carriers enhance the bioavailability and brain cancer inhibitory efficacy of curcumin both in vitro and in vivo. *Drug Deliv.* 2016;23:1383–1392.

Chinsriwongkul A, Opanasopit P, Ngawhirunpat T, Rojanarata T, Sila-On W, Ruktanonchai U. Oleic acid enhances all-trans retinoic acid loading in nano-lipid emulsions. *PDA J Pharm Sci Technol.* 2010;64:113–123.

Constantinides PP, Tustian A, Kessler DR. Tocol emulsions for drug solubilization and parenteral delivery. *Adv Drug Deliv Rev.* 2004;56:1243–1255. doi: 10.1016/j.addr.2003.12.005.

Date AA, Desai N, Dixit R, Nagarsenker M. Self-nanoemulsifying drug delivery systems: Formulation insights, applications and advances. *Nanomedicine.* 2010;5:1595–1616.

Decuzzi P, Ferrari M. The receptor-mediated endocytosis of nonspherical particles. *Biophys J Elsevier.* 2008;94:3790–3797.

de Moura LD, Ribeiro LN, de Carvalho FV, Rodrigues da Silva GH, Lima Fernandes PC, Brunetto SQ, Ramos CD, Velloso LA, de Araújo DR, de Paula E. Docetaxel and lidocaine co-loaded (NLC-in-hydrogel) hybrid system designed for the treatment of melanoma. *Pharmaceutics.* 2021;13(10):1552.

Devaraj GN, Parakh S, Devraj R, Apte S, Rao B, Rambhau D. Release studies on niosomes containing fatty alcohols as bilayer stabilizers instead of cholesterol. *J. Colloid Interface Sci.* 2002;251:360–365.

Dolatabadi S, Karimi M, Nasirizadeh S, Hatamipour M, Golmohammadzadeh S, Jaafari MR. Preparation, characterization and in vivo pharmacokinetic evaluation of curcuminoids-loaded Solid Lipid Nanoparticles (SLNs) and Nanostructured Lipid Carriers (NLCs). *J Drug Deliv Sci Technol.* 2021;62:102352.

Dong X, Wang W, Qu H, Han D, Zheng J, Sun G. Targeted delivery of doxorubicin and vincristine to lymph cancer: Evaluation of novel nanostructured lipid carriers in vitro and in vivo. *Drug Deliv.* 2016;23:1374–1378. doi: 10.3109/10717544.2015.1041580.

Efficacy and safety of liposomal lactoferrin in COVID-19 patients with mild-to-moderate disease and in COVID-19 asymptomatic patients (NCT04475120) [Internet]. Sponsored by University of Rome Tor Vergata; 2020. Available from: https://clinicaltrials.gov/ct2/show/NCT04475120?

Ekambaram P, Sathali A, Priyanka K. Solid lipid nanoparticles: A review. *Sci Rev Chem Commun.* 2012;2(1):80–102.

Ezzat HM, Elnaggar YSR, Abdallah OY. Improved oral bioavailability of the anticancer drug catechin using chitosomes: Design, in-vitro appraisal and in-vivo studies. *Int J Pharm.* 2019;565:488–498.

Fang Y, Xue J, Gao S, et al. Cleavable PEGylation: A strategy for overcoming the "PEG dilemma" in efficient drug delivery. *Drug Deliv.* 2017;24(Suppl 1):22–32.

Fornasier M, Pireddu R, Del Giudice A, Sinico C, Nylander T, Schillén K, Galantini L, Murgia S. Tuning lipid structure by bile salts: Hexosomes for topical administration of catechin. *Colloids Surf B Biointerfaces.* 2021;199:111564.

Germain M, Caputo F, Metcalfe S, Tosi G, Spring K, Aslund AKO, et al. Delivering the power of nanomedicine to patients today. *J Control Release.* 2020;326:164–171. doi: 10.1016/j.jconrel.2020.07.007.

Gonçalves RF, Martins JT, Abrunhosa L, Baixinho J, Matias AA, Vicente AA, Pinheiro AC. Lipid-based nanostructures as a strategy to enhance curcumin bioaccessibility: Behavior under digestion and cytotoxicity assessment. *Food Res Int.* 2021;143:110278.

Gopi S, Amalraj A, Jacob J, Kalarikkal N, Thomas S, Guo Q. Preparation, characterization and in vitro study of liposomal curcumin powder by cost effective nanofiber weaving technology. *New J Chem.* 2018;42:5117–5127.

Gref R, Domb A, Quellec P, et al. The controlled intravenous delivery of drugs using PEG-coated sterically stabilized nanospheres. *Adv Drug Deliv Rev.* 1995;16:215–232.

Guimarães D, Cavaco-Paulo A, Nogueira E. Design of liposomes as drug delivery system for therapeutic applications. *Int J Pharm.* 2021;601:120571. doi: 10.1016/j.ijpharm.2021.120571.

Guo Y, Mao X, Zhang J, Sun P, Wang H, Zhang Y, et al. Oral delivery of lycopene-loaded microemulsion for brain-targeting: Preparation, characterization, pharmacokinetic evaluation and tissue distribution. *Drug Deliv Taylor & Francis.* 2019;26:1191–1205.

Haider M, Abdin SM, Kamal L, Orive G. Nanostructured lipid carriers for delivery of chemotherapeutics: A review. *Pharmaceutics.* 2020;12:288. doi: 10.3390/pharmaceutics12030288.

Hızır-Kadı I, Gültekin-Özgüven M, Altin G, Demircan E, Özçelik B. Liposomal nanodelivery systems generated from proliposomes for pollen extract with improved solubility and in vitro bioaccessibility. *Heliyon.* 2020;6:05030.

How CW, Rasedee A, Manickam S, Rosli R. Tamoxifen-loaded nanostructured lipid carrier as a drug delivery system: Characterization, stability assessment and cytotoxicity. *Colloids Surf B Biointerfaces.* 2013;112:393–399. doi: 10.1016/j.colsurfb.2013.08.009.

Huang J, Wang Q, Chu L, Xia Q. Liposome-chitosan hydrogel bead delivery system for the encapsulation of linseed oil and quercetin: Preparation and in vitro characterization studies. *LWT.* 2020;117:108615.

Ipar VS, Dsouza A, Devarajan, PV. Enhancing curcumin oral bioavailability through nanoformulations. *Eur. J. Drug Metab. Pharmacokinet.* 2019;44:459–480.

Jagdevappa P, Prashant G, Ravindra K, Sachin J, Satish M, Meghanath S. Applications of solid lipid nanoparticle in novel drug delivery system. *Br Biomed Bull.* 2013;1(2):103–118.

Jagwani S, Jalalpure SS, Dhamecha D, Jadhav K, Bohara RA. Pharmacokinetic and pharmacodynamic evaluation of resveratrol loaded cationic liposomes for targeting hepatocellular carcinoma. *ACS Biomater. Sci. Eng.* 2020;6:4969–4984.

Jahangir MA, Taleuzzaman M, Kala C, Gilani SJ. Advancements in polymer and lipid-based nanotherapeutics for cancer drug targeting. *Curr Pharm Des.* 2020;26:5119–5127. doi: 10.2174/1381612826999200820173253.

Jaiswal P, Gidwani B, Vyas A. Nanostructured lipid carriers and their current application in targeted drug delivery. *Artif Cells Nanomed Biotechnol.* 2016;44(1):27–40. doi: 10.3109/21691401.2014.909822.

Jenkins SI, Weinberg D, Al-Shakli AF, Fernandes AR, Yiu HHP, Telling ND, et al. "Stealth" nanoparticles evade neural-immune cells but also evade major brain cell populations: Implications for PEG-based neurotherapeutics. *J Controlled Release.* 2016;224:136–145.

Jiang H, Pei L, Liu N, Li J, Li Z, Zhang S. Etoposide-loaded nanostructured lipid carriers for gastric cancer therapy. *Drug Deliv.* 2016;23:1379–1382. doi: 10.3109/10717544.2015.1048491.

Khan AW, Kotta S, Ansari SH, Sharma RK, Ali J. Potentials and challenges in self-nanoemulsifying drug delivery systems. *Expert Opin Drug Deliv.* 2012;9:1305–1317.

Lázaro I, Haddad S, Sacca S, Orellana-Tavra C, Fairen-Jimenez D, Forgan RS. Selective surface PEGylation of UiO-66 nanoparticles for enhanced stability, cell uptake, and pH-responsive drug delivery. *Chem.* 2017;2:561–578.

Lee MH, Lee IY, Chun YG, Kim BK. Formulation and characterization of β-caryophellene-loaded lipid nanocarriers with different carrier lipids for food processing applications. *LWT.* 2021;149:111805. doi: 10.1016/j.lwt.2021.111805.

Li ZL, Peng S, Chen X, Zhu YQ, Zou LQ, Liu W, Liu CM. Pluronics modified liposomes for curcumin encapsulation: Sustained release, stability and bioaccessibility. *Food Res Int.* 2018;108:246–253.

Liang R, Chen L, Yokoyama W, Williams PA, Zhong F. Niosomes consisting of tween-60 and cholesterol improve the chemical stability and antioxidant activity of (-)-epigallocatechin gallate under intestinal tract conditions. *J Agric Food Chem.* 2016;64:9180–9188.

Liu CS, Chen L, Changshun L, Dai JL, Ma B, Tang QF, Tan XM. Self-microemulsifying drug delivery system for improved oral delivery and hypnotic efficacy of ferulic acid. *Int J Nanomed.* 2020a;15:2059–2070.

Liu D, Liu Z, Wang L, Zhang C, Zhang N. Nanostructured lipid carriers as novel carrier for parenteral delivery of docetaxel. *Colloids Surf B Biointerfaces.* 2011;85:262–269. doi: 10.1016/j.colsurfb.2011.02.038.

Liu X, Wang P, Zou YX, Luo ZG, Tamer TM. Co-encapsulation of vitamin C and β-carotene in liposomes: Storage stability, antioxidant activity, and in vitro gastrointestinal digestion. *Food Res Int.* 2020;136:109587.

McClements DJ. Nanoemulsions versus microemulsions: Terminology, differences, and similarities. *Soft Matter.* 2012;8:1719–1729.

Mehnert W, Mäder K. Solid lipid nanoparticles: Production, characterization and applications. *Adv Drug Deliv Rev.* 2001;47:165–196. doi: 10.1016/S0169-409X(01)00105-3.

Mukherjee S, Ray S, Thakur RS. Solid lipid nanoparticles: A modern formulation approach in drug delivery system. *Indian J Pharm Sci.* 2009;71(4):349–358. doi: 10.4103/0250-474X.57282.

Müller RH, Mäder K, Gohla S. Solid lipid nanoparticles (SLN) for controlled drug delivery—a review of the state of the art. *Eur J Pharm Biopharm.* 2000;50:161–177. doi: 10.1016/S0939-6411(00)00087-4.

Müller RH, Radtke M, Wissing SA. Nanostructured lipid matrices for improved microencapsulation of drugs. *Int J Pharm.* 2002;242:121–128. doi: 10.1016/S0378-5173(02)00180-1.

Müller RH, Shegokar R, Keck C. 20 years of lipid nanoparticles (SLN and NLC): Present state of development and industrial applications. *Curr Drug Discov Technol.* 2011;8:207–227. doi: 10.2174/157016311796799062.

Noor NM, Sheikh K, Somavarapu S, Taylor KMG. Preparation and characterization of dutasteride-loaded nanostructured lipid carriers coated with stearic acid-chitosan oligomer for topical delivery. *Eur J Pharm Biopharm.* 2017;117:372–384. doi: 10.1016/j.ejpb.2017.04.012.

Ong YS, Yazan LS, Ng WK, Abdullah R, Mustapha NM, Sapuan S, Foo JB, Tor YS, How CW, Rahman NA, et al. Thymoquinone loaded in nanostructured lipid carrier showed enhanced anticancer activity in 4T1 tumor-bearing mice. *Nanomedicine.* 2018;13:1567–1582. doi: 10.2217/nnm-2017-0322.

Pardeike J, Hommoss A, Muller RH. Lipid nanoparticles (SLN, NLC) in cosmetic and pharmaceutical dermal products. *Int J Pharm.* 2009;366:170–184.

Patil PH, Belgamwar VS, Patil PR, Surana SJ. Solubility enhancement of raloxifene using inclusion complexes and cogrinding method. *J Pharm.* 2013;2013:527380. doi: 10.1155/2013/527380.

Patwekar S, Gattani S, Giri R, Bade A, Sangewar B, Raut V. Review on nanoparticles used in cosmetics and dermal products. *World J Pharm Pharm Sci.* 2014;3(8):1407–1421.

Perche F, Torchilin VP. Recent trends in multifunctional liposomal nanocarriers for enhanced tumor targeting. *J Drug Delivery.* 2013;201:3705265.

Prakash B, Kujur A, Yadav A, Kumar A, Singh PP, Dubey NK. Nanoencapsulation: An efficient technology to boost the antimicrobial potential of plant essential oils in food system. *Food Control.* 2018;89:1–11. doi: 10.1016/j.foodcont.2018.01.018.

Pu C, Tang W, Li X, Li M, Sun Q. Stability enhancement efficiency of surface decoration on curcumin-loaded liposomes: Comparison of guar gum and its cationic counterpart. *Food Hydrocoll.* 2019;87:29–37.

Puri A, Loomis K, Smith B, Lee JH, Yavlovich A, Heldman E, et al. Lipid-based nanoparticles as pharmaceutical drug carriers: From concepts to clinic. *Critic Rev Therap Drug Carrier Syst.* 2009;26:6. doi: 10.1615/CritRevTherDrugCarrierSyst.v26.i6.10.

Ragelle H, Danhier F, Préat V, Langer R, Anderson DG. Nanoparticle-based drug delivery systems: A commercial and regulatory outlook as the field matures. *Expert Opin Drug Deliv.* 2017;14:851–864.

Reddy AS, Lakshmi BA, Kim S, Kim J. Synthesis and characterization of acetyl curcumin-loaded core/shell liposome nanoparticles via an electrospray process for drug delivery, and theranostic applications. *Eur J Pharm Biopharm.* 2019;142:518–530.

Rein MJ, Renouf M, Cruz-Hernandez C, Actis-Goretta L, Thakkar SK, Da Silva Pinto M. Bioavailability of bioactive food compounds: A challenging journey to bioefficacy. *Br J Clin Pharmacol.* 2013;75:588–602.

Rezaei A, Fathi M, Jafari SM. Nanoencapsulation of hydrophobic and low-soluble food bioactive compounds within different nanocarriers. *Food Hydrocoll.* 2019;88:146–162.

Roger E, Lagarce F, Benoit JP. Development and characterization of a novel lipid nanocapsule formulation of Sn38 for oral administration. *Eur J Pharm Biopharm.* Elsevier B.V. 2011;79:181–188.

Rostami E, Kashanian S, Azandaryani AH, Faramarzi H, Dolatabadi JEN, Omidfar K. Drug targeting using solid lipid nanoparticles. *Chem Phys Lipids.* 2014;181:56–61. doi: 10.1016/j.chemphyslip.2014.03.006.

Sabzichi M, Samadi N, Mohammadian J, Hamishehkar H, Akbarzadeh M, Molavi O. Sustained release of melatonin: A novel approach in elevating efficacy of tamoxifen in breast cancer treatment. *Colloids Surf B Biointerfaces.* 2016;145:64–71. doi: 10.1016/j.colsurfb.2016.04.042.

Sang KK, Eun HL, Vaishali B, Lee S, Lee YK, Kim CY, et al. Tricaprylin microemulsion for oral delivery of low molecular weight heparin conjugates. *J Control Release.* 2005;105:32–42.

Sarathchandiran I. A review on nanotechnology in solid lipid nanoparticles. *Int J Pharm Deve Tech.* 2012;2(1):45–61.

Schellekens H, Hennink WE, Brinks V. The immunogenicity of polyethylen glycol: Facts and fiction. *Pharm Res.* 2013;30:1729–1734.

Sedaghat Doost A, Afghari N, Abbasi H, Nikbakht Nasrabadi M, Dewettinck K, Van der Meeren P. Nano-lipid carriers stabilized by hydrophobically modified starch or sucrose stearate for the delivery of lutein as a nutraceutical beverage model. *Colloids Surf A Physicochem Eng Aspects.* 2020;605:125349. doi: 10.1016/j.colsurfa.2020.125349.

Semple SC, Chonn A, Cullis PR. Interactions of liposomes and lipid-based carrier systems with blood proteins: Relation to clearance behavior in vivo. *Adv Drug Del Rev.* 1998;32:3–17.

Shah NV, Seth AK, Balaraman R, Aundhia CJ, Maheshwari RA, Parmar GR. Nanostructured lipid carriers for oral bioavailability enhancement of raloxifene: Design and in vivo study. *J Adv Res.* 2016;7:423–434. doi: 10.1016/j.jare.2016.03.002.

Stolnik S, Illum L, Davis SS. Long circulating microparticulate drug carriers. *Adv Drug Deliv Rev.* 1995;16:195–214.

Suk JS, Xu Q, Kim N, Hanes J, Ensign LM. PEGylation as a strategy for improving nanoparticle-based drug and gene delivery. *Adv Drug Delivery Rev.* 2016;99:28–51.

Tavano L, Muzzalupo R, Picci N, de Cindio B. Co-encapsulation of antioxidants into niosomal carriers: Gastrointestinal release studies for nutraceutical applications. *Colloids Surfaces B Biointerfaces.* 2014;114:82–88.

Tikekar RV, Nitin N. Distribution of encapsulated materials in colloidal particles and its impact on oxidative stability of encapsulated materials. *Langmuir.* 2012;28:9233–9243. doi: 10.1021/la301435k.

Uner M, Yener G. Importance of solid lipid nanoparticles (SLN) in various administration routes and future perspectives. *Int J Nanomedicine.* 2007;2(3):289–300.

Vanaja K, Wahl M, Bukarica L, Heinle H. Liposomes as carriers of the lipid soluble antioxidant resveratrol: Evaluation of amelioration of oxidative stress by additional antioxidant vitamin. *Life Sci.* 2013;93:917–923. doi: 10.1016/j.lfs.2013.10.019.

Verhoef JJ, Anchordoquy TJ. Questioning the use of PEGylation for drug delivery. *Drug Delivery and Translational Research.*2013;3(6):499–503. doi: 10.1007/s13346-013-0176-5.

Voon SH, Tiew SX, Kue CS, Lee HB, Kiew LV, Misran M, et al. Chitosan-coated poly(lactic-co-glycolic acid)-diiodinated boron-dipyrromethene nanoparticles improve tumor selectivity and stealth properties in photodynamic cancer therapy. *J Biomed Nanotechnol.* 2016;12:1431–1452.

Wacker M. Nanocarriers for intravenous injection—The long hard road to the market. *Int J Pharm.* 2013;457:50–62. doi: 10.1016/j.ijpharm.2013.08.079.

Wang Q, Huang J, Hu C, Xia N, Li T, Xia Q. Stabilization of a non-aqueous self-double-emulsifying delivery system of rutin by fat crystals and nonionic surfactants: Preparation and bioavailability study. *Food Funct.* 2017;8:2512–2522.

Wang Y, Zhang H, Hao J, Li B, Li M, Xiuwen W. Lung cancer combination therapy: Co-delivery of paclitaxel and doxorubicin by nanostructured lipid carriers for synergistic effect. *Drug Deliv.* 2016;23:1398–1403. doi: 10.3109/10717544.2015.1055619.

Wong HL, Bendayan R, Rauth AM, Li Y, Wu XY. Chemotherapy with anticancer drugs encapsulated in solid lipid nanoparticles. *Adv Drug Deliv Rev.* 2007;59:491–504. doi: 10.1016/j.addr.2007.04.008.

Wu Y, Mou B, Song S, Tan CP, Lai, OM, Shen C, Cheong LZ. Curcumin-loaded liposomes prepared from bovine milk and krill phospholipids: Effects of chemical composition on storage stability, in-vitro digestibility and anti-hyperglycemic properties. *Food Res Int.* 2020;136:109301.

Xu J, Gattacceca F, Amiji M. Biodistribution and pharmacokinetics of EGFR-targeted thiolated gelatin nanoparticles following systemic administration in pancreatic tumour-bearing mice. *Mol Pharm.* 2013;10:2031–2044.

Yan X, Scherphof GL, Kamps JAAM. Liposome opsonization. *J Liposome Res.* 2005;15:109–139.

Yang J, Ju Z, Dong S. Cisplatin and paclitaxel co-delivered by folate-decorated lipid carriers for the treatment of head and neck cancer. *Drug Deliv.* 2017;241:792–799. doi: 10.1080/10717544.2016.1236849.

Yang X, Li Y, Li M, et al. Hyaluronic acid-coated nanostructured lipid carriers for targeting paclitaxel to cancer. *Cancer Lett.* 2012;334:338–345.

Yingchoncharoen P, Kalinowski DS, Richardson DR. Lipid-based drug delivery systems in cancer therapy: What is available and what is yet to come. *Pharmacol Rev.* 2016;68:701–787. doi: 10.1124/pr.115.012070.

Yuan H, Wang LL, Du YZ, You J, Hu FQ, Zeng S. Preparation and characteristics of nanostructured lipid carriers for control-releasing progesterone by melt-emulsification. *Colloids Surf B Biointerf.* 2007;60:174–179. doi: 10.1016/j.colsurfb.2007.06.011.

Zhang F, Liu M, Wan H. Discussion about several potential drawbacks of PEGylated therapeutic proteins. *Biol Pharm Bull.* 2014;37:335–339.

Zhang MW, Corona PT, Ruocco N, Alvarez D, de Molina PM, Mitragotri S, et al. Controlling complex nanoemulsion morphology using asymmetric cosurfactants for the preparation of polymer nanocapsules. *Langmuir.* 2018;34:978–990. doi: 10.1021/acs.langmuir.7b02843.

Zhang XG, Miao J, Dai YQ, Du YZ, Yuan H, Hu FQ. Reversal activity of nanostructured lipid carriers loading cytotoxic drug in multi-drug resistant cancer cells. *Int J Pharm.* 2008;361:239–244. doi: 10.1016/j.ijpharm.2008.06.002.

Zhao YN, Cao Y, Sun J, Liang Z, Wu Q, Cui S, Zhi DF, Guo S, Zhen Y, Zhang S. Anti-breast cancer activity of resveratrol encapsulated in liposomes. *J Mater ChemB.* 2020;8:27–37.

8 The Use of Bioadhesive, Thermoresponsive Nanostructured Systems for the Delivery of Phytopharmaceutical Constituents

Curcumin, Gallic Acid, p-Coumaric Acid and Propolis

Sabrina Barbosa de Souza Ferreira, Rafaela Said dos Santos, Camila Felix Vecchi, and Marcos Luciano Bruschi

8.1 INTRODUCTION

Phytopharmaceuticals have demonstrated a huge importance in the field of development of drug delivery systems due to their wide biological activities. These compounds are derivatives of the secondary metabolism of plants. Among them, polyphenols have shown versatile activities, including antibacterial, antiviral, antihelminthic, anti-inflammatory, anticancer, and antioxidant. More than 8,000 polyphenols have been identified in nature (Wang et al. 2020; Pandey and Rizvi 2009).

Some of the most reported polyphenols are curcumin, p-coumaric acid, gallic acid and the complex matrix rich in polyphenols: propolis. Due to their interesting biological activities, it is important to improve their utilization by incorporating in drug delivery systems (Wang et al. 2020).

The use of formulations which enable the availability of the phytopharmaceutical in the body and improve its safety and efficacy by modifying the drug release are well known as drug delivery systems (Bruschi et al. 2017). During the development of these systems, the biological aspects related to the application site (e.g., mucosal surfaces or skin) should be considered, including good bioadhesion properties to improve the residence time in the administration site and the availability of the phytopharmaceutical agent (Gowthamarajan et al. 2012; Meher et al. 2013). Thus, the development of drug delivery systems using materials with bioadhesive properties may address these aspects.

Thermoresponsive characteristics of formulations could be used to improve the administration of the compounds. Since most of the materials responsive to temperature are copolymers with hydrophilic and hydrophobic segments, such as poloxamer 407 and poly(N-propylacrylamide), they can allow the solubilization of hydrophilic and hydrophobic phytopharmaceuticals (Matanović, Kristl, and Grabnar 2014). In addition, these polymers may facilitate the administration in body surfaces

DOI: 10.1201/9781003300588-8

and cavities, such as skin, vaginal mucosa, oral mucosa, oral cavity, and intestinal mucosa (Pereira et al. 2013; Borghi-Pangoni et al. 2016; Rosseto et al. 2021; Souza Ferreira et al. 2020).

The aim of this chapter is to discuss definitions, strategies, and technologies involved in the development of systems consisting of polyphenolic compounds, including curcumin, propolis, gallic acid, and p-coumaric acid and their bioadhesive and/or thermoresponsive properties. The fundamentals regarding bioadhesive and/or thermoresponsive systems and bioactive agents of natural origin are described. The bioadhesive, thermoresponsive formulations containing bioactive agents of natural source are considered in terms of strategies, technology, and types of systems. In addition, the safety and toxicity of these formulations are also addressed, as well as their patents.

8.2 DEFINITIONS AND FUNDAMENTALS

8.2.1 Bioadhesive, Thermoresponsive, Nanostructured Systems

Polymers are macromolecules characterized by their size, chemical structure, and intra and intermolecular interactions. They have chemical units that are joined by covalent bonds, which are repeated along the chain. They can be of natural or synthetic origin (Elias 1997). Initially, they can be classified into homopolymers and copolymers. Homopolymers come from a single monomeric unit, and copolymers have two or more different monomeric units; that is, they are obtained using two or more monomers (Paoli 2009). According to its structure and functionalities, it is possible to acquire important properties, including bioadhesion and thermoresponsiveness, and they may be useful in the development of nanostructured systems (Bruschi et al. 2017).

Bioadhesive polymers can be designated as long molecules of monomers (identical or similar units) linked by covalent bonds, forming adhesive interactions with biological substrates, such as skin or mucosa. When these polymers are capable of adhering with the mucosa, they are called mucoadhesive polymers. Several advantages have been reported in the literature regarding these materials, such as prolonged residence time, administration in specific sites, improvement of absorption and modified drug release (Andrews, Laverty, and Jones 2009; Wypych 2009; Carvalho et al. 2010; Almeida et al. 2013; Gupta et al. 2013; Pereira et al. 2013). Their physicochemical characteristics such as viscoelastic properties, presence of groups that form hydrogen bonds, degree of hydration and swelling, pH, charge, length, weight, and conformation of the chain make them differentiated from other polymers (Pliszczak et al. 2012; Almeidaet al. 2013; Carvalho, Chorilli, and Gremião 2014; Andrews, Laverty, and Jones 2009; Wypych 2009). The polymer's swelling capacity, the capacity to absorb aqueous liquid and swell, is extremely important, as this characteristic can affect both its adhesion properties and drug release control properties (Perioli et al. 2004). The ideal properties of these polymers are described in Figure 8.1 (Sudhakar, Kuotsu, and Bandyopadhyay 2006; Bruschi et al. 2017; Andrews, Laverty, and Jones 2009)

In addition, the formulations prepared with these bio-/mucoadhesive polymers may improve the drug permeation and availability for local and systemic effects. The type of the polymer may change the bioadhesive properties of the formulation due to the adhesive interaction. Diverse materials have been used in the development of these nanostructured systems, including poly(acrylic acid) derivatives, cellulose derivatives, alginates, gums, gelatin, and chitosan (Andrews, Laverty, and Jones 2009; Carvalho et al. 2010; Carvalho, Chorilli, and Gremião 2014)

Thermoresponsive polymers respond to temperature variations and change their conformational structure through the reversible *in situ* gelling phenomenon (Qiu and Park 2001; Almeida et al. 2013; Gupta, Samanta, and Raichur 2010). The ideal thermoresponsive polymer is a solution in environmental temperatures and, after administration, exhibits gelling properties (Matanović, Kristl, and Grabnar 2014). Many advantages have been reported in the literature for these systems, including easy administration, good spreadability over the skin or mucosa, and low clearance in the mucosa due to the gelation temperature (Liu et al. 2009; Dumortier et al. 2006; Qiu and Park 2001). Besides, these systems can dissolve hydrophilic and hydrophobic drugs, are good drug release, and

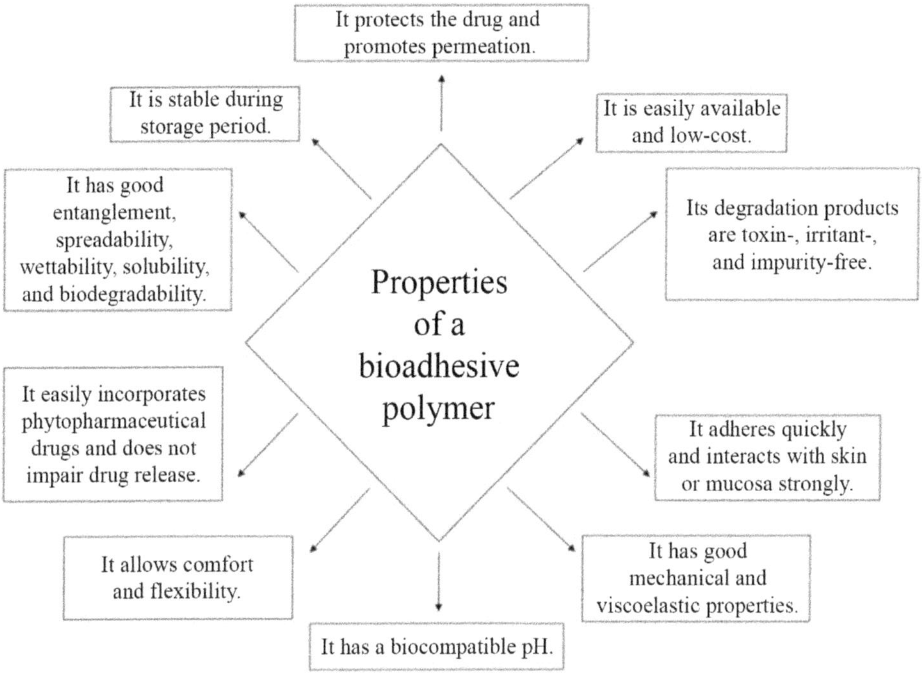

FIGURE 8.1 Ideal properties of bioadhesive polymers.

are nontoxic to the skin and mucosa, which make them safe materials in the pharmaceutical industry (Mayol et al. 2008; Kikuchi and Okano 2005; Dumortier et al. 2006; Liu et al. 2009; Bruschi et al. 2017).

These thermoresponsive polymers may present hydrophobic groups, including methyl, ethyl, and propyl (James et al. 2014), and the most commonly used polymers are poloxamers, xyloglucan, and poly(N-isopropylacrylamide).

8.2.1.1 Poly(Acrylic Acid) Derivatives

Most of them are known as carbomers, and they belong to a high-molecular-weight synthetic polymeric class containing acrylic acid chains cross-linked by allylsaccharose, pentaerythrol allyl ethers or divinyl glycol (Tamburic and Craig 1995; Gómez-Carracedo et al. 2004; Bonacucina, Martelli, and Palmieri 2004; Chawla and Saraf 2012; Berginc et al. 2014; Carvalho et al. 2013). The US Food and Drug Administration (FDA) has classified polymers derived from poly(acrylic acid) (PAA) as GRAS (generally considered safe) for oral administration. PAA polymers (Figure 8.2) are transparent, nontoxic, and non-irritating (Ugwoke et al. 2000; Rosseto et al. 2021).

The ability to absorb water, hydrate, and swell while maintaining the cross-linked structure and water insolubility makes poly(acrylic acid) derivatives excellent bio-/mucoadhesive systems,

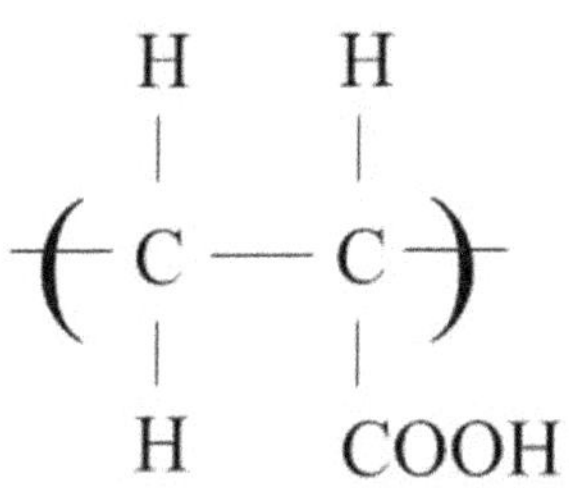

FIGURE 8.2 Chemical structure of polymer poly(acrylic acid).

mainly due to the hydrogen bonds between the carboxylic acid groups and the carboxylic acid groups belonging to the sialic acid of the mucin glycoprotein (Guo 2008; Tamburic and Craig 1995; Berginc et al. 2014). Bioadhesive formulations containing carbomers can be used in the topical, oral, ophthalmic, intestinal, nasal, vaginal, and rectal routes (Islam et al. 2004; Bonacucina, Martelli, and Palmieri 2004; Guo 2008). The carbomers, formed by cross-linked acrylic acid monomers, are differentiated by the degree of cross-linking and manufacturing conditions into Carbopol 934P®, 971P®, and 974P® (Bonacucina, Martelli, and Palmieri 2004).

The benzene-polymerized C934P was developed for the pharmaceutical industries with a high degree of purity in the 1960s and, to date, has been widely used. This polymer is highly cross-linked with excellent stability at high viscosities, which confers excellent bioadhesive properties (Hosmani, Thorat, and Kasture 2006; Lubrizol 2013; Pereira et al. 2013).

C971P and C974P, polymerized in ethyl acetate, differ by the cross-linking agent, although they have the same amount of carboxyl groups. What occurs with C971P is that it exhibits a higher concentration of carboxyls available to perform weak interactions, thus allowing for greater intumescence. In addition, C971P exhibits lower cross-linking with higher deformation capacity, which results in a honey-like consistency in a semisolid formulation (Souza Ferreira et al. 2015; Bonacucina, Martelli, and Palmieri 2004; Hosmani, Thorat, and Kasture 2006; Carvalho et al. 2013; Rosseto et al. 2021).

C974P, on the other hand, is highly cross-linked and forms highly viscous gels, with lower deformation capacity, resulting in a mayonnaise-like consistency. Thus, it is referred to as a viscoelastic fluid, which means it behaves as an elastic solid under low shear stress and starts to flow when the shear stress exceeds the yield values (Efentakis, Koutlis, and Vlachou 2000; Mezreb et al. 2004; Bonacucina, Martelli, and Palmieri 2004; Hosmani, Thorat, and Kasture 2006; Carvalho et al. 2013; Souza Ferreira et al. 2017; Rosseto et al. 2021).

8.2.1.2 Polycarbophil

Polycarbophil, in turn, is a homopolymer of acrylic acid cross-linked with divinyl glycol, reported having excellent viscosity properties and bioadhesive strength, as well as great swelling and wetting values (Kerec et al. 2002; Zhu et al. 2013). This polymer is used extensively to control the release of various drugs into mucous membranes and can be used in the topical, buccal, vaginal, ophthalmic, and nasal routes. Like the carbomers, they are anionic polymers and may gel at pH around 7.0 due to the repulsion between the ionized chains, allowing the formation of high levels of interlacing between the polymeric chains and the mucous barrier, which results in high mucoadhesive force (Han et al. 2006; Ivarsson and Wahlgren 2012; Tamburic and Craig 1995; Kerec et al. 2002; Gómez-Carracedo et al. 2004; Zhu et al. 2013; Almeida et al. 2013).

8.2.1.3 Poloxamers

Poloxamer is a linear tri-block copolymer consisting of hydrophobic central polypropylene oxide (PPO) groups and hydrophilic polyethylene oxide (PEO) end groups (Figure 8.3) (Singh et al. 2013; Pires et al. 2009). Also known as Pluronic, it is commonly used as a nonionic surfactant

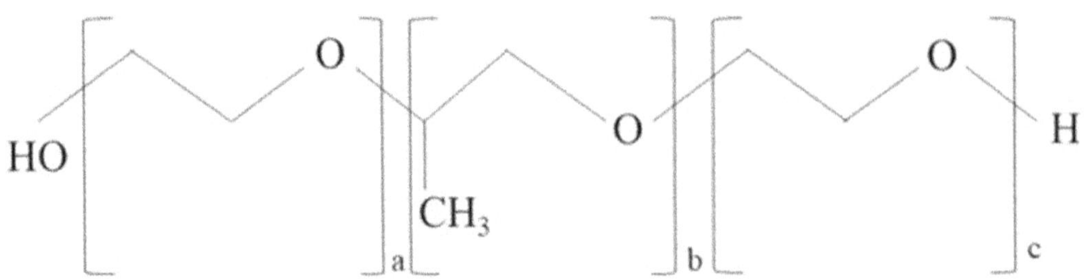

FIGURE 8.3 Chemical structure of the tri-block copolymer poloxamer.

and thermosetting gelling agent. It is extensively used by pharmaceutical industries, for the development of new dosage forms, for its unique gelling ability and nontoxic nature (Qiu, Hamilton, and Temenoff 2011). Various types of poloxamers are available, depending on the polymer chain composition and length, such as poloxamers 407 (Pluronic F-127), 188 (Pluronic F-68), and 181 (Pluronic L61) (Alexandridis, Holzwarth, and Hatton 1994).

Among the different types of poloxamer, the poloxamer 407 (MW = 12.000), with a PEO/PPO ratio of 2:1, is a widely used gelling agent in pharmaceutical developments due to its ability to form a gel at comparatively low concentrations at body temperature in suitable concentrations (20–30%, w/w of poloxamer 407) (Wang et al. 2013). This phenomenon happens due to the self-assembly of micelles with dehydration of hydrophobic PPO core and hydrophilic PEO shell. However, as the temperature increases, they immediately form a gel. Such a gel is unsuitable for a drug carrier system because it loses its structural integrity after adding an aqueous solution (Lee et al. 2004; Qiu, Hamilton, and Temenoff 2011). In order to counteract its weak bio-/mucoadhesive properties, poor mechanical properties, and short residence time, P407 may be used with bioadhesive polymers (cellulose derivatives, carbomers, chitosan) or other poloxamers (poloxamer 188) for the desired properties.

8.2.1.4 Polyvinyl Alcohol

Polyvinyl alcohol (PVA) is a semi-crystalline polymer that has properties such as water solubility, adhesive nature, and film-forming ability; it is used in the biomedical field, is nontoxic and biocompatible, has excellent chemical resistance and good mechanical properties, and is a biodegradable synthetic polymer. It can also be used as an emulsion stabilizer and coating agent, being used in transdermal and ophthalmic pharmaceutical forms (Rodrigues 2006). The properties of PVA are controlled by molecular weight and degree of hydrolysis (GH) (Kumar et al. 2007).

PVA is widely used in the food, pharmaceutical, and cosmetic industries due to its barrier properties against moisture, oxygen, and other external agents. It is an excellent adhesive highly resistant to solvents, oils, and greases; forms clear, rigid films with high resistance to traction and abrasion; and acts as a barrier to oxygen, being superior to any other polymer; however, it must be protected from moisture, which increases the rate of transmission of gases through it (Wainwright and Pr 1998).

Another characteristic that makes it a very interesting polymer is the fact that PVA is one of the few fully biodegradable synthetic polymers, making its applications extend to the area of controlled drug release due to its low toxicity and high biocompatibility (Farias et al. 2012).

8.2.1.5 Polyvinylpyrrolidone

PVP, also known as polyvinylpyrrolidone, is a straight-chain homopolymer of 1-vinyl-2 pyrrolidone groups with different polymerization indices that result in polymers with different molecular weights. It is an amorphous polymeric material that is hygroscopic in nature and capable of complex formation. The presence of PVP in a binary system guarantees an increase in the permeability of the drug in relation to the homopolymer, and therefore, it has great importance in the pharmaceutical area (Risbud et al. 2000). PVP is a nontoxic material and also has applications in the biomedical

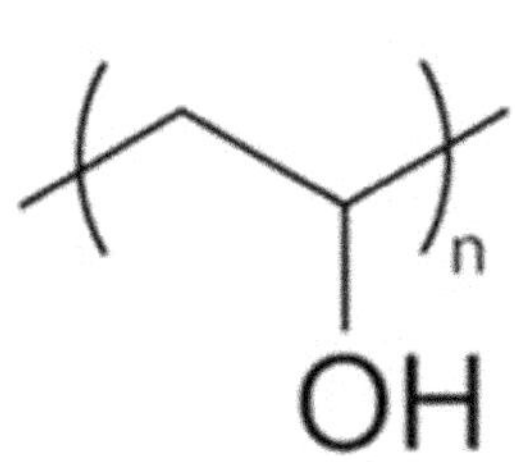

FIGURE 8.4 Chemical structure of polyvinyl alcohol.

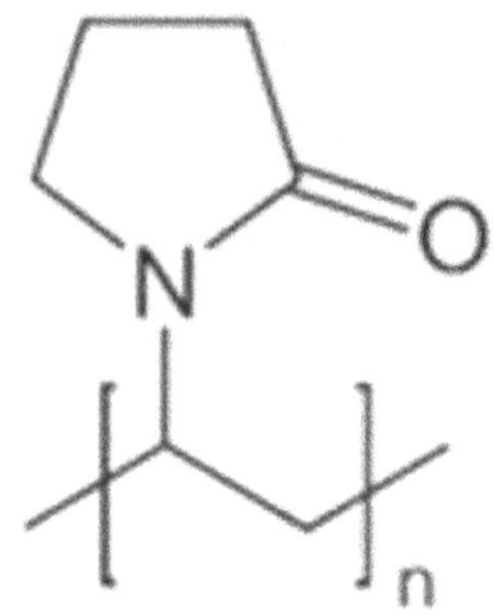

FIGURE 8.5 Chemical structure of polyvinylpyrrolidone.

field. It presents characteristics such as adhesion, excellent physiological compatibility, and reasonable solubility in water and most organic solvents (Rodrigues 2006).

8.2.2 Bioactive Agents of Natural Origin

Phytopharmaceutical constituents, also known as primary or secondary metabolites of plants, are widely produced during the life cycle of plants. The primary metabolites are associated with photosynthesis, cellular respiration, and plant growth and development, including carbohydrates, lipids, amino acids, and nucleotides. The secondary metabolites are produced due to the adaptation to environmental conditions (Nascimento and Fett-Neto 2010).

Among these secondary metabolites, polyphenols are widely found in nature with around 8,000 compounds described in literature and have distinct chemical structures, which range from simple structure, like phenolic acid, to highly polymerized structures, like condensed tannins (Manach et al. 2004; Pandey and Rizvi 2009). The main types of polyphenols are phenolic acids, flavonoids, stilbenes, and lignins (Pinto, 2016). In addition, these constituents display many biological activities, including antioxidant activity due to its ability to scavenge free radicals and delay lipid autoxidation. Consequently, they may prevent several serious disorders, including cancer, atherosclerosis, inflammatory, and coronary heart diseases (Chaillou and Nazareno 2006). In this chapter, the polyphenolic compounds studied are curcumin, gallic acid, p-coumaric acid (Table 8.1), and the complex matrix propolis (rich in polyphenolic compounds).

8.2.2.1 Curcumin

Curcumin (CUR) is a substance isolated from *Curcuma longa* Linn., also known as turmeric, which belongs to the Zingiberaceae family. The (1E,6E)-1,7-bis(4-hydroxy-3-methoxyphenyl)-1,6-heptadiene-3,5-dione, popularly called CUR, is present in curry and saffron, very common in typical Indian and Chinese dishes. Curcuminoids are found in the rhizome of the plant and can be divided into three compounds according to their abundance: CUR, demethoxycurcumin, and bisdemethoxycurcumin (Figure 8.6), corresponding to 77%, 17%, and 3% of the active, respectively (Mazzarino et al. 2011; Goel and Aggarwal 2010; Ji and Shen 2014).

Its polyphenolic structure ($C_{12}H_{20}O_6$; MW 368,385 g/mol) has two methyl ether-substituted meta phenolic rings, tautomeric keto-enolic groups, and an unsaturated carbon chain in its structural formula, contributing to lipophilicity (log P = 3.29) (Alves et al. 2018; Hussain et al. 2017). The tautomeric keto-enolic groups provide three possible forms: a ß-diketonic and two isomers in an equilibrating keto-enol tautomeric form. The enol form is more stable and predominant in pH levels around 3 and 7 because of the planar conformation, which is also dependent of temperature, solvent, and polarity (Araiza-Calahorra, Akhtar, and Sarkar 2018; Jovanovic et al. 1999). It exhibits

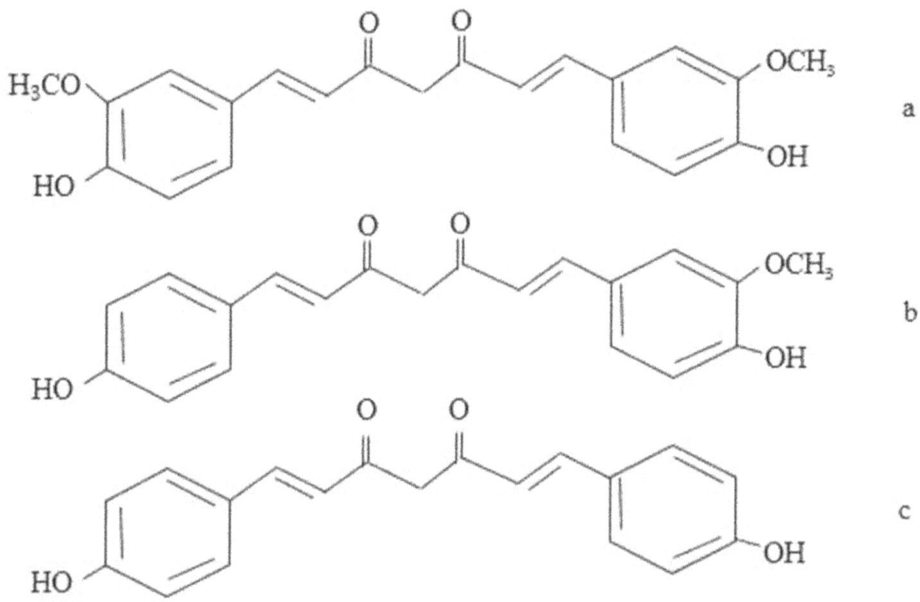

FIGURE 8.6 Curcumin molecules: (a) CUR, (b) desmethoxycurcumin, (c) bisdesmethoxycurcumin.

low solubility in water (3.2 mg/L at 25°C) but is soluble in acetone, dimethyl sulfoxide (DMSO), and ethanol. CUR has three ionization constants (pKa), 8.38, 9.88, and 10.51, which give it a basic character variable according to the determination technique (Dias Pereira et al. 2022). It is stable at acidic pH (pH = 1 to 6) and unstable at neutral and basic pH, degrading more than 90%, due to the formation of phenylate anion, increasing the production of CUR radicals (Naksuriya et al. 2014; Sueth-Santiago et al. 2015; Goel, Kunnumakkara, and Aggarwal 2008; Siviero et al. 2015; Alves et al. 2018; Araiza-Calahorra, Akhtar, and Sarkar 2018). Besides the high lipophilicity and instability in neutral pH, CUR has demonstrated photosensitivity, which hinder its clinical applications (Alves et al. 2018; Araiza-Calahorra, Akhtar, and Sarkar 2018).

Despite the unfavorable chemical properties previously discussed, CUR is an active agent widely studied in treating and preventing various diseases such as cystic fibrosis, psoriasis, depression, asthma, arthritis, brain injury, and diabetes. It has anti-inflammatory, antioxidant, and antimicrobial action and helps prevent cardiovascular diseases, besides antineoplastic properties (Mazzarino et al. 2011; Sueth-Santiago et al. 2015; Goel and Aggarwal 2010; Ji and Shen 2014).

It has been observed a huge necessity to develop drug delivery systems to improve its solubility, bioavailability, and clinical efficacy. Huge effort has been done to improve its chemical properties by incorporating the active pharmaceutical ingredients in diverse formulations, including nanocrystals, liquid crystals, solid lipid nanoparticles, transdermal films, micro-/nanospheres, nanoemulsions, phospholipid complexes, and bioadhesive, thermoresponsive, nanostructured systems (Samanta and Roccatano 2013; Zhao et al. 2012; Altunbas et al. 2011; Ferreira et al. 2019; Das, Kasoju, and Bora 2010; Sun et al. 2014; Said dos Santos et al. 2021; Naksuriya et al. 2014; Ferreira et al. 2020).

8.2.2.2 Gallic Acid

Gallic acid is a trihydroxybenzoic acid where the hydroxy groups are in positions 3, 4, and 5, being a conjugate acid of a gallate with the formula $C_6H_2(OH)_3CO_2H$ classified as a phenolic acid. It has application as an astringent and antineoplastic. As a biological function, it plays the role of cyclooxygenase 2 inhibitor, it is a metabolite of plant origin with antioxidant activity and the ability to donate a hydrogen to a receptor, it induces apoptosis, in addition to being a human xenobiotic metabolite and inhibiting EC.13.11. 33 (arachidonate 15-lipoxygenase) (Pubchem).

Gallic acid is an odorless white solid; some samples may be brown due to partial oxidation (Guard 1999). It is soluble in water, hot water, alcohol, ether, glycerol, and acetone and is practically insoluble in benzene, chloroform, and petroleum ether (Budavari et al. 1996). Studies in the literature have shown therapeutic use in veterinary medicine, such as being used as an intestinal astringent and antidiarrheal agent (Budavari et al. 1996; Pelletier and Phaneuf 1976). It can be absorbed in the gastrointestinal tract and is largely excreted in the urine (Gosselin et al. 1984; Parke 1968).

The compound is used in food additives as a flavoring agent in addition to being used in the cosmetic area as an antioxidant agent, obtained through chemical or enzymatic hydrolysis of tannins.

8.2.2.3 p-Coumaric Acid

p-Coumaric acid, also known as 4-hydroxycinnamic acid, is a coumaric acid where the hydro substituent is located at the phenyl ring position C-4, being derived from cinnamic acid hydroxylate. There are three coumaric acid isomers, ortho-, meta-, and para-coumaric, which differ by the hydroxyl position of the phenyl group. p-Coumaric acid is the most abundant of the isomers in nature and can exist in the form of trans-p-coumaric acid or cis-p-coumaric acid (PubChem).

It appears as a solid and crystalline, is slightly soluble in water, and is very soluble in ethanol and diethyl ether. Its chemical function is to be a hydrogen donor, and its biological function is a plant metabolite produced during a reaction in plants, such as in flowers, conifers, or gymnosperms.

Studies in the literature have shown its pharmacological use as an antioxidant, free radical scavenger, and anti-infective agent, in addition to being used as a male contraceptive agent. It is also used in the cosmetic area for skin conditioning.

TABLE 8.1
Physical and Chemical characteristics of curcumin, gallic acid and p-coumaric acid

	Curcumin	Gallic Acid	p-Coumaric Acid
Structure		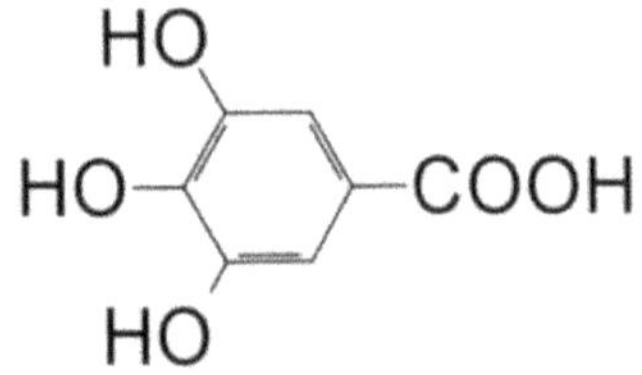 A chemical structure where six carbon atoms are connected to three hydroxide groups and one carboxyl group.	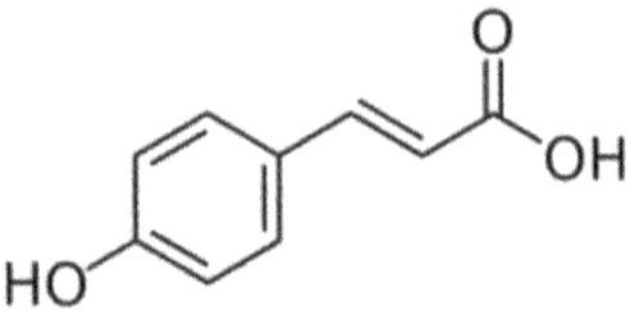 A chemical structure of trans-4-(Trifluoromethyl)cinnamic acid that has two hydroxide groups attached to it.
Formula	$C_{21}H_{20}O_6$	$C_6H_2(OH)_3CO_2H$	$C_9H_8O_3$
Molecular weight	368.4	170.12	164.16
Log P	3.29	0.7	1.5
pKa	8.38, 9.88, 10.51	4.40	Not present
Melting point	180 to 183 °C	~210 °C	211.5 °C
Relative density (water = 1)	0.9348 g/mL at 15°C	1.7 g/mL	Not present

8.2.2.4 Propolis

Propolis (PRP) is produced by bees of the genus *Apis mellifera* L., from the collection of resins, waxes, and gums from petals, flowers, stems, and tree bark around the apiary. This adhesive resin is a product of the salivary enzymes of the bee (Funari and Ferro 2006). This gum resin (Figure 8.7)

FIGURE 8.7 Macroscopic appearance of propolis *in natura*.

is used by bees to seal the gaps in the hive, protecting the hive against wind, insects, and dirt. PRP varies in color from yellow-green to dark brown, depending on its origin and harvesting time, with a characteristic odor (Ghisalberti 2016; Moreira 1986; Marcucci 1995; Burdock 1998). The PRP *in natura* is processed, and its extract is used as a final or intermediate pharmaceutical form (Bruschi et al. 2006; Said dos Santos et al. 2020; Burdock 1998).

Its chemical composition is complex and is related to the existing flora around the apiary, which can vary from region to region and the time of year it is collected (Bankova 2005; Bruschi, Franco, and Gremião 2003; Popova et al. 2004). PRP, in general, consists of approximately 50–60% resinous compounds, 30–40% waxes, 5–10% volatile oils and aromatic acids, 5% balsams and pollen grains, and 5% various other substances, such as polyphenols (flavonoids and phenolic acids), vitamins, mineral salts, and impurities (Park et al. 2002; Francisco, Cerquetani, and Bruschi 2013; Bruschi, Franco, and Gremião 2003). Polyphenols are one of the most important compounds occurring in plants, with flavonoids and phenolic acids being the most interesting constituents of the polyphenol class and, consequently, the most important class of pharmacologically active compounds. One of the main phenolic compounds found in propolis is artepelin C, being reported as the most abundant compound isolated in Brazilian PRP (Funari, Ferro, and Mathor 2007; Buriol et al. 2009; Miguel et al. 2010; Salatino et al. 2005; Bruschi et al. 2006; Lee et al. 2004).

PRP has antibacterial, fungicidal, antioxidant, antiviral, anti-inflammatory, immunostimulant, anticancer, anti-ulcer, hypotensive, and cytostatic properties (Bruschi et al. 2002; Cabral et al. 2012; Burdock 1998; Bruschi, Franco, and Gremião 2003), which, consequently, lead to important pharmacological activities and pharmaceutical uses. There are also reports of tissue-regenerating and healing action in ulcers and anesthetic action (Burdock 1998; Francisco, Cerquetani, and Bruschi 2013). For the improvement of health, they are used in the prevention of inflammatory diseases, heart diseases, cancer, and diabetes, mainly through their antioxidant activity (Sulaiman et al. 2011; Francisco et al. 2018b; Pereira et al. 2013).

8.3 BIOADHESIVE, THERMORESPONSIVE SYSTEMS CONTAINING THE NATURAL BIOACTIVE AGENTS

8.3.1 Strategies

As previously described, polyphenolic compounds, like curcumin, gallic acid, p-coumaric acid, and the natural product propolis, show a wide range of structures and pharmacological activities. However, they show diverse limitations, such as low bioavailability due to limited stability in light, moisture, pH, temperature, oxygen concentration, and low water solubility (Paini et al. 2015; Das, Kasoju, and Bora 2010). It is important to develop strategies, methods, and systems to exploit and improve their therapeutic potential. The bioadhesive properties of formulations containing these polyphenols and propolis have been explored, such as strategies using isolated synthetic or natural bioadhesive polymers, chemical modification of bioadhesive polymers, nanotechnology, multiparticulates, physicochemical structuring, or even as polymer blends with thermoresponsive materials. The thermoresponsive properties have also been used as a strategy by chemical modification of the thermoresponsive polymer, nanotechnology, or physicochemical structuring systems.

Bioadhesive polymers are materials capable of bonding to biological substrates in two ways: by adhering to the cell membrane (bioadhesive) or the mucous layer (mucoadhesive). Bioadhesion is a property exploited as a strategy to increase the length of time a drug remains in the body. As a result, the bioavailability of drugs can be improved, and the number of doses taken daily can be reduced. Another advantage is that bioadhesion can be modulated to occur at specific sites, decreasing toxicity and increasing the therapeutic effectiveness of drugs (Andrews, Laverty, and Jones 2009)(Asane et al. 2008; Villanova, Oréfice, and Cunha 2010; Andrews, Laverty, and Jones 2009).

In the literature, it is possible to find works with this type of system, such as the semisolid mucoadhesive system for oral application, containing propolis (Bruschi et al. 2007); bioadhesive emulgels, containing propolis (Said dos Santos et al. 2020; Dano et al. 2021); and bioadhesive emulgels, containing propolis and/or curcumin (Said dos Santos et al. 2021), among others.

Some authors have discussed the importance of natural source bioadhesive polymers, like biocompatibility, nontoxicity, biodegradability, pollution-free processing, and easy availability (Ifat, Karsten, and Fredmen 2001). Cashew nut tree gum, which has a polysaccharide complex, has been used in curcumin tablets due to its adhesive properties for buccal applications (Gowthamarajan et al. 2012). In addition, composites containing gallic acid, gelatin, ZnO nanoparticles, and L-DOPA inspired additives (such as catechol) have been developed for their antimicrobial activity. L-DOPA-inspired additives are amino acids that have been reported as responsible for present adhesion characteristics (Ramirez-Barron et al. 2021). Marine invertebrates are sources of bioadhesives with biomimetic interest. They have been used for the development of formulations containing gallic acid using protein-based adhesives, like tunicates (sea squirts). These tunicates show a body wall composed of a cellulose fiber (tunicin) and proteins (DOPA and 3,4,5-trihydroxyphenylalanine—TOPA), which contribute to adhesion (Almeida, Reis, and Silva 2020). Nanofibers and solutions using these compounds and gallic acid have been developed (Prajatelistia et al. 2016).

The functionalization of synthetic or natural bioadhesive polymers has been performed to improve bioadhesion and performance in the application site. Most of these chemical modifications have been developed with chitosan (CS), which is a strong bioadhesive polymer. This strategy has been carried out between CS and N-acetyl-L-cysteine to carry curcumin in nanostructured lipid carriers for ocular application (Liu et al. 2016). In addition, CS has been grafted with gallic acid to obtain a hydrogel for wound healing applications (Sun et al. 2021). Modifications with hexanoyl glycol and chitosan provided thermosensitive properties and were used to carry gallic acid in hydrogel formulations (Park et al. 2021).

Multiparticulate pellets may be used as a strategy to improve mucoadhesion, residence time, and localized delivery of the drug at the target site and improve the therapeutic effect. Thus, formulations containing curcumin were developed to increase these characteristics (Desai and Momin 2020).

Thermoresponsive properties are commonly provided from copolymers with the self-assembly properties, facilitating administration, solubility of hydrophilic and hydrophobic compounds like polyphenols and propolis, and consequently therapeutic efficacy. The most commonly used polymers are poloxamers, which contain poly(ethylene) oxide and poly(propylene) oxide segments, and pNIPAM. These polymers may be isolated, chemically modified, blended with polymers, or found in nature. Polymeric micelles containing curcumin and poloxamer 407 and Pluronic P123 were developed to improve the unfavorable physicochemical characteristics and could improve curcumin solubility (Braga et al. 2020).

Chemically modified thermoresponsive polymers have been used as a strategy to improve the solubility of curcumin using poloxamer 407 and poly(N-isopropylacrylamide) (pNIPAM). Poloxamer 407-formulations were prepared with the polyelectrolyte hydrophilic monomer and

2-acrylamido-2-methylpropane sulfonic acid to improve the mechanical properties of these formulations (Khan et al. 2021). In addition, pNIPAM was grafted with sodium alginate for the delivery of curcumin. This strategy promotes higher structural integrity and better gelling properties (Zakerikhoob et al. 2021). Natural molecules may be chemically modified; for example, coiled proteins may be assembled into functional, physically cross-linked hydrogels with a tunable phase behavior (Hill et al. 2019).

The structuring physicochemical technique is commonly used in liquid crystals, liquid crystalline precursors with unique properties, to increase the viscosity and improve the administration performance. Bioadhesive or thermoresponsive polymers have also been incorporated in some formulations as a strategy to improve their clinical efficacy. However, the bioadhesive property may also be provided by the structure of these liquid crystals. The phytopharmaceutical compounds incorporated in these formulations are p-coumaric acid and curcumin (Ferreira et al. 2021; Fonseca-Santos et al. 2016; Santezi et al. 2021).

During the last decades, the nanotechnology has been widely used as a strategy in the fabrication of formulations containing phytopharmaceutical compounds with bioadhesive and/or thermoresponsive properties (Oliveira et al. 2018). This approach may offer specific targeting for the treatment of diseases, improve physicochemical properties of drugs, like polyphenols, promote convenient administration, and consequently, availability and better clinical use. Nanotechnology may also decrease drug dose and toxicity (Bruschi, De Francisco, and Nova 2019). When bioadhesive or thermoresponsive polymers are included in these formulations their beneficial properties are increased. Considering the diverse formulations prepared using this strategy, a wide range of materials have been used to provide these characteristics, like polymers, lipids, surfactants, and metals. This strategy is mostly used for curcumin due to its insolubility in a hydrophilic environment, photosensitivity, and pH-sensitivity, which has been incorporated in polymeric nanoparticles, liposomes, microspheres, nanotubes, silver nanoparticles, nanogels, microgels, magnetic nanocarriers, micelles, and nanopolymers with bioadhesives and/or thermoresponsive characteristics (Ali et al. 2014; Ternullo et al. 2020; Sanoj Rejinold et al. 2011; Soto-Quintero et al. 2019; Luckanagul et al. 2018; Hu et al. 2014; Kulkarni et al. 2021; Babaei et al. 2020; Ferreira et al. 2019; Oliveira et al. 2018; Berginc et al. 2014). But gallic acid also has been incorporated in microemulsions and ZnO nanoparticles (Salat et al. 2018; Mitsou et al. 2020; Hoyo et al. 2019).

In this context, polymer blends are another strategy widely used to combine bioadhesive, thermoresponsive properties to formulations containing phytopharmaceutical constituents, like curcumin, gallic acid, p-coumaric acid and propolis (Douglas et al. 2015; Mayol et al. 2008). This strategy is commonly faster and lower-cost than developing new polymerization routes and/or new monomers. In addition, it is possible to obtain new and improved functionalities, such as improved mechanical characteristics, facilitating administration and performance, new release mechanisms, and biocompatibility (Lefnaoui and Moulai-Mostefa 2014; Liu et al. 2005). Improved adhesion and drug delivery performance *in vitro* and *in vivo* have been reported in literature for these formulations on the mucosal surfaces (Bhowmik et al. 2013; Chang et al. 2002; Han et al. 2006). Using this strategy, it is not only possible to create liquid and semisolid systems (gels or emulgels, liquid precursor crystal systems) but also solid systems (film-forming formulations, films, patches). Polymer blends are widely used for propolis and curcumin in a variety of polymers, including poloxamer 407, carbomers (Carbopol 934P, Carbopol 971P, Carbopol 974P), polycarbophil, chitosan, xanthan gum, and hydroxypropyl methylcellulose (Bruschi et al. 2007; Fabri et al. 2011; Rahman et al. 2021; Nasra et al. 2017; Kawathe, Nanadgude, and Poddar 2017; Bonfim et al. 2020; Pereira et al. 2013; Ferreira et al. 2019, 2020; Santezi et al. 2021; Rosseto et al. 2021). Interestingly, this strategy may be used in the skin, oral mucosa, and vaginal mucosa, increasing the residence time and allowing comfortable administration (Bruschi, De Souza Ferreira, and da Silva 2020).

Thus, many strategies are utilized to improve the use of these phytopharmaceutical compounds by combining polymers, improve physical or chemical properties, provide structure for different sources (synthetic or natural), and allow for different application sites. However, it is also important to consider the technology involved in the fabrication of these formulations and types of systems and their possible utilization.

8.3.2 Technology

The techniques involved in the preparation of the formulations containing curcumin, gallic acid, p-coumaric acid, or propolis may show a significant impact on their performance, concerning drug release, adhesion, administration, availability, and clinical use (Russo et al. 2016). This is particularly true for solid systems like films and nanotechnology-based formulations, including microemulsions, nanoparticles, micro-/nanogels. Moreover, according to the solubility of the drug in a semisolid system, the order addition of the compounds may change and influence in its stability (Ferreira et al. 2020).

8.3.2.1 Films

The polymers used for the preparation of pharmaceutical films are normally water-soluble but not restricted to them. The pharmaceutical film-coating technology is based on the use of polymers dispersed in an aqueous medium or in organic solvents (Bunhak 2007).

The preparation of oral films can be carried out, mainly, using three techniques: hot melt extrusion, print technologies and solvent casting. Although different, all these processes aim to provide a more homogeneous pharmaceutical form visually possible (Fernandes 2017). In the solvent casting method, the drug, polymers, and other excipients are mixed and dispersed in suitable solvents until a homogeneous mixture is obtained. This mixture is then poured into suitable containers and dried in an oven at room temperature or lyophilized (Morales and McConville 2011;Neelagiri, Reddy, and Rao 2013; Madhavi et al. 2013). The solution obtained after the drying process is cut to the desired size (Bala et al. 2013).

Hot melt extrusion (HME) is a technique that does not use solvents, so the process becomes faster and safer. The decrease in the number of steps allows the process of obtaining it to be continuous, which makes this technique viable for a large-scale film production (Jones et al. 2009). In the HME technique, the raw materials must be able to deform easily inside the extruder, by the action of heat, and solidify after exiting.

Printing technologies use inkjets and 3D printing. Most of the time, the bioactive agent is dissolved in a matrix with suitable characteristics of viscosity and surface tension necessary for printing using inkjet, extrusion, or pressure-assisted microsyringe, for example. One of the advantages of this technology is that the bioactive agents are often protected from thermal degradation (Fernandes 2017). This methodology has the flexibility to print exact amounts of the drug on each base film, which enable a more individualized therapy (Preis, Breitkreutz, and Sandler 2015).

8.3.2.2 Gel and Emulgel

The pharmaceutical technology for preparing these systems is based on the use of polymers dispersed in an aqueous medium or gelling polymers—viscosity increasers, whereby the emulgel contains an immiscible liquid (oil), stabilized by the emulsifying agent (polymer) (Villanova, Oréfice, and Cunha 2010).

The gels, for the most part, are prepared by dispersing poly(acrylic acid)-derived, cellulose-derived, or natural polymers in a purified aqueous phase with stirring. Polymers with thermoresponsive characteristics are added after this first stage of dispersion of other polymer types and left in the refrigerator for 24–48 hours to ensure complete polymer wetting. Subsequently, the polymeric systems are again stirred to complete dissolution and are neutralized with triethanolamine or a basic agent. Finally, in these systems, the active agents are added (Bassi da Silva et al. 2017; Borghi-Pangoni et al. 2017; Bruschi et al. 2007; Souza Ferreira et al. 2015).

According to the solubility of the phytopharmaceutical compound and the polymers involved in the formulation, it is possible to use a different approach during the fabrication of these formulations.

For example, if this formulation comprises a hydrophobic compound like curcumin and a co-block polymer with a hydrophobic region, it is possible to prepare this gel by a solid dispersion method to promote intimate contact between the hydrophobic compound and the hydrophobic site of this copolymer avoiding its destabilization. By this method, the copolymer may be solubilized in ethanol with the hydrophobic compounds and subsequent elimination of the solvent by a rotary evaporator. Then this solid dispersion may be dispersed in the aqueous environment finalizing the gel (Souza Ferreira et al. 2020).

Emulgels are characterized as dispersed systems, consisting of two immiscible liquids stabilized by an emulsifying agent, usually a polymer (Aulton and Taylor 2013; Allen, Popovich, and Ansel 2012; Villanova, Oréfice, and Cunha 2010). They are prepared by dispersing the polymer in a purified aqueous phase with agitation. After the complete dispersion of the polymer, the pH is neutralized, and oil and active agents are added (Francisco et al. 2019; Said dos Santos et al. 2020, 2021). Besides this way, they can be prepared in a more traditional way, where two phases are prepared, one aqueous and one oily, according to the characteristics of each component and at a suitable temperature, which finally one of these phases will be added to the other, under constant agitation (Yadav et al. 2016).

Thermoresponsive polymers, like poloxamer 407, show a copolymeric structure with hydrophilic and hydrophobic segments, which may be considered during the preparation of emulgels. Since they may act as surfactants, so they can become gelling agents, stabilizing these formulations. Their preparation may be based on the incorporation of the hydrophobic constituent into the oil phase with the subsequent addition to the polymeric dispersion containing poloxamer 407. In case of a polymer blend strategy, the bioadhesive polymer may be added later (Ferreira et al. 2020).

8.3.2.3 Liquid Crystalline Systems

Liquid crystal precursor systems are basically composed of oil phase, surfactants, and aqueous phase. The preparation of these formulations is based on the dispersion and mixture of the compounds using stirrers. The order of each step depends on the drug solubility; for example, if the drug is hydrophobic, this is incorporated into the oily phase. The surfactant may be added to this mixture of the drug and the oily phase, with subsequent addition of the aqueous phase. The bioadhesive and/or thermoresponsive polymers are dispersed in the aqueous phase. Despite this simple preparation in laboratory scale, the scale-up production of such liquid crystals is not well established and is a concern of researchers (Toledo, Rosseto, and Bruschi 2018; Santezi et al. 2021; Ferreira et al. 2021; Fonseca-Santos et al. 2016).

8.3.2.4 Multiparticulate Pellets

The fabrication of multiparticulate pellets is simpler than nanocarriers due to the improved flowability of these larger pellets and the possibility of using easier standardized equipment for scale-up. These systems are prepared by the extrusion/spheronization method with different equipment. Initially, the excipient powders are sifted and blended in a mortar to facilitate the mixture. The powder blends are wetted with a granulating liquid (purified water) and the wet mass is extruded to form the extrudate using a screw axial extruder. These extrudates are then spheronized with different parameters in a spheronizer to form pellets, which are dried in tray dryer and fluidized bed dryer. The dried pellets are sifted for separation of fines. The mucoadhesive properties are achieved by its enteric coating with fluidized bed coating with a pH-sensitive polymer, anti-tacking agent, plasticizer, and solvent to disperse these agents (Desai and Momin 2020).

8.3.2.5 Nanotechnology

Nanotechnology may be a useful strategy for development of improved delivery systems containing phytopharmaceuticals. The most utilized systems are polymeric nanoparticles, lipid nanoparticles, magnetic nanoparticles, micro-/nanogels, liposomes, polymeric micelles, nanotubes, and niosomes, which have been used to carry propolis, curcumin, gallic acid, and p-coumaric acid, showing different composition and structuration. Their structure impacts the preparation methods and techniques. Even the same type of nanoparticles shows different preparation methods according to materials and drugs used (Russo et al. 2016).

Diverse techniques and technologies have been carried out to fabricate polymeric nanoparticles containing propolis, curcumin, gallic acid, and p-coumaric acid, such as spray drying (Villa Nova et al. 2019; Francisco et al. 2019; Oliveira et al. 2018), ionic cross-linking method (Chuah et al. 2013; Sanoj Rejinold et al. 2011), electrostatic layer-by-layer technique (Paini et al. 2015), chemical reaction (Bielska et al. 2013), and microwave-assisted synthesis (Xu et al. 2020). Spray drying is one of the techniques used to dry and/or obtain micro-/nanoparticles due to the atomization of the mixture of the aqueous or organic solution of polymers and drugs. The micro-/nanoparticles are produced by the contact of the droplets with the hot air and their consequent drying (Villa Nova et al. 2019). The cross-linking method is prepared by physical or chemical techniques. The physical cross-linking is also known as ionic cross-linking, the method used to prepare curcumin nanoparticles with chitosan, and chemically modified chitosan was formed by dropping the cross-linking agent under stirring without chemical reaction (Chuah et al. 2013; Sanoj Rejinold et al. 2011). Ionic cross-linking can be prepared by using chemically modified polymers with opposite charges, which interact and precipitate when they get make contact (Bielska et al. 2013). Another technique used to prepare microcapsules containing polyphenolic compounds is the electrostatic layer-by-layer self-assembly method. Usually, carbonate particles are used as templates to allow the alternate adsorption of opposite charges of polyelectrolytes, creating the capsule shell. Once this multilayer shell is finalized, the templates are eliminated by dissolution (chelating agents or acidic medium). This method allows the creation of variable shell thickness. The incorporation of the drug, like polyphenols, is performed by increasing the shell permeability using a solution of ethanol, purified water, and the drug (Paini et al. 2015).

Lipid nanoparticles also show diverse preparation techniques. Specifically for lipid nanoparticles containing propolis, curcumin, gallic acid, or p-coumaric acid, the literature reports the use of stirring and ultrasonication methods for the preparation of propolis extract lipid nanoparticles. An aqueous phase is incorporated into an oily phase with subsequent high shear dispersion and sonication step. Propolis extract was added into the oily phase (Rosseto et al. 2017).

The composition of magnetic nanoparticles is also variable with silver, iron, and others. The bioadhesive and/or thermoresponsive characteristics can be promoted by polymeric coating. The fabrication methods reported for micro-/nanoparticles containing curcumin, propolis, gallic acid, or p-coumaric acid were coprecipitation (Toledo, Rosseto, and Bruschi 2018), chemical (Soto-Quintero et al. 2019), and polyol methods (Imanifard et al. 2017). The coprecipitation method is carried out using iron salts dissolved in water, which are precipitated by sodium hydroxide (Toledo, Rosseto, and Bruschi 2018). Chemical synthesis was performed by the oxidation-reduction of sodium citrate, curcumin, and silver in water solution. These silver nanoparticles were coated with a thermoresponsive polymer using the free radical precipitation (Soto-Quintero et al. 2019). For the preparation of magnetic nanoparticles using the polyol method, iron was slowly heated with triethylene glycol under nitrogen atmosphere and until the reflux temperature of 285°C was reached. This step allows the formation of the magnetic nanoparticles (Imanifard et al. 2017).

Micro-/nanogels, composed of a cross-linked network of polymers, was used as a strategy to incorporate curcumin. The preparation technique is coupling reaction, where thermoresponsive polymers (poloxamer 407 or pNIPAM) are chemically modified or grafted with other polymers like chitosan. Subsequently, the micro-/nanogels are prepared by chemical synthesis or simple sonication method in aqueous media. The phytopharmaceutical loading may occur through the incubation method (Kulkarni et al. 2021; Luckanagul et al. 2018).

Microemulsions, reported in literature with the phytochemical compounds, consist of the oil phase (extra virgin olive oil and sunflower oil), surfactants (Tween 80, monoglycerides of vegetable fatty acids), and chitosan solution. The formulations were prepared by a gentle mixture of the aqueous phase and the dispersion of surfactants and oil phase until an optically transparent system was achieved (Mitsou et al. 2020).

Liposomes are usually prepared through the film hydration method. Initially, a solvent, phospholipids (phosphatidylcholine), the drug, and surfactants (polysorbate 20) are mixed and dried in a rota evaporator. Subsequently, the thin film is hydrated in a phosphate buffer. The bioadhesive characteristics are achieved by coating the liposomes with polymers, including chitosan or Carbopol 974P, using a drop-wise method, with subsequent storage in a refrigerator. The incorporation of liposomes in a gel with bioadhesive characteristics is the other option (Berginc et al. 2014; Ternullo et al. 2020).

According to the composition, polymeric micelles are prepared by the solid dispersion method, also known as solvent evaporation/membrane rehydration method. The solid dispersion method is quite similar to the liposome preparation technique using solvents and rota evaporator, but replacing the oily phase and surfactants to well-described polymers in literature (Braga et al. 2020). On the other hand, the thermoresponsive polymers can be prepared by atom transfer radical polymerization, providing the coupling of a central polymer chain with two lateral thermoresponsive blocks. Subsequently, the polymeric micelles are fabricated using solid dispersion method as previously reported (Hu et al. 2015).

Niosomes also show a similar preparation method of liposomes and polymeric micelles using the solvent, a mixture of surfactants, and the oil phase. This mixture is submitted to a rotary evaporator with posterior rehydration with a phosphate buffer. Subsequently, these niosomes are sonicated with a microtip probe sonicator to achieve the niosomal dispersion (Manosroi et al. 2011).

Therefore, multiple technologies were developed to facilitate and innovate the preparation of these different systems. These modern technologies enable drug performance, stability, comfortable administration, and suitable therapeutic response (Souza Ferreira et al. 2020).

8.3.3 Type of Systems

Despite the well-defined characteristics of drug delivery systems containing bioadhesive or/and thermoresponsive properties, they may generate diverse types of systems (Bruschi et al. 2017). These systems include films, film-forming systems, gels, emulgels, organogels, liquid crystalline systems, multiparticulate systems, and diverse systems prepared using the nanotechnology strategy (polymeric nanoparticles, lipid nanoparticles, magnetic nanoparticles, liposomes, niosomes, micro-/nanogels, microemulsions, and polymeric micelles).

8.3.3.1 Films and Film-Forming Systems

Films are thin-thick polymeric matrices that are obtained from different sources and have different chemical characteristics. They can be formed by polymers of synthetic, semi-synthetic, or natural origin, from, for example, polysaccharides, proteins, lipids, and derivatives (Batista, Tanada-palmu, and Grosso 2005). Films can be obtained from various materials, such as fatty acids, aromatic hydrocarbon derivatives, porphyrins, phthalocyanines, tetracyanoquinodimethane, biomolecules (phospholipids, proteins, pigments, lignin, etc.), conventional polymers, copolymers, polymeric liquid crystals, and conductive polymers (Ulman 2013).

When compared to other more traditional dosage forms, such as long lengths and capsules, the buccal films mainly show better mechanical resistance, greater precision, and better flexibility in relation to the administered dose. In addition, they can act as a modified release system for biologically active agents, which can be easily removed when necessary (Avachat, Gujar, and Wagh 2013).

Pharmaceutical films have a wide range of applications, being used from polymeric coatings for tablets to final pharmaceutical forms. Soluble films that exhibit rapid disintegration and dissolution can be administered via the buccal, sublingual, ophthalmic, vaginal, or topical routes (Bunhak 2007).

The films can disintegrate immediately in the oral cavity, making the drug available for local action or for absorption in much shorter times, an advantage over conventional dosage forms that require time for disintegration in gastric fluids (Dixit and Puthli 2009). Another feature is that the use of oral films may be preferred over tablets in terms of flexibility and comfort (Mohamed, Haider, and Ali 2011). Furthermore, compared to other release methods, such as oral gels, orally disintegrating films have the advantage of increasing the residence time of the active compound in the oral cavity (Boddupalli 2010). Films for oral administration can be administered to the mucosa, the gums, or the periodontal pocket or sublingually.

Generally speaking, polymers, plasticizers, and active compounds are used for the composition of oral administration films, which can be of natural and synthetic origin, as well as other components, such as saliva stimulating agents, surfactants, sweeteners, and flavoring agents (Aggarwal et al. 2011).

Some studies in the literature have already been carried out using films with bioadhesive and thermoresponsive properties and also for the binding of phytopharmaceutical constituents, such as propolis, for example. Studies using films with ethylcellulose and gelatin were proposed for the incorporation of residues from the preparation of propolis extract to evaluate the development and characterization of the obtained structures (Toledo et al. 2015). Film-forming systems with propolis were also developed for topical use using polymers derived from acrylic acid, such as Carbopol 971P, Carbopol 974P, and polycarbophil, in addition to the use of thermoresponsive polymer, such as poloxamer 407 (P407) (Rosseto et al. 2021).

Films containing bioadhesive polymers, such as polyvinyl alcohol (PVA) and polyvinyl pyrrolidone (PCP), and thermoresponsive ones, such as poloxamer 407, were developed and characterized for the release of synthetic actives, such as metronidazole and methylene blue, this one for the treatment of infections caused by *Candida albicans*. The formulations containing metronidazole showed desirable characteristics for film application on the oral mucosa for fast drug release, enabling the local drug action or absorption during an extended period of time, considering the mucoadhesive properties. The films with methylene blue enabled the active availability for inhibition of the growth of *C. albicans* (Vecchi et al. 2021; Vecchi and Bruschi 2019; Vecchi et al. 2020).

8.3.3.2 Gels and Emulgels

Gels consist of a liquid portion confined within a polymeric matrix. What characterizes the gel itself is this matrix formed by gelling substances. In addition, gels can evaporate and form a thin adhesive film (Prista 2008; Allen, Popovich, and Ansel 2012; Aulton and Taylor 2013). Emulsions are preparations of semisolid consistency, with different viscosities, intended to be applied to the skin or mucous membranes to exert a local action or to provide cutaneous penetration of active ingredients. It is a system of at least two phases: one is dispersed, in the form of small droplets (internal or dispersed phase), and the other is spread (external or continuous phase) (Said dos Santos et al. 2020). They induce greater emolliency and consequent hydration due to one of their phases being oil, besides favoring the absorption of the drug (Prista 2008; Allen, Popovich, and Ansel 2012; Aulton and Taylor 2013).

When emulsions and gels are used in the combined form, they are called emulgels. This gelling ability allows the formation of more stable emulsions by decreasing interfacial tension and increasing the viscosity of the aqueous phase (Khullar et al. 2012). Being intermediate preparations between gels and emulsions, they overcome some disadvantages, such as having thixotropic properties, being easy to spread and easily removable (because they are not greasy), being emollient and water-soluble, having a long shelf life and pleasant appearance, and being pleasant to the touch (Singla et al. 2012).

Semisolid systems are attractive pharmaceutical forms to be used as a modified release system, extending the residence time of the device and improving bioavailability. Some studies in the literature have already been performed using different polymers with bioadhesive and thermoresponsive properties in the development of these systems containing propolis extract and/or curcumin, such as Carbopol 934P® and 974P polymers (Dano et al. 2021; Said dos Santos et al. 2021), Carbopol 971 (Dano et al. 2021), poloxamer 188 (Francisco et al. 2019), poloxamer 407 and/or polycarbophil (PCB) (Souza Ferreira et al. 2020; Said dos Santos et al. 2021), poloxamer 407 and C974 (Ferreira et al. 2020), and poloxamer 407 and Carbopol 934P (Mazia et al. 2016; Pereira et al. 2013), which have bioadhesive and thermoresponsive properties in modifying the release of active agents (Rowe, Sheskey, and Quinn 2009; Bruschi et al. 2007; Souza Ferreira et al. 2015). These systems provide gels, emulgels, and organogels.

8.3.3.3 Liquid Crystalline Systems

Liquid crystals (LC) are systems with properties of both solid crystals and conventional liquids. They show structural order, defined bonds, and rigidity and, at the same time, mobility, fluidity, and disorganized regions. These LC are composed by the mixture of oil phase, surfactant, and aqueous phase. When mixed, they show ordered mesophases with lamellar, hexagonal, or cubic shapes. These mesophases provide interesting characteristics to these drug delivery systems, including bioadhesion, ability to modify the drug release, loading of hydrophobic and hydrophilic drugs, and application in mucosal surfaces, like the oral cavity and vaginal mucosa. Bioadhesive and/or thermoresponsive polymers, such as poloxamer 407 and Carbopol 974P, are added in the aqueous phase (Fonseca-Santos et al. 2016; Ferreira et al. 2021). The oily phase is commonly composed of oleic acid and cholesterol, while the surfactants usually used are polyoxypropylene (5) polyoxyethylene (20) cetyl alcohol and triethanolamine oleate (Santezi et al. 2021; Ferreira et al. 2021; Fonseca-Santos et al. 2016; Toledo, Rosseto, and Bruschi 2018). According to small angle X-ray analysis and polarized light microscopy, it is possible to determine the type of mesophases and determine how the structures interact with the phytochemical constituents, as well as their bioadhesive properties (Fonseca-Santos et al. 2016; Ferreira et al. 2021).

8.3.3.4 Multiparticulate Systems

Multiparticulate systems, such as pellets, have been used mainly for intestinal delivery. They show diverse advantages, including better distribution into the intestinal wall, higher total surface for adhesive bonds, slower transit to colon, and lower risk of systemic drug toxicity due to the absence of internalization by epithelial cells (Desai and Momin 2020). When these pellets are coated with a mucoadhesive layer, their residence time in colonic transit increases their therapeutic efficacy. Desai and Momin (2020) prepared pellets containing curcumin and Carbopol 940 and hydroxypropyl cellulose as bioadhesive polymers. These materials promoted great swelling ability, strong adhesion, improved permeability, and sustained release pattern (Desai and Momin 2020).

8.3.3.5 Nanotechnology

Diverse systems were prepared using nanotechnology to facilitate the therapeutic utilization of curcumin, propolis, gallic acid, and p-coumaric acid. These systems include micro-/nanoparticles with diverse composition, liposomes and niosomes, micro-/nanogels, and microemulsions (Rosseto et al. 2017; Francisco et al. 2019; Villa Nova et al. 2019; Chuah et al. 2013; Sanoj Rejinold et al. 2011; Oliveira et al. 2018; Toledo, Rosseto, and Bruschi 2018; Berginc et al. 2014; Manosroi et al. 2011; Kulkarni et al. 2021; Mitsou et al. 2020). These systems fabricated by the nanotechnology approach are displayed in Figure 8.8.

According to micro-/nanoparticle composition, they are classified in polymeric, solid lipid, and magnetic. Polymeric micro-/nanoparticles have been used to carry propolis and curcumin (Belloto de Francisco et al. 2016; Oliveira et al. 2018). They are defined as solid drug carriers, biodegradable or not depending on the used polymer or natural material used to prepare these systems. For example, propolis by-product was already used as an excipient in nanoparticles and also shown antioxidant and immunostimulant activity (Belloto de Francisco et al. 2016; Villa Nova et al. 2019). Among the advantages, the small size has been reported to facilitate the passage through biological membranes (endothelium, tumors, epithelium of gastrointestinal tract). They are considered the most suitable systems for injectable administration and their biodegradability may prolong the drug release (Bruschi, De Francisco, and Villa Nova 2019).

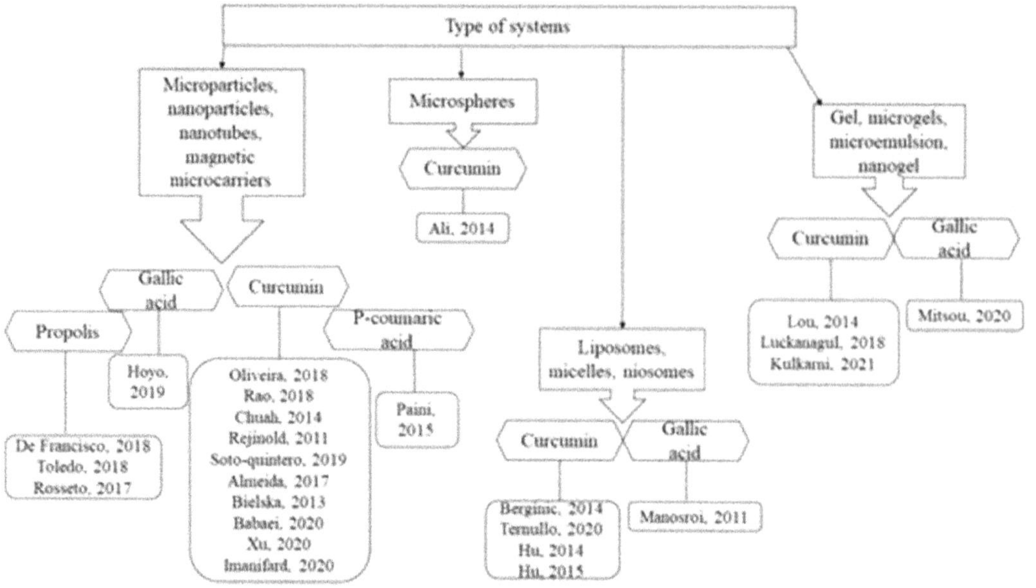

FIGURE 8.8 Main types of systems prepared by nanotechnological approach.

On the other hand, solid lipid nanoparticles consist of emulsifiers and a solid lipid or a mixture of solid lipids, which improves the kinetic release and stability due to the solid state. The literature has reported the use of compatible lipids with the human body to improve safety (Bruschi, Rosseto, and de Toledo 2019). Rosseto and colleagues (2017) prepared solid lipid nanoparticles containing propolis by-product and propolis extract. The presence of this propolis blend significantly affected the properties of these formulations, regarding drug loading and release, size, and morphology (Rosseto et al. 2017). In addition, magnetic nanoparticles are prepared with diverse metal oxides, like nickel, cobalt, and copper and are affected by an external magnetic field. They show good biocompatibility with the human body. Iron oxide magnetic nanoparticles have been used to carry propolis and propolis by-products. These systems showed great physicochemical properties and slow release of propolis (Toledo, Rosseto, and Bruschi 2018). Two studies reported the use of curcumin incorporated in magnetic nanoparticles with thermoresponsive shells. Imanifard and colleagues (2017 investigated the use of iron nanoparticles coated with poly(N-isopropylacrylamide) and found good encapsulation efficiency (83%), temperature-dependent drug release for cancer treatment (Imanifard et al. 2017). Soto-Quintero et al. (2019) incorporated curcumin in silver nanoparticles coated with thermoresponsive polymer. These formulations were easily fabricated and showed promising biomedical applications, mainly antimicrobials (Soto-Quintero et al. 2019).

Because of their lipophilic characteristics, liposomes are usually used to incorporate poorly soluble drugs like curcumin. Liposomes increase their solubility, allow sustained drug release, enable administration in aqueous medium. Ternullo and colleagues (2020) prepared liposomes containing curcumin, which were incorporated in chitosan hydrogels. The positive charges of liposomes helped in the bioadhesive characteristics of chitosan and in the *ex vivo* permeability tests (Ternullo et al. 2020). Berginic et al. (2014) studied the preparation of liposomes containing curcumin coated with chitosan or Carbopol 974P for vaginal administration. These formulations demonstrated superior mucoadhesion, drug delivery and permeability (Berginc et al. 2014).

Niosomes are surfactant-based systems structured by the self-assembly in bilayer vesicles. They are able to carry both hydrophilic and lipophilic substances with higher stability and are cheaper than liposomes. Manosroi (2011) reported the encapsulation of gallic acid in niosomes, which were incorporated in Carbopol 980-gel for antioxidant and anti-wrinkle activity and showed interesting characteristics for skin aging (Manosroi et al. 2011).

Micro-/nanogels are hydrogels confined into a micro-/nanoscopic size, presenting good permeability, including blood-brain barrier permeability, controlled release of drugs, high drug loading, and biocompatibility (Bruschi et al. 2019). These systems were used to incorporate curcumin with thermoresponsive and/or bioadhesive properties. Kulkarni and co-authors (2021) prepared poloxamer-based microgels with curcumin for breast cancer. These studies showed modified release of curcumin and promising properties as carriers for curcumin in breast cancer (Kulkarni et al. 2021). Luckanagul and colleagues (2018) developed nanogels containing chitosan and poly(N-isopropylacrylamide) to carry curcumin with potential delivery for low-soluble drugs, with safety and activity (Luckanagul et al. 2018).

Microemulsions consist of a thermodynamically stable system containing water, oil phase, and a mixture of surfactants and co-surfactants. Their characteristics enable the application in diverse sites, including nasal, ocular, dermal, and oral routes. Mitsou and co-authors (2020) investigated microemulsions containing gallic acid as nasal carriers and achieved mucoadhesion, including chitosan, in the aqueous phase. These systems showed biocompatibility, improved permeability, and maintained their antioxidant activity even after the incorporation in microemulsions (Mitsou et al. 2020).

Polymeric micelles containing amphiphilic block copolymers are widely used because of stable storage, increased therapeutic efficacy, and lower side effects. Due to the hydrophobic segments of these block copolymers, curcumin has been incorporated into well-studied polymers like poloxamer 407 but also in new modified synthetic polymers. These systems showed improved stability due to lower critical micelle concentration, good drug loading, and temperature-dependent release (Braga et al. 2020; Hu et al. 2014).

8.4 TOXICITY AND ACTIVITY

Polyphenols are secondary metabolic compounds of plants, and due to their wide range of structures, they also show diverse biological activities (e.g., antioxidant, antimicrobial, and anti-inflammatory). Considering the antioxidant activity, these compounds show potential properties to combat cardiovascular disorders, cancer, and diabetes due to the activity against reactive oxygen species from endogenous or exogenous metabolism and limiting oxidative stress (Paini et al. 2015). Consequently, it is important to investigate these properties and the safety when the polyphenols are incorporated in the formulations with bioadhesive and/or thermoresponsive properties.

8.4.1 In Vivo Evaluation

Many *in vivo* studies have been conducted with polyphenolic compounds, and its natural products incorporated in thermoresponsive and/or bioadhesive formulations. Most of them include curcumin (Lou et al. 2014; Fonseca-Santos et al. 2016; Desai and Momin 2020; Xu et al. 2020), propolis (Bonfim et al. 2020; Rosseto et al. 2017), gallic acid (Manosroi et al. 2011), and p-coumaric acid (Ferreira et al. 2021).

The *in vivo* studies containing propolis extract and/or by-product consisted of a preclinical vulvovaginal candidiasis in a murine model and wound closure test in a mouse model (Bonfim et al. 2020; Rosseto et al. 2017). Bonfim and co-authors tested a bioadhesive, thermoresponsive formulation containing propolis extract, poloxamer 407, and Carbopol 934P against vulvovaginal candidiasis. Their results support the use of propolis formulation as a potential agent to combat vulvovaginal candidiasis with superior results than nystatin (Bonfim et al. 2020). Rosseto and colleagues (2017) evaluated the wound closure of mice using a solid lipid nanoparticle formulation containing propolis and propolis by-product, and their system could improve the wound healing process (Rosseto et al. 2017).

Curcumin shows a wide range of biological properties. Consequently some *in vivo* studies were found in literature of formulations containing bioadhesive and/or thermoresponsive systems and curcumin. These *in vivo* studies include inflammatory, vulvovaginal candidiasis, ulcerative colitis, ocular irritation, and cancer models (Fonseca-Santos et al. 2016; Desai and Momin 2020; Lou et al. 2014; Xu et al. 2020). Fonseca-Santos et al. (2016) developed liquid crystal systems containing curcumin and investigated its anti-inflammatory effects *in vivo* in a Swiss mice model. It was found to have important and comparable anti-inflammatory activity with dexamethasone-commercial cream, which demonstrates the utility of incorporating curcumin in formulations for skin diseases (Fonseca-Santos et al. 2016). Desai and Momin (2020) developed multiparticulate systems containing curcumin and cyclosporine with Carbopol 940 and hydroxypropyl cellulose as mucoadhesive polymers for ulcerative colitis in male Wistar rat models. The outcomes revealed that the two drugs could show a synergism, and it was necessary to use lower doses to achieve the therapeutic effects in comparison with the administration of individual drugs (Desai and Momin 2020). Lou and co-authors (2014) studied nanoparticles consisting of curcumin with thermoresponsive characteristics (poloxamer 407) for diabetic retinopathy. They performed eye irritation tests in New Zealand white rabbits, the formulations were considered safe for eye administration, and the formulation improved the bioavailability of curcumin in rabbits' eyes (Lou et al. 2014). Xu et al. (2020) developed thermoresponsive nanoparticles containing curcumin and tested their efficacy in tumor models in mice. This study showed promising results for tumor application (Xu et al. 2020).

The *in vivo* study involving gallic acid was performed in rabbits. Manosroi and colleagues (2011) prepared gels containing niosomes incorporated with gallic acid for skin anti-aging. The irritation test was performed in rabbits through the closet patch test, and no skin irritation was observed (Manosroi et al. 2011).

Liquid crystalline systems containing p-coumaric acid were developed for treatment of vulvogavinal candidiasis and the *in vivo* activity was evaluated in an experimental mouse model. The test revealed lower fungal infection and clinical symptoms of vaginal yeast infection (Ferreira et al. 2021).

8.4.2 In Vitro Evaluation

Diverse *in vitro* studies were carried out in phytopharmaceutical compounds, especially propolis, curcumin, and gallic acid, which were incorporated into formulations with bioadhesive, thermoresponsive characteristics. Most of them involve cytotoxicity in healthy and diseased cell lines (intestinal cancer, breast cancer, oral cancer, prostate cancer), antioxidants, *Artemia salina* assay, susceptibility tests against *Candida albicans*, and *Streptococcus mutans* biofilm formation. This data is displayed in Table 8.2.

8.4.3 Clinical Trials

Diverse biological activities and a huge number of formulations were developed with curcumin, propolis, p-coumaric acid, and gallic acid, with bioadhesive, thermoresponsive characteristics (Ferreira et al. 2020; Francisco et al. 2019; Rosseto et al. 2017; Oliveira et al. 2018; Mitsou et al. 2020; Bonfim et al. 2020; Hoyo et al. 2019; Santezi et al. 2021). However, few of them were evaluated by *in vivo* tests, and to our knowledge, only one formulation with bioadhesive and/or thermoresponsive properties was tested in clinical trials. For example, Manosroi and colleagues (2011) studied gels containing niosomes incorporated with gallic acid for skin anti-aging. The anti-aging was evaluated by measuring skin elasticity, skin surface microstructure, skin hydration and skin erythema and pigmentation in 31 human volunteers and showed improvement on skin elastic and roughness and could be applied as an anti-wrinkle product (Manosroi et al. 2011).

TABLE 8.2
***In vitro* Studies of Propolis, Curcumin, and Gallic Acid Incorporated in Bioadhesive and/or Thermoresponsive Formulations**

System	Phytopharmaceutical	Test Type	Reference(s)
Microparticles	Propolis by-product	Antioxidant (DPPH, FRAP, ABTS) Cell culture assays (Caco-2 and HT29-MTX cell lines)	(Francisco et al. 2018a)
Magnetic nanoparticles	Propolis extract Propolis by-product	Cytotoxicity study (*Artemia salina*) Cytotoxicity study using fibroblasts	(Toledo, Rosseto, and Bruschi 2018)
Nanostructured lipid systems	Propolis extract Propolis by-product	Cellular viability (keratinocytes—HaCaT cells)	(Rosseto et al. 2017)
Microparticles	Curcumin	Antioxidant (ABTS) cytotoxicity study (*Artemia salina*, fibroblasts)	(Oliveira et al. 2018)
Nanoparticles	Curcumin	Cytotoxicity study (MCF-7 cancer cells)	(Rao et al. 2018)
Nanoparticles	Curcumin	MTT cell viability (human colorectal adenocarcinoma—HT-29)	(Chuah et al. 2013)
Nanoparticles	Gallic acid	Cytotoxicity study (fibroblasts)	(Salat et al. 2018)
Microemulsion	Gallic acid	Antioxidant (galvinoxyl free radical) MTT cell viability (RPMI 2650—CCL-30)	(Mitsou et al. 2020)
ZnO Nanoparticles	Gallic acid	Antioxidant (DPPH) Cytotoxicity Study (fibroblasts and keratinocytes—HaCat cells)	(Hoyo et al. 2019)
Niosomes	Gallic acid	Antioxidant (DPPH) cytotoxicity study (fibroblasts)	(Manosroi et al. 2011)
Nanoparticles	Curcumin	Cytotoxicity study (L929, MCF7, PC3, and KB)	(Sanoj Rejinold et al. 2011)
Gel	Curcumin	Cytotoxicity study via MTT assay (MDA-MB-231 cell line)	(Kulkarni et al. 2021)
Nanogel	Curcumin	Cytotoxicity study via MTT assay (MDA-MB-231, Caco-2, HepG2, HT-29)	(Luckanagul et al. 2018)
Micelles	Curcumin	Cytotoxicity study via MTT assay (murine fibrosarcoma—L929)	(Hu et al. 2014)
Magnetic microcarriers	Curcumin	Cytotoxicity study from the sulforhodamine B method (VERO cells and Caco-2 cells)	(Almeida et al. 2017)
Magnetic nanoparticles	Curcumin	Cytotoxicity Study via MTT assay (Human breast cancer cells, MCF-7)	(Imanifard et al. 2017)
Nanoparticles	Curcumin	Cytotoxicity study via MTT assay (Eca-109, HepG2 and LoVo cells)	(Xu et al. 2020)
Gel	Propolis extract	Susceptibility test	(Bonfim et al. 2020)
Gel	Curcumin	Cytotoxicity study via MTT assay (FNB6, FaDu and Cal27)	(Ferreira et al. 2019)
Emulgel	Curcumin	Cytotoxicity study via MTT assay (FNB6, FaDu and Cal27)—monolayer cultures Cytotoxicity study via MTT assay (FNB6 and Cal27)—tissue-engineered models	(Ferreira et al. 2020)
Liquid crystalline films	Curcumin	Biofilm formation	(Santezi et al. 2021)

8.5 PATENTS

Patents were reviewed and categorized, focusing the applications and pharmaceutical dosage forms using this innovative strategy of bioadhesive or thermoresponsive systems containing one of the phytochemicals compounds.

Table 8.3 considers mostly the patent literature on the application of the phytochemicals compounds, related to thermoresponsive and mucoadhesive systems for pharmaceutical and biomedical products.

TABLE 8.3
Patents of Formulations Containing Phytopharmaceuticals with Bioadhesive and/or Thermoresponsive Properties

Patent Number	Title	Major Invention
Curcumin		
US20170216221A1 (Sabliov et al. 2017)	Delivery of bioactive, nanoencapsulated antioxidants	Nanoparticles containing a protein such as zein or a polymer such as poly(lactic-co-glycolic acid) to deliver lutein or another antioxidant, such as CUR, to target tissues, such as the eye, in bioactive form being released for a long time while protecting the antioxidant from degradation.
EP2814332B1 (Salman, Azcárate, and Catalán 2017)	Nanoparticles comprising a vegetable hydrophobic protein and a water miscible non-volatile organic solvent and uses thereof	The invention relates to a biocompatible nanoparticulate delivery system comprising nanoparticles based on vegetable hydrophobic proteins, particularly zein, and water-miscible non-volatile organic solvents having high mucosal bioadhesion capacity. The invention also relates to an *in situ* self-assembly method that allows the spontaneous formation of said nanoparticles and to the uses and applications thereof.
Gallic Acid		
EP2625966A1 (Salman, Azcàrate, and Catalán 2018)	Nanoparticles comprising a vegetable hydrophobic protein and a water miscible non-volatile organic solvent and uses thereof	The invention relates to a biocompatible nanoparticulate delivery system comprising nanoparticles based on vegetable hydrophobic proteins, particularly zein, and water-miscible non-volatile organic solvents having high mucosal bioadhesion capacity. The invention also relates to an *in situ* self-assembly method that allows the spontaneous formation of said nanoparticles and to the uses and applications thereof.
CN106511259A US9592295B2 (EN) (Lu, Lu, and Liu 2016)	Reverse temperature-sensitive reversible hydrogel composition	The invention relates to a reverse temperature-sensitive reversible hydrogel composition comprising a water-soluble block copolymer and at least one associated gel adjuvant. The hydrogel composition shows improved gel efficiency and improvement on solubleness and/or stability of a slightly-water-soluble and/or non-water-soluble medicinal preparation(s). The hydrogel composition can be applied in various medicines and cosmetic products and applications such as the treatment of esophageal, ear, vaginal, rectal, ophthalmic, and dermatological diseases and deficiencies.

TABLE 8.3 *(Continued)*
Patents of Formulations Containing Phytopharmaceuticals with Bioadhesive and/or Thermoresponsive Properties

Patent Number	Title	Major Invention
p-Coumaric Acid		
ES2298618T3 (Andruskova et al. 2008)	Pharmaceutical compositions that include flavonoids and menthol	A pharmaceutical composition comprising (1) one or more purified flavonoids, such as p-coumaric acid; (2) purified menthol; and (3) pharmaceutically acceptable excipients.
CA2948357A1 (Murphy et al. 2015)	Bioadhesive compounds and methods of synthesis and use	Synthesis methods for creating polymeric compounds comprising phenyl derivatives (PD), such as p-coumaric acid, or PDp i.e., polymers modified with PD, with desired surface-active effects are described. The polymer backbone of PDp has structural or performance features that can be tailored to control physical properties of PDp, allowing it to be useful for different applications, i.e., tissue adhesives or sealants, adhesion-promoting coatings, and antifouling coatings.
Propolis		
BR102019007253A2 (Luciana et al. 2020)	Process of production of thermoresponsive photo hydrogel of phthalocyanes for topical application in antimicrobial photodynamic therapy and their hydrogel	The present invention consists of the development (processing and production) of formulations based on photoactive hydrogel with mucoadhesive-thermoresponsive phthalocyanines, using techniques for the preparation of gels containing propolis.
JP2015013889A EP 2395997 A1 20111221 (EN) (Attali et al. 2009)	Treating inflammation and inflammatory pain in mucosa using mucosal prolonged-release bioadhesive therapeutic carriers	Provided is a pharmaceutical composition for the prevention or treatment of inflammatory pains and diseases in a mucosa of the oral cavity, pharynx, and/or larynx, comprising an alpha-2 adrenergic receptor agonist, and propolis is a preservative that can be used in the present invention.

8.6 CONCLUDING REMARKS

The bioadhesive and/or thermoresponsive characteristics of formulations demonstrate huge importance to increase the residence time, bioavailability of drugs with low stability, solubility, and difficult administration, such as curcumin, gallic acid, p-coumaric acid and propolis. Therefore, the development of strategies combining these characteristics or using specific drug delivery systems, such as nanoparticles, films, microemulsions, and gels is quite important. New technologies may facilitate the scale up process and the market of these products, and they may result in more efficient, safer and individualized therapies. However, much has to be done in terms of *in vivo* biological tests, clinical trials to improve the number of patents, their commercialization, and the emergence of new products on the market.

REFERENCES

Aggarwal J, Singh G, Saini S, and Rana A C. Fast Dissolving Films: A Novel Approach to Oral Drug Delivery. *International Research Journal of Pharmacy*. 2011, 2 (12): 69–71.

Alexandridis P, Holzwarth J F, and Hatton T A. Micellization of Poly (Ethylene Oxide)-Poly (Propylene Oxide)-Poly (Ethylene Oxide) Triblock Copolymers in Aqueous Solutions: Thermodynamics of Copolymer Association Solutions: Thermodynamics of Copolymer Association. *Macromolecules*. 1994, 27: 2414–2425. https://doi.org/10.1021/ma00087a009.

Ali M S, Pandit V, Jain M, and Dhar K L. Mucoadhesive Microparticulate Drug Delivery System of Curcumin against Helicobacter Pylori Infection: Design, Development and Optimization. *Journal of Advanced Pharmaceutical Technology and Research*. 2014, 5 (1): 48–56. https://doi.org/10.4103/2231-4040.126996.

Allen L V, Popovich N G, and Ansel H C. 2012. Ansel's Pharmaceutical Dosage Forms and Drug Delivery Systems. In *Ansel's Pharmaceutical Dosage Forms and Drug Delivery Systems: Ninth Edition*. 9th ed. https://doi.org/10.1177/875512251002600315.

Almeida E A M S, Bellettini I C, Garcia F P, Farinácio M T, Nakamura C V, Rubira A F, Martins A F, and Muniz E C. Curcumin-Loaded Dual pH- and Thermo-Responsive Magnetic Microcarriers Based on Pectin Maleate for Drug Delivery. *Carbohydrate Polymers*. 2017, 171: 259–266. https://doi.org/10.1016/j.carbpol.2017.05.034.

Almeida H, Amaral M H, Lobão P, and Lobo J M S. Applications of Poloxamers in Ophthalmic Pharmaceutical Formulations: An Overview. *Expert Opinion on Drug Delivery*. 2013, 10 (9): 1223–1237. https://doi.org/10.1517/17425247.2013.796360.

Almeida H, Amaral M H, Lobão P, and Lobo J M S. In Situ Gelling Systems: A Strategy to Improve the Bioavailability of Ophthalmic Pharmaceutical Formulations. *Drug Discovery Today*. 2013, 00 (00). https://doi.org/10.1016/j.drudis.2013.10.001.

Almeida M, Reis R L, and Silva T H. Marine Invertebrates Are a Source of Bioadhesives with Biomimetic Interest. *Materials Science and Engineering C*. 2020, 108 (September 2019): 110467. https://doi.org/10.1016/j.msec.2019.110467.

Altunbas A, Lee S J, Rajasekaran S A, Schneider J P, and Pochan D J. Encapsulation of Curcumin in Self-Assembling Peptide Hydrogels as Injectable Drug Delivery Vehicles. *Biomaterials*. 2011, 32 (25): 5906–5914. https://doi.org/10.1016/j.biomaterials.2011.04.069.

Alves R C, Fernandes R P, Fonseca-Santos B, Victorelli F D, and Chorilli M. A Critical Review of the Properties and Analytical Methods for the Determination of Curcumin in Biological and Pharmaceutical Matrices. *Critical Reviews in Analytical Chemistry*. 2018. https://doi.org/10.1080/10408347.2018.1489216.

Andrews G P, Laverty T P, and Jones D S. Mucoadhesive Polymeric Platforms for Controlled Drug Delivery. *European Journal of Pharmaceutics and Biopharmaceutics*. 2009, 71 (3): 505–518. https://doi.org/10.1016/j.ejpb.2008.09.028.

Andruskova V, Uhlar J, Lehocky P, and Horak J. 2008. ES2298618T3—Composiciones farmacéuticas que comprenden flavonoides y mentol. ES 2 298 618 T3, issued 2008.

Araiza-Calahorra A, Akhtar M, and Sarkar A. Recent Advances in Emulsion-Based Delivery Approaches for Curcumin: From Encapsulation to Bioaccessibility. *Trends in Food Science and Technology*. 2018, 71: 155–169. https://doi.org/10.1016/j.tifs.2017.11.009.

Asane G S, Nirmal S A, Rasal K B, Naik A A, Mahadik M S, and Rao Y M. Polymers for Mucoadhesive Drug Delivery System: A Current Status. *Drug Development and Industrial Pharmacy*. 2008, 34 (11): 1246–1266. https://doi.org/10.1080/03639040802026012.

Attali P, Zakin C L, Roulet V, Scheuir C, and Zakin L. 2009. Treating of inflammation and inflammatory pain in mucosa using mucosal prolonged release bioadhesive therapeutic carriers, issued 2009.

Aulton M, and Taylor K M G. 2013. Aulton Delineamento de Formas Farmacêuticas. In *Pharmaceutics: The Science of Dosage Form Design*. 4th ed. https://doi.org/10.1016/0168-3659(89)90050-3.

Avachat A M, Gujar K N, and Wagh K V. Development and Evaluation of Tamarind Seed Xyloglucan-Based Mucoadhesive Buccal Films of Rizatriptan Benzoate. *Carbohydrate Polymers*. 2013, 91 (2): 537–542. https://doi.org/10.1016/j.carbpol.2012.08.062.

Babaei M, Davoodi J, Dehghan R, Zahiri M, Abnous K, Taghdisi M S, Ramezani M, and Alibolandi M. Thermosensitive Composite Hydrogel Incorporated with Curcumin-Loaded Nanopolymersomes for Prolonged and Localized Treatment of Glioma. *Journal of Drug Delivery Science and Technology*. 2020, 59 (July): 101885. https://doi.org/10.1016/j.jddst.2020.101885.

Bala R, Khanna S, Pawar P, and Arora S. Orally Dissolving Strips: A New Approach to Oral Drug Delivery System. *International Journal of Pharmaceutical Investigation*. 2013, 3 (2): 67. https://doi.org/10.4103/2230-973x.114897.

Bankova V. Chemical Diversity of Propolis and the Problem of Standardization. *Journal of Ethnopharmacology*. 2005, 100 (1–2): 114–117. https://doi.org/10.1016/j.jep.2005.05.004.

Bassi da Silva J, Khutoryanskiy V V, Bruschi M L, and Cook M T. A Mucosa-Mimetic Material for the Mucoadhesion Testing of Thermogelling Semi-Solids. *International Journal of Pharmaceutics*. 2017, 528 (1–2): 586–594. https://doi.org/10.1016/j.ijpharm.2017.06.025.

Batista J A, Tanada-palmu P S, and Grosso C R F. Efeito da adição de ácidos graxos em filmes à base de pectina1.Pdf. *Ciência e Tecnologia Alimentos*. 2005, 25 (4): 781–788.

Belloto de Francisco L M, Costa Y T C, Outuki P M, Souza R P, Mendonca P S B, Novello C R, Consolaro M E L, and Bruschi M L. Nanoparticles of Waste Material of Propolis and Gelatin as a Novel System for Delivery of L-Ascorbic Acid. *Current Drug Delivery*. 2016, 14 (7). https://doi.org/10.2174/1570159x14666160322144315.

Berginc K, Suljaković S, Škalko-Basnet N, and Kristl A. Mucoadhesive Liposomes as New Formulation for Vaginal Delivery of Curcumin. *European Journal of Pharmaceutics and Biopharmaceutics*. 2014, 87 (1): 40–46. https://doi.org/10.1016/j.ejpb.2014.02.006.

Bhowmik M, Kumari P, Sarkar G, Bain M K, Bhowmick B, Mollick M D R, Mondal D, et al. Effect of Xanthan Gum and Guar Gum on In Situ Gelling Ophthalmic Drug Delivery System Based on Poloxamer-407. *International Journal of Biological Macromolecules*. 2013, 62 (November): 117–123. https://doi.org/10.1016/J.IJBIOMAC.2013.08.024.

Bielska D, Karewicz A, Kamiński K, Kiełkowicz I, Lachowicz T, Szczubiałka K, and Nowakowska M. Self-Organized Thermo-Responsive Hydroxypropyl Cellulose Nanoparticles for Curcumin Delivery. *European Polymer Journal*. 2013, 49 (9): 2485–2494. https://doi.org/10.1016/j.eurpolymj.2013.02.012.

Boddupalli B M. Mucoadhesive Drug Delivery System: An Overview. *Journal of Advenced Pharmaceutical Technology and Research*. 2010, 1. https://doi.org/10.4103/0110-5558.76436.

Bonacucina G, Martelli S, and Palmieri G F. Rheological, Mucoadhesive and Release Properties of Carbopol Gels in Hydrophilic Cosolvents. *International Journal of Pharmaceutics*. 2004, 282 (1–2): 115–130. https://doi.org/10.1016/j.ijpharm.2004.06.012.

Bonfim A P, Sakita K M, Faria D R, Arita G S, Vendramini F A V R, Capoci I R G, Braga A G, et al. Preclinical Approaches in Vulvovaginal Candidiasis Treatment with Mucoadhesive Thermoresponsive Systems Containing Propolis. *PLoS One*. 2020, 15 (12 December): 1–20. https://doi.org/10.1371/journal.pone.0243197.

Borghi-Pangoni F B, Junqueira M V, de Souza Ferreira S B, Silva L L, Rabello B R, de Castro L V, Baesso M L, Diniz A, Caetano W, and Bruschi M L. Preparation and Characterization of Bioadhesive System Containing Hypericin for Local Photodynamic Therapy. *Photodiagnosis and Photodynamic Therapy*. 2017, 19. https://doi.org/10.1016/j.pdpdt.2017.06.016.

Borghi-Pangoni F B, Junqueira M V, Ferreira S B S, Silva L L, Rabello B R, Caetano W, Diniz A, and Bruschi M L. Screening and In Vitro Evaluation of Mucoadhesive Thermoresponsive System Containing Methylene Blue for Local Photodynamic Therapy of Colorectal Cancer. *Pharmaceutical Research*. 2016, 33 (3): 776–791. https://doi.org/10.1007/s11095-015-1826-8.

Braga G, Campanholi K S S, Ferreira S B S, Calori I R, Oliveira J H, Vanzin D, Bruschi M L, et al. Tautomeric and Aggregational Dynamics of Curcumin-Supersaturated Pluronic Nanocarriers. *ACS Applied Polymer Materials*. 2020, 2 (11): 4493–511. https://doi.org/10.1021/acsapm.0c00589.

Bruschi M L, Borghi-Pangoni F B, da Silva J B, Junqueira M V, and Ferreira S B S. 2019. Nanogels: Preparation, Applications, and Clinical Considerations. In *Nanodispersions for Drug Delivery*, edited by Raj K. Reservani and Anil K. Sharma, 1st ed., 277–316. Oakville: Apple Academic Press.

Bruschi M L, Borghi-Pangoni F B, Junqueira M V, and Ferreira S B S. 2017. Nanostructured Therapeutic Systems with Bioadhesive and Thermoresponsive Properties. In *Nanostructures for Noveltherapy*, 313–342. Amsterdam: Elsevier.

Bruschi M L, De Francisco L M B, and Villa Nova M. Polymeric Nanoparticles: General Features, Polymers, and Formulation Aspects. *Nanoparticulate Drug Delivery Systems*, 2019, 3–36. https://doi.org/10.1201/9781351137263-1.

Bruschi M L, Ferreira S B S, and da Silva J B. 2020. Mucoadhesive and Mucus-Penetrating Polymers for Drug Delivery. In *Nanotechnology for Oral Drug Delivery: From Concept to Applications*, edited by João Pedro Martins and Hélder A. Santos, 1st ed., 77–123. London: Academic Press.

Bruschi M L, Franco S L, and Gremião M P D. Application of an HPLC Method for Analysis of Propolis Extract. *Journal of Liquid Chromatography & Related Technologies*. 2003, 26 (14): 2399–2409. https://doi.org/10.1081/JLC-120023254.

Bruschi M L, Jones D S, Panzeri H, Gremião M P D, de Freitas O, and Lara E H G. Semisolid Systems Containing Propolis for the Treatment of Periodontal Disease: In Vitro Release Kinetics, Syringeability, Rheological, Textural, and Mucoadhesive Properties. *Journal of Pharmaeutical Sciences*. 2007, 96 (8): 2074–2089. https://doi.org/10.1002/jps.

Bruschi M L, Klein T, Lopes R S, Franco S L, and Gremião M P D. Contribuição Ao Protocolo de Controle de Qualidade Da Própolis e de Seus Extratos. *Revista de Ciencias Farmaceuticas*. 2002, 23 (2): 289–306.

Bruschi M L, Lara E H G, Martins C H G, Vinholis A H C, Casemiro L A, Panzeri H, and Gremião M P D. Preparation and Antimicrobial Activity of Gelatin Microparticles Containing Propolis against

Oral Pathogens. *Drug Development and Industrial Pharmacy*. 2006, 32 (2): 229–238. https://doi.org/10.1080/03639040500466312.

Bruschi M L, Panzeri H, De Freitas O, Lara E H G, and Gremião M P D. Sistemas de Liberação de Fármaco Intrabolsa Periodontal. *Revista Brasileira de Ciências Farmacêuticas*. 2006, 42 (1): 29–47. https://doi.org/10.1590/S1516-93322006000100004.

Bruschi M L, Rosseto H C, and Toledo L A S. 2019. Solid Lipid Nanoparticles: General Aspects, Preparation Methods, and Applications in Drug Delivery. In *Nanoparticulate Drug Delivery Systems*, edited by Raj K. Keservani and Anil K. Sharma, 1st ed., 315–335. Oakville: Apple Academic Press.

Budavari S, O'neil M J, Smith A, Heckelman P E, Obenchain Jr J R, Gallipeau J A R, and D'Arecea M A. 1996. *The Merck Index: An Encyclopedia of Chemicals, Drugs, and Biologicals*, Vol. 450. Whitehouse Station, NJ: Merck & Co. Inc.

Bunhak E J. Influencia Do Sulfato de Condroitina Na Formação de Filmes Isolados de Polimetacriato: Avaliação Do Índice de Intumescimento e Permeabilidade Ao Capor d'água. *Química Nova*. 2007, 30 (2): 312–317.

Burdock G A. Review of the Biological Properties and Toxicity of Bee Propolis (Propolis). *Food and Chemical Toxicology*. 1998, 36 (4): 347–363. https://doi.org/10.1016/S0278-6915(97)00145-2.

Buriol L, Finger D, Schmidt E M, dos Santos J M T, da Rosa M R, Quináia S P, Torres Y R, Santa H S D, Pessoa C, and de Moraes M O. Composição Química e Atividade Biológica de Extrato Oleoso de Própolis: Uma Alternativa Ao Extrato Etanólico. *Química Nova*. 2009, 32: 296–302.

Cabral I S R, Oldoni T L C, de Alencar S M, Rosalen P L, and Ikegaki M. The Correlation between the Phenolic Composition and Biological Activities of Two Varieties of Brazilian Propolis (G6 and G12). *Brazilian Journal of Pharmaceutical Sciences*. 2012, 48 (3): 557–564. https://doi.org/10.1590/S1984-82502012000300023.

Carvalho F C, Calixto G, Hatakeyama I N, Luz G M, Gremião M P D, and Chorilli M. Rheological, Mechanical, and Bioadhesive Behavior of Hydrogels to Optimize Skin Delivery Systems. *Drug Development and Industrial Pharmacy*. 2013, 29 (11): 1750–1757. https://doi.org/10.3109/03639045.2012.734510.

Carvalho F C, Chorilli M, and Gremião M P D. Plataformas Bio(Muco) Adesivas Poliméricas Baseadas Em Nanotecnologia Para Liberação Controlada de Fármacos—Propriedades, Metodologias e Aplicações. *Polimeros*. 2014, 24 (2): 203–213. https://doi.org/10.4322/polimeros.2014.043.

Carvalho F C, Evagelista R C, Gremião M P D, and Bruschi M L. Mucoadhesive Drug Delivery Systems. *Brazilian Journal of Pharmaceutical Sciences*. 2010, 46 (1): 1–18. https://doi.org/10.3109/03639045.2011.623355.

Chaillou L L, and Nazareno M A. New Method to Determine Antioxidant Activity of Polyphenols. *Journal of Agricultural and Food Chemistry*. 2006, 54 (22): 8397–8402. https://doi.org/10.1021/jf061729f.

Chang Y, Oh J Y K, Kong H S, Kim E J, Jang D D, Nam K T, and Kim C-K. Prolonged Antifungal Effects of Clotrimazole-Containing Mucoadhesive Thermosensitive Gels on Vaginitis. *Journal of Controlled Release*. 2002, 82 (1): 39–50. https://doi.org/10.1016/S0168-3659(02)00086-X.

Chawla V, and Saraf S A. Rheological Studies on Solid Lipid Nanoparticle Based Carbopol Gels of Aceclofenac. *Colloids and Surfaces B: Biointerfaces*. 2012, 92: 293–298. https://doi.org/10.1016/j.colsurfb.2011.12.006.

Chuah L H, Billa N, Roberts C J, Burley J C, and Manickam S. Curcumin-Containing Chitosan Nanoparticles as a Potential Mucoadhesive Delivery System to the Colon. *Pharmaceutical Development and Technology*. 2013, 18 (3): 591–599. https://doi.org/10.3109/10837450.2011.640688.

Dano M E L, Said dos Santos R, Bassi da Silva J, Junqueira M V, de Souza Ferreira S B, and Bruschi M L. Design of Emulgel Platforms for Local Propolis Delivery: The Influence of Type and Concentration of Carbomer. *Journal of Molecular Liquids*. 2021, 334: 116025. https://doi.org/10.1016/j.molliq.2021.116025.

Das R K, Kasoju N, and Bora U. Encapsulation of Curcumin in Alginate-Chitosan-Pluronic Composite Nanoparticles for Delivery to Cancer Cells. *Nanomedicine: Nanotechnology,Biology,and Medicine*. 2010, 6 (1): 153–160. https://doi.org/10.1016/j.nano.2009.05.009.

Desai N, and Momin M. Colon Targeted Bioadhesive Pellets of Curcumin and Cyclosporine for Improved Management of Inflammatory Bowel Disease. *Drug Delivery and Translational Research*. 2020, 10 (5): 1288–1301. https://doi.org/10.1007/s13346-020-00756-x.

Dias Pereira C I, de Freitas C F, Braga T L, Braga G, Gonçalves R S, Tessaro A L, Mikcha J M G, Hioka N, and Caetano W. Concise Behavior of Curcumin in Water-Ethanol: Critical Water Aggregation Percentage and Multivariate Analysis of Protolytic Equilibria. *Dyes and Pigments*. 2022, 197 (May 2021). https://doi.org/10.1016/j.dyepig.2021.109887.

Dixit R P, and Puthli S P. Oral Strip Technology: Overview and Future Potential. *Journal of Controlled Release*. 2009, 139 (2): 94–107. https://doi.org/10.1016/j.jconrel.2009.06.014.

Douglas P, Albadarin A B, Al-Muhtaseb A H, Mangwandi C, and Walker G M. Thermo-Mechanical Properties of Poly ε-Caprolactone/Poly l-Lactic Acid Blends: Addition of Nalidixic Acid and Polyethylene Glycol Additives. *Journal of the Mechanical Behavior of Biomedical Materials*. 2015, 45: 154–165. https://doi.org/10.1016/j.jmbbm.2015.01.022.

Dumortier G, Grossiord J L, Agnely F, and Chaumeil J C. A Review of Poloxamer 407 Pharmaceutical and Pharmacological Characteristics. *Pharmaceutical Research*. 2006, 23 (12): 2709–2728. https://doi.org/10.1007/s11095-006-9104-4.

Efentakis M, Koutlis A, and Vlachou M. Development and Evaluation of Oral Multiple-Unit and Single-Unit Hydrophilic Controlled-Release Systems. *AAPS PharmSciTech*. 2000, 1 (4): 1–9. https://doi.org/10.1208/pt010434.

Elias, H-G. 1997. *An Introduction to Polymer Science*., 1st ed. Hobocken: VCH.

Fabri F V, Cupertino R R, Hidalgo M M, Weffort de Oliveira R M M, and Bruschi M L. Preparation and Characterization of Bioadhesive Systems Containing Propolis or Sildenafil for Dental Pulp Protection. *Drug Development and Industrial Pharmacy*. 2011, 37 (12): 1446–1454. https://doi.org/10.3109/03639045.2011.584387.

Farias M G, de Carvalho C W P, Takeiti C Y, and Ascheri J L R. 2012. O Efeito Da Permeabilidade Ao Vapor de Água, Atividade de Água, Molhabilidade e Solubilidade Em Água Em Filmes de Amido e Polpa de Acerola. In *Embrapa Agroindústria de Alimentos-Artigo Em Anais de Congresso(ALICE)*. In: WORKSHOP DA REDE DE NANOTECNOLOGIA APLICADA AO AGRONEGÓCIO, 6, 2012. . . .

Fernandes F P. 2017. Desenvolvimento e Caracterização de Filmes e Comprimidos Bucais a Base de Pectina e Goma Gelana Para Liberação Tópica de Triancinolona. Universidade de São Paulo.

Ferreira P S, Victorelli F D, Rodero C F, Fortunato G C, Araújo V H S, Fonseca-Santos B, Bauab T M, Dijck V M, and Chorilli M. P-Coumaric Acid Loaded into Liquid Crystalline Systems as a Novel Strategy to the Treatment of Vulvovaginal Candidiasis. *International Journal of Pharmaceutics*. 2021, 603 (May): 120658. https://doi.org/10.1016/j.ijpharm.2021.120658.

Ferreira S B S, Braga G, Oliveira E L, Bassi da Silva J, Rosseto H C, de Castro Hoshino L V, Baesso M L, et al. Design of a Nanostructured Mucoadhesive System Containing Curcumin for Buccal Application: From Physicochemical to Biological Aspects. *Beilstein Journal of Nanotechnology*. 2019, 10: 2304–2328. https://doi.org/10.3762/bjnano.10.222.

Ferreira S B S, Slowik K M, Castro Hoshino L V, Baesso M L, Murdoch C, Colley H E, and Bruschi M L. Mucoadhesive Emulgel Systems Containing Curcumin for Oral Squamous Cell Carcinoma Treatment: From Pre-Formulation to Cytotoxicity in Tissue-Engineering Oral Mucosa. *European Journal of Pharmaceutical Sciences*. 2020, 151 (March): 105372. https://doi.org/10.1016/j.ejps.2020.105372.

Fonseca-Santos B, dos Santos A M, Rodero C F, Gremião M P D, and Chorilli M. Design, Characterization, and Biological Evaluation of Curcumin-Loaded Surfactant-Based Systems for Topical Drug Delivery. *International Journal of Nanomedicine*. 2016, 11: 4553–4562. https://doi.org/10.2147/ijn.s108675.

Francisco L M B, Cerquetani J A, and Bruschi M L. Development and Characterization of Gelatin and Ethylcellulose Microparticles Designed as Platforms to Delivery Fluoride. *Drug Development and Industrial Pharmacy*. 2013, 39 (11): 1644–1650. https://doi.org/10.3109/03639045.2012.728610.

Francisco L M B, Pinto D, Rosseto H C, Toledo L A S, Said Dos Santos R, Costa P, Rodrigues F, Oliveira M B P P, Sarmento B, and Bruschi M L. Development of a Microparticulate System Containing Brazilian Propolis By-Product and Gelatine for Ascorbic Acid Delivery: Evaluation of Intestinal Cell Viability and Radical Scavenging Activity. *Food and Function*. 2018a, 9 (8): 4194–4206. https://doi.org/10.1039/c8fo00863a.

Francisco L M B, Pinto D, Rosseto H C, Toledo L A S, Santos R S, Tobaldini-Valério F, Svidzinski T, et al. Evaluation of Radical Scavenging Activity, Intestinal Cell Viability and Antifungal Activity of Brazilian Propolis By-Product. *Food Research International*. 2018b, 105: 537–547. https://doi.org/10.1016/j.foodres.2017.11.046.

Francisco L M B, Rosseto H C, Toledo L A S, dos Santos R S, de Souza Ferreira S B, and Bruschi M L. Organogel Composed of Poloxamer 188 and Passion Fruit Oil: Sol-Gel Transition, Rheology, and Mechanical Properties. *Journal of Molecular Liquids*. 2019, 289. https://doi.org/10.1016/j.molliq.2019.111170.

Funari C S, and Ferro V O. Análise de Própolis. *Ciencia e Tecnologia de Alimentos*. 2006, 26 (1): 171–178. https://doi.org/10.1590/S0101-20612006000100028.

Funari C S, Ferro V O, and Mathor M B. Analysis of Propolis from Baccharis Dracunculifolia DC. (Compositae) and Its Effects on Mouse Fibroblasts. *Journal of Ethnopharmacology*. 2007, 111 (2): 206–212. https://doi.org/10.1016/j.jep.2006.11.032.

Ghisalberti E L. Propolis: A Review. *Bee World*. 2016, no. July: 59–84. https://doi.org/10.1080/0005772X.1979.11097738.

Goel A, and Aggarwal B A. Curcumin, the Golden Spice from Indian Saffron, Is a Chemosensitizer and Radiosensitizer for Tumors and Chemoprotector and Radioprotector for Normal Organs. *Nutrition and Cancer*. 2010, 62 (7): 919–930. https://doi.org/10.1080/01635581.2010.509835.

Goel A, Kunnumakkara A B, and Aggarwal B B. Curcumin as 'Curecumin': From Kitchen to Clinic. *Biochemical Pharmacology*. 2008, 75 (4): 787–809. https://doi.org/10.1016/j.bcp.2007.08.016.

Gómez-Carracedo A, Alvarez-Lorenzo C, Gómez-Amoza J L, and Concheiro A. 2004. Glass Transitions and Viscoelastic Properties of Carbopol® and Noveon® Compacts. *International Journal of Pharmaceutics*. https://doi.org/10.1016/j.ijpharm.2004.01.023.

Gosselin R E, Smith R P, Hodge H C, and Braddock J E. 1984. *Clinical Toxicology of Commercial Products*, Vol. 1085. Baltimore: Williams &Wilkins.

Gowthamarajan K, Jawahar N, Wake P, Jain K, and Sood S. Development of Buccal Tablets for Curcumin Using Anacardium Occidentale Gum. *Carbohydrate Polymers*. 2012, 88 (4): 1177–1183. https://doi.org/10.1016/j.carbpol.2012.01.072.

Guard U S C. Chemical Hazard Response Information System (CHRIS)-Hazardous Chemical Data. *Commandant Instruction*. 1999: 16465.

Guo J-H. Carbopol ® Polymers for Pharmaceutical Drug Delivery Applications. *Drug Development and Delivery*. 2008, 3 (6): 1–3.

Gupta N V, Natasha S, Getyala A, and Bhat R S. Bioadhesive Vaginal Tablets Containing Spray Dried Microspheres Loaded with Clotrimazole for Treatment of Vaginal Candidiasis. *Acta Pharmaceutica*. 2008, 63: 359–372. https://doi.org/10.2478/acph-2013-0027.

Gupta S, Samanta M K, and Raichur A M. Dual-Drug Delivery System Based on In Situ Gel-Forming Nanosuspension of Forskolin to Enhance Antiglaucoma Efficacy. *AAPS PharmSciTech*. 2010, 11 (1): 322–325.

Han I K, Kim Y B, Kang H S, Sul D, Jung W W, Cho H J, and Oh Y K. Thermosensitive and Mucoadhesive Delivery Systems of Mucosal Vaccines. *Methods*. 2006, 38: 106–111. https://doi.org/10.1016/j.ymeth.2005.10.003.

Hill L K, Meleties M, Katyal P, Xie X, Delgado-Fukushima E, Jihad T, Liu C F, et al. Thermoresponsive Protein-Engineered Coiled-Coil Hydrogel for Sustained Small Molecule Release. *Biomacromolecules*. 2019, 20 (9): 3340–3351. https://doi.org/10.1021/acs.biomac.9b00107.

Hosmani A, Thorat Y S, and Kasture P V. Carbopol and Its Pharmaceutical Significance: A Review. *Pharmainfo. Net*. 2006, 1: 1–10.

Hoyo J, Ivanova K, Guaus E, and Tzanov T. Multifunctional ZnO NPs-Chitosan-Gallic Acid Hybrid Nanocoating to Overcome Contact Lenses Associated Conditions and Discomfort. *Journal of Colloid and Interface Science*. 2019, 543: 114–121. https://doi.org/10.1016/j.jcis.2019.02.043.

Hu Y, Darcos V, Monge S, and Li S. Thermo-Responsive Drug Release from Self-Assembled Micelles of Brush-like PLA/PEG Analogues Block Copolymers. *International Journal of Pharmaceutics*. 2015, 491 (1–2): 152–161. https://doi.org/10.1016/j.ijpharm.2015.06.020.

Hu Y, Darcos V, Monge S, Li S, Zhou Y, and Su F. Thermo-Responsive Release of Curcumin from Micelles Prepared by Self-Assembly of Amphiphilic P(NIPAAm-Co-DMAAm)-b-PLLA-b-P(NIPAAm-Co-DMAAm) Triblock Copolymers. *International Journal of Pharmaceutics*. 2014, 476 (1): 31–40. https://doi.org/10.1016/j.ijpharm.2014.09.029.

Hussain Z, Thu H E, Amjad M W, Hussain F, Ahmed T A, and Khan S. Exploring Recent Developments to Improve Antioxidant, Anti-Inflammatory and Antimicrobial Efficacy of Curcumin: A Review of New Trends and Future Perspectives. *Materials Science and Engineering C*. 2017, 77: 1316–1326. https://doi.org/10.1016/j.msec.2017.03.226.

Ifat K, Karsten M, and Fredmen M. Correlation between Drug Release Kinetics from Proteineous Matrices and Protein Folding: Elasticity and Compressibility Study. *Journal of Controlled Release*. 2001, 67: 261–274.

Imanifard S, Zarrabi A, Zarepour A, Jafari M, Khosravi A, and Razmjou A. Nanoengineered Thermoresponsive Magnetic Nanoparticles for Drug Controlled Release. *Macromolecular Chemistry and Physics*. 2017, 218 (23): 1–8. https://doi.org/10.1002/macp.201700350.

Islam M T, Rodríguez-Hornedo N, Ciotti S, and Ackermann C. Rheological Characterization of Topical Carbomer Gels Neutralized to Different pH. *Pharmaceutical Research*. 2004, 21 (7): 1192–1199. https://doi.org/10.1023/B:PHAM.0000033006.11619.07.

Ivarsson D, and Wahlgren M. Comparison of In Vitro Methods of Measuring Mucoadhesion: Ellipsometry, Tensile Strength and Rheological Measurements. *Colloids and Surfaces B: Biointerfaces*. 2012, 92: 353–359. https://doi.org/10.1016/j.colsurfb.2011.12.020.

Ji H-F, and Shen L. Can Improving Bioavailability Improve the Bioactivity of Curcumin? *Trends in Pharmacological Sciences*. 2014, 35 (6): 265–266. https://doi.org/10.1016/j.tips.2014.04.001.

Jones D S, Bruschi M L, de Freitas O, Gremião M P D, Lara E H G, and Andrews G P. Rheological, Mechanical and Mucoadhesive Properties of Thermoresponsive, Bioadhesive Binary Mixtures Composed of Poloxamer 407 and Carbopol 974P Designed as Platforms for Implantable Drug Delivery Systems for Use in the Oral Cavity. *International Journal of Pharmaceutics*. 2009, 372 (1–2): 49–58. https://doi.org/10.1016/j.ijpharm.2009.01.006.

Jovanovic S V, Steenken S, Boone C W, and Simic M G. H-Atom Transfer Is a Preferred Antioxidant Mechanism of Curcumin. *Journal of the American Chemical Society*. 1999, 121 (41): 9677–9681. https://doi.org/10.1021/ja991446m.

Kawathe N K, Nanadgude T D, and Poddar S S. Comparative Study of Mucoadhesive Vaginal Film and Tablet of Curcumin. *International Journal of Drug Delivery Technology*. 2017, 7 (3): 146–147. https://doi.org/10.25258/ijddt.v7i03.9558.

Kerec M, Bogataj M, Mugerle B, Gašperlin M, and Mrhar A. Mucoadhesion on Pig Vesical Mucosa: Influence of Polycarbophil/Calcium Interactions. *International Journal of Pharmaceutics*. 2002, 241: 135–143. https://doi.org/10.1016/S0378-5173(02)00231-4.

Khan S, Akhtar N, Minhas M U, Shah H, Khan K U, and Thakur R S S. A Difunctional Pluronic®127-Based In Situ Formed Injectable Thermogels as Prolonged and Controlled Curcumin Depot, Fabrication, In Vitro Characterization and In Vivo Safety Evaluation. *Journal of Biomaterials Science, Polymer Edition*. 2021, 32 (3): 281–319. https://doi.org/10.1080/09205063.2020.1829324.

Khullar R, Kumar D, Seth N, and Saini S. Formulation and Evaluation of Mefenamic Acid Emulgel for Topical Delivery. *Saudi Pharmaceutical Journal*. 2012, 20 (1): 63–67. https://doi.org/10.1016/j.jsps.2011.08.001.

Kikuchi A, and Okano T. 2005. Hydrogels – Stimuli Sensitive Hydrogels. In *Polymeric Drug Delivery Systems*, edited by Glen S. Kwon, 145th ed., 275–322. Boca Raton: Taylor&Francis Group.

Kulkarni A S, Tapase S R, Kodam K M, and Shinde V S. Thermoresponsive Pluronic Based Microgels for Controlled Release of Curcumin against Breast Cancer Cell Line. *Colloids and Surfaces B: Biointerfaces*. 2021, 205 (April): 111834. https://doi.org/10.1016/j.colsurfb.2021.111834.

Kumar R S, Suresh C G, Brannigan J A, Dodson G G, and Gaikwad S M. Bile Salt Hydrolase, the Member of Ntn-hydrolase Family: Differential Modes of Structural and Functional Transitions during Denaturation. *IUBMB Life*. 2007, 59 (2): 118–125.

Lee J B, Yoon J J, Lee D S, and Park T G. Photo-Crosslinkable, Thermo-Sensitive and Biodegradable Pluronic Hydrogels for Sustained Release of Protein. *Journal of Biomaterials Science, Polymer Edition*. 2004, 15 (12): 1571–1583. https://doi.org/10.1163/1568562042459751.

Lefnaoui S, and Moulai-Mostefa N. Investigation and Optimization of Formulation Factors of a Hydrogel Network Based on Kappa Carrageenan-Pregelatinized Starch Blend Using an Experimental Design. *Colloids and Surfaces A: Physicochemical and Engineering Aspects*. 2014, 458: 117–125. https://doi.org/10.1016/j.colsurfa.2014.01.007.

Liu D, Li J, Pan H, He F, Liu Z, Wu Q, Bai C, Yu S, and Yang X. Potential Advantages of a Novel Chitosan-N-Acetylcysteine Surface Modified Nanostructured Lipid Carrier on the Performance of Ophthalmic Delivery of Curcumin. *ScientificReports*. 2016, 6 (January): 1–14. https://doi.org/10.1038/srep28796.

Liu J, Lin S, Li L, and Liu E. Release of Theophylline from Polymer Blend Hydrogels. *International Journal of Pharmaceutics*. 2005, 298: 117–125. https://doi.org/10.1016/j.ijpharm.2005.04.006.

Liu Y, Zhu Y Y, Wei G, and Lu W Y. Effect of Carrageenan on Poloxamer-Based In Situ Gel for Vaginal Use: Improved In Vitro and In Vivo Sustained-Release Properties. *European Journal of Pharmaceutical Sciences*. 2009, 37 (3–4): 306–312. https://doi.org/10.1016/j.ejps.2009.02.022.

Lou J, Hu W, Tian R, Zhang H, Jia Y, Zhang J, and Zhang L. Optimization and Evaluation of a Thermoresponsive Ophthalmic In Situ Gel Containing Curcumin-Loaded Albumin Nanoparticles. *International Journal of Nanomedicine*. 2014, 9 (1): 2517–2525. https://doi.org/10.2147/IJN.S60270.

Lu S X, Lu J, and Liu L. 2016. Reversely thermo-reversible hydrogel compositions—US009592295B2. US 9 592 295 B2, issued 2016.

Lubrizol. 2013. *Pharmaceutical Polymers Typical Properties and Specifications*. Cleveland, OH: Lubrizol Advanced Materials.

Luciana S, Rosa F, Lonardoni M V C, Sakita K M, Sato F, Martha J, et al. 2020. Process of production of thermorresponsive photo hydrogel of phthalocyanes for topical application in antimicrobial photodynamic therapy and their hydrogel, issued 2020.

Luckanagul J A, Pitakchatwong C, Bhuket P R N, Muangnoi C, Rojsitthisak P, Chirachanchai S, Wang Q, and Rojsitthisak P. Chitosan-Based Polymer Hybrids for Thermo-Responsive Nanogel Delivery of Curcumin. *Carbohydrate Polymers*. 2018, 181 (April 2017): 1119–1127. https://doi.org/10.1016/j.carbpol.2017.11.027.

Madhavi B R, Murthy V S N, Rani A P, and Kumar G D. Buccal Film Drug Delivery System-An Innovative and Emerging Technology. *Journal of Molecular Pharmaceutics & Organic Process Research.* 2013, 1 (3): 2–6.

Manach C, Scalbert A, Morand C, Rémésy C, and Jiménez L. Polyphenols: Food Sources and Bioavailability. *American Journal of Clinical Nutrition.* 2004, 79 (5): 727–747. https://doi.org/10.1093/ajcn/79.5.727.

Manosroi A, Jantrawut P, Akihisa T, Manosroi W, and Manosroi J. In Vitro and In Vivo Skin Anti-Aging Evaluation of Gel Containing Niosomes Loaded with a Semi-Purified Fraction Containing Gallic Acid from Terminalia Chebula Galls. *Pharmaceutical Biology.* 2011, 49 (11): 1190–1203. https://doi.org/10.3109/13880209.2011.576347.

Marcucci M C C. Propolis: Chemical Composition, Biological Properties and Therapeutic Activity. *Apidologie.* 1995, 26 (2): 83–99. https://doi.org/10.1051/apido:19950202.

Matanović M R, Kristl J, and Grabnar P A. Thermoresponsive Polymers: Insights into Decisive Hydrogel Characteristics, Mechanisms of Gelation, and Promising Biomedical Applications. *International Journal of Pharmaceutics.* 2014, 472 (1–2): 262–275. https://doi.org/10.1016/j.ijpharm.2014.06.029.

Mayol L, Quaglia F, Borzacchiello A, Ambrosio A, and La Rotonda M I. A Novel Poloxamers/Hyaluronic Acid In Situ Forming Hydrogel for Drug Delivery: Rheological, Mucoadhesive and In Vitro Release Properties. *European Journal of Pharmaceutics and Biopharmaceutics.* 2008, 70 (1): 199–206. https://doi.org/10.1016/J.EJPB.2008.04.025.

Mazia R S, Pereira R R A, de Francisco L M B, Natali M R M, Dias Filho B P, Nakamura C V, Bruschi M L, and Ueda-Nakamura T. Formulation and Evaluation of a Mucoadhesive Thermoresponsive System Containing Brazilian Green Propolis for the Treatment of Lesions Caused by Herpes Simplex Type I. *Journal of Pharmaceutical Sciences.* 2016, 105 (1): 113–121. https://doi.org/10.1016/j.xphs.2015.11.016.

Mazzarino L S L F C, Curta J C, Licínio M A, Costa A, Pacheco L K, Siqueira J M, et al. Curcumin-Loaded Lipid and Polymeric Nanocapsules Stabilized by Nonionic Surfactants: An In Vitro and In Vivo Antitumor Activity on B16-F10 Melanoma and Macrophage Uptake Comparative Study. *Journal of Biomedical Nanotechnology.* 2011, 7 (3): 406–414. https://doi.org/10.1166/jbn.2011.1296.

Meher J G, Tarai M, Yadav N P, Patnaik A, Mishra A, and Yadav K S. Development and Characterization of Cellulose-Polymethacrylate Mucoadhesive Film for Buccal Delivery of Carvedilol. *Carbohydrate Polymers.* 2013, 96 (1): 172–180. https://doi.org/10.1016/j.carbpol.2013.03.076.

Mezreb N, Charrueau C, Boy P, Allain P, and Chaumeil J C. Production of Carbopol 974P and Carbopol 971P Pellets by Extrusion-Spheronization: Optimization of the Processing Parameters and Water Content. *Drug Development and Industrial Pharmacy.* 2004, 30 (5): 481–490. https://doi.org/10.1081/DDC-120037476.

Miguel M G, Nunes S, Dandlen S A, Cavaco A M, and Antunes M D. Phenols and Antioxidant Activity of Hydro-Alcoholic Extracts of Propolis from Algarve, South of Portugal. *Food and Chemical Toxicology.* 2010, 48 (12): 3418–3423. https://doi.org/10.1016/j.fct.2010.09.014.

Mitsou E, Pletsa V, Sotiroudis G T, Panine P, Zoumpanioti M, and Xenakis A. Development of a Microemulsion for Encapsulation and Delivery of Gallic Acid. The Role of Chitosan. *Colloids and Surfaces B: Biointerfaces.* 2020, 190 (March): 110974. https://doi.org/10.1016/j.colsurfb.2020.110974.

Mohamed M I, Haider M, and Ali M A M. Buccal Mucoadhesive Films Containing Antihypertensive Drug. *Journal of Chemical and Pharmaceutical Research.* 2011, 3 (6): 665–686.

Morales J O, and McConville J T. Manufacture and Characterization of Mucoadhesive Buccal Films. *European Journal of Pharmaceutics and Biopharmaceutics.* 2011, 77 (2): 187–199. https://doi.org/10.1016/j.ejpb.2010.11.023.

Moreira T F. Composição Química Da Própolis: Vitaminas e Aminoácidos. *Revista Brasileira de Farmacognosia.* 1986, 1 (1): 12–19. https://doi.org/10.1590/s0102-695x1986000100003.

Murphy J L, Radano C P, Dalsin J L, and Koepsel J T. 2015. Bioadhesive compounds and methods of synthesis and use, issued 2015.

Naksuriya O, Okonogi S, Schiffelers R M, and Hennink W E. Curcumin Nanoformulations: A Review of Pharmaceutical Properties and Preclinical Studies and Clinical Data Related to Cancer Treatment. *Biomaterials.* 2014, 35 (10): 3365–3383. https://doi.org/10.1016/j.biomaterials.2013.12.090.

Nascimento N C, and Fett-Neto A G. Plant Secondary Metabolism and Challenges in Modifying Its Operation: An Overview. *Methods in Molecular Biology.* 2010, 643: 1–13. https://doi.org/10.1007/978-1-60761-723-5.

Nasra M M A, Khiri H M, Hazzah H A, and Abdallah O Y. Formulation, In-Vitro Characterization and Clinical Evaluation of Curcumin In-Situ Gel for Treatment of Periodontitis. *Drug Delivery.* 2017, 24 (1): 133–142. https://doi.org/10.1080/10717544.2016.1233591.

Neelagiri R, Reddy M, and Rao N. Buccal Patch as Drug Delivery System: An Overview. *International Journal of Pharmaceutical Current Research*. 2013, 5 (2): 40–47.

Oliveira M B, da Silva J B, Montanha M C, Kimura E, Diniz A, and Bruschi M L. Design and Characterization of Mucoadhesive Gelatin-Ethylcellulose Microparticles for the Delivery of Curcumin to the Bladder. *Current Drug Delivery*. 2018. https://doi.org/10.2174/1567201815666180503121043.

Paini M, Aliakbarian B, Casazza A A, Perego P, Ruggiero C, and Pastorino L. Chitosan/Dextran Multilayer Microcapsules for Polyphenol Co-Delivery. *Materials Science and Engineering C*. 2015, 46: 374–380. https://doi.org/10.1016/j.msec.2014.10.047.

Pandey K B, and Rizvi S I. Plant Polyphenols as Dietary Antioxidants in Human Health and Disease. *Oxidative Medicine and Cellular Longevity*. 2009, 2 (5): 270–278. https://doi.org/10.4161/oxim.2.5.9498.

Paoli M A. 2009. *Degradação e Estabilização de Polímeros*. São Paulo: Artliber.

Park S G, Li M X, Cho W K, Joung Y K, and Huh K M. Thermosensitive Gallic Acid-Conjugated Hexanoyl Glycol Chitosan as a Novel Wound Healing Biomaterial. *Carbohydrate Polymers*. 2021, 260 (January): 117808. https://doi.org/10.1016/j.carbpol.2021.117808.

Park Y K, Alencar S M, Scamparini A R P, and Aguiar C L. Própolis Produzida No Sul Do Brasil, Argentina e Uruguai: Evidências Fitoquímicas de Sua Origem Vegetal. *Ciência Rural*. 2002, 32 (6): 997–1003. https://doi.org/10.1590/s0103-84782002000600013.

Parke D V. 1968. *The Biochemistry of Foreign Compounds*, 1 st ed. Pergamon

Pelletier R, and Phaneuf L P. Handbook of Veterinary Drugs: A Compendium for Research and Clinical Use. *The Canadian Veterinary Journal*. 1976, 17 (9): 247.

Pereira R R A, Godoy J S R, Svidzinski T I E, and Bruschi M L. Preparation and Characterization of Mucoadhesive Thermoresponsive Systems Containing Propolis for the Treatment of Vulvovaginal Candidiasis. *Journal of Pharmaeutical Sciences*. 2013, 102 (4): 1222–1234. https://doi.org/10.1002/jps.

Perioli L, Ambrogi V, Angelici F, Ricci M, Giovagnoli S, Capuccella M, and Rossi C. Development of Mucoadhesive Patches for Buccal Administration of Ibuprofen. *Journal of Controlled Release*. 2004, 99: 73–82.

Pinto E. 2016. "Métodos Analíticos Na Determinação de Polifenóis Em Frutos." Faculdade de Farmácia da Universidade de Porto, Portugal. Available at: <http://tmeventos.com.br/frut2016/pdfs/C7EDGARPINTO.pdf>

Pires A, Fortuna A, Alves G, and Falcão A. Intranasal Drug Delivery: How, Why and What For? *Journal of Pharmaceutical Sciences*. 2009, 12 (3): 288–311. https://doi.org/10.18433/j3nc79.

Pliszczak D, Bordes C, Bourgeois S, Marote P, Zahouani H, Tupin S, Mattei C P, and Lantéri P. Mucoadhesion Evaluation of Polysaccharide Gels for Vaginal Application by Using Rheological and Indentation Measurements. *Colloids and Surfaces B: Biointerfaces*. 2012, 92: 168–174. https://doi.org/10.1016/j.colsurfb.2011.11.039.

Popova M, Bankova V, Butovska D, Petkov V, Nikolova-Damyanova B, Sabatini A G, Marcazzan G L, and Bogdanov S. Validated Methods for the Quantification of Biologically Active Constituents of Poplar-Type Propolis. *Phytochemical Analysis*. 2004, 15 (4): 235–240. https://doi.org/10.1002/pca.777.

Prajatelistia E, Ju S W, Sanandiya N D, Jun S H, Ahn J S, and Hwang H D. Tunicate-Inspired Gallic Acid/Metal Ion Complex for Instant and Efficient Treatment of Dentin Hypersensitivity. *Advanced Healthcare Materials*. 2016, 5 (8): 919–927. https://doi.org/10.1002/adhm.201500878.

Preis M, Breitkreutz J, and Sandler N. Perspective: Concepts of Printing Technologies for Oral Film Formulations. *International Journal of Pharmaceutics*. 2015, 494 (2): 578–584.

Prista L N. 2008. Tecnologia Farmacêutica I. Sétima ed. Lisboa: Fundação Calouste Gulbenkian.

James H P, John R, Alex A, and Anoop K R. Smart Polymers for the Controlled Delivery of Drugs—A Concise Overview. *Acta Pharmaceutica Sinica B*. 2014, 4 (2): 120–127. https://doi.org/10.1016/j.apsb.2014.02.005.

Qiu Y, Hamilton S K, and Temenoff J. 4—Improving Mechanical Properties of Injectable Polymers and Composites. *Injectable Biomaterials Science and Applications*. 2011: 61–91. https://doi.org/10.1533/9780857091376.1.61.

Qiu Y, and Park K. Environment-Sensitive Hydrogels for Drug Delivery. *Advanced Drug Delivery Reviews*. 2001, 53: 321–339. https://doi.org/10.1016/j.addr.2012.09.024.

Rahman L, Lembang R S, Lallo S, Usmanengsi S R H, and Permana A D. Bioadhesive Dermal Patch as Promising Approach for Improved Antibacterial Activity of Bioactive Compound of Zingiber Cassumunar Roxb in Ex Vivo Staphylococcus Aureus Skin Infection Model. *Journal of Drug Delivery Science and Technology*. 2021, 63 (September 2020): 102522. https://doi.org/10.1016/j.jddst.2021.102522.

Ramirez-Barron S N, Sanchez-Valdes S, Betancourt R, Gallardo C A, Puente-Urbina B, Rodriguez-Fernández O S, Carneiro-da-Cunha M G, Santos-Correia M T, and Sanchez-Martinez Z V. Preparation and Characterization of Gelatin-Gallic Acid/ZnO Nanocomposite with Antibacterial Properties as a Promising Multi-Functional Bioadhesive for Wound Dressing Applications. *International Journal of Adhesion and Adhesives*. 2021, 104 (September 2020). https://doi.org/10.1016/j.ijadhadh.2020.102749.

Rao K M, Kumar A, Suneetha M, and Han S S. pH and Near-Infrared Active; Chitosan-Coated Halloysite Nanotubes Loaded with Curcumin-Au Hybrid Nanoparticles for Cancer Drug Delivery. *International Journal of Biological Macromolecules*. 2018, 112: 119–125. https://doi.org/10.1016/j.ijbiomac.2018.01.163.

Risbud M V, Hardikar A A, Bhat S V, and Bhonde R R. pH-Sensitive Freeze-Dried Chitosan–Polyvinyl Pyrrolidone Hydrogels as Controlled Release System for Antibiotic Delivery. *Journal of Controlled Release*. 2000, 68 (1): 23–30.

Rodrigues I R. 2006. Síntese e Caracterização de Redes Poliméricas à Base de Quitosana Com PVP e PVA Para Aplicação Na Liberação Controlada de Fármacos. Master Degree Dissertation. State Univesity of Rio Grande do Sul. 107 p. Available at: <https://lume.ufrgs.br/handle/10183/8349>.

Rosseto H C, Toledo L A S, Francisco L M B, Esposito E, Lim Y, Valacchi G, Cortesi R, and Bruschi M L. Nanostructured Lipid Systems Modified with Waste Material of Propolis for Wound Healing: Design, In Vitro and In Vivo Evaluation. *Colloids and Surfaces B: Biointerfaces*. 2017, 158: 441–452. https://doi.org/10.1016/j.colsurfb.2017.07.029.

Rosseto H C, Toledo L A S, Said dos Santos R, de Francisco L M B, Vecchi C F, Esposito E, Cortesi R, and Bruschi M L. Design of Propolis-Loaded Film Forming Systems for Topical Administration: The Effect of Acrylic Acid Derivative Polymers. *Journal of Molecular Liquids*. 2021, 322 (xxxx): 114514. https://doi.org/10.1016/j.molliq.2020.114514.

Rowe R C, Sheskey P J, and Quinn M E. 2009. *Handbook Pharmaceutical Excipients*. 6th ed. London: Pharmaceutical Press.

Russo E, Selmin F, Baldassari S, Gennari C G M, Caviglioli G, Cilurzo F, Minghetti P, and Parodi B. A Focus on Mucoadhesive Polymers and Their Application in Buccal Dosage Forms. *Journal of Drug Delivery Science and Technology*. 2016, 32: 113–125. https://doi.org/10.1016/j.jddst.2015.06.016.

Sabliov C M, Astete C E, Bodoki E, Vostinaru O, Samoila O C, and Dinte E. 2017. Delivery of bioactive, nano-encapsulated antoxdants. US 2017/0216221 A1, issued 2017.

Said dos Santos R, Bassi da Silva J, Rosseto H C, Vecchi C F, Campanholi K S S, Caetano W, and Bruschi M L. Emulgels Containing Propolis and Curcumin: The Effect of Type of Vegetable Oil, Poly(Acrylic Acid) and Bioactive Agent on Physicochemical Stability, Mechanical and Rheological Properties. *Gels*. 2021, 7 (3). https://doi.org/10.3390/gels7030120.

Said dos Santos R, Rosseto H C, Bassi da Silva J, Vecchi C F, Caetano W, and Bruschi M L. The Effect of Carbomer 934P and Different Vegetable Oils on Physical Stability, Mechanical and Rheological Properties of Emulsion-Based Systems Containing Propolis. *Journal of Molecular Liquids*. 2020, 307. https://doi.org/10.1016/j.molliq.2020.112969.

Said dos Santos R, Vecchi C F, Rosseto H C, Bassi da Silva J, Dano M E L, Castro-Hoshino L V, Baesso M L, and Bruschi M L. Emulgels Containing Carbopol 934P and Different Vegetable Oils for Topical Propolis Delivery: Bioadhesion, Drug Release Profile, and Ex Vivo Skin Permeation Studies. *AAPS PharmSciTech*. 2021, 21 (6): 1–10. https://doi.org/10.1208/s12249-020-01748-3.

Salat M, Petkova P, Hoyo J, Perelshtein I, Gedanken A, and Tzanov T. Durable Antimicrobial Cotton Textiles Coated Sonochemically with ZnO Nanoparticles Embedded in an In-Situ Enzymatically Generated Bioadhesive. *Carbohydrate Polymers*. 2018, 189 (November 2017): 198–203. https://doi.org/10.1016/j.carbpol.2018.02.033.

Salatino A, Teixeira E W, Negri G, and Message D. Origin and Chemical Variation of Brazilian Propolis. *Evidence-Based Complementary and Alternative Medicine*. 2005, 2 (1): 33–38. https://doi.org/10.1093/ecam/neh060.

Salman H H A, Azcàrate I G, and Catalán I E. 2018. Nanoparticles comprising a vegetable hydrophobic protein and a water miscible non-volatile organic solvent and uses thereof, issued 2018.

Salman H H A, Azcárate I G, and Catalán I E. 2017. EP 2 814 332 B1—Nanoparticles comprising a vegetable hydrophobic protein and a water miscible non-volatile organic solvent and uses thereof. EP2013/052795, issued 2017.

Samanta S, and Roccatano D. 2013. Interaction of Curcumin with PEO – PPO – PEO Block Copolymers: A Molecular Dynamics Study. *The Journal of Physical Chemistry B*. 2013, 117: 3250–3257.

Sanoj Rejinold N, Muthunarayanan M, Divyarani V V, Sreerekha P R, Chennazhi K P, Nair S V, Tamura H, and Jayakumar R. Curcumin-Loaded Biocompatible Thermoresponsive Polymeric Nanoparticles for Cancer Drug Delivery. *Journal of Colloid and Interface Science*. 2011, 360 (1): 39–51. https://doi.org/10.1016/j.jcis.2011.04.006.

Santezi C, Reina B D, Annunzio S R, Calixto G, Chorilli M, and Dovigo L N. Photodynamic Potential of Curcumin in Bioadhesive Formulations: Optical Characteristics and Antimicrobial Effect against Biofilms. *Photodiagnosis and Photodynamic Therapy*. 2021, 35 (April). https://doi.org/10.1016/j.pdpdt.2021.102416.

Singh R, Tønnesen H H, Kristensen S, and Berg K. The Influence of Pluronics® on Dark Cytotoxicity, Photocytotoxicity, Localization and Uptake of Curcumin in Cancer Cells: Studies of Curcumin and Curcuminoids XLIX. *Photochemical and Photobiological Sciences*. 2013, 12 (3): 559–575. https://doi.org/10.1039/c2pp25249j.

Singla V, Saini S, Joshi B, and Rana A C. Emulgel: A New Platform for Topical Drug Delivery. *International Journal of Pharma and Bio Sciences*. 2012, 3 (1): 485–498.

Siviero A, Gallo E, Maggini V, Gori L, Mugelli A, Firenzuoli F, and Vannacci A. Curcumin, a Golden Spice with a Low Bioavailability. *Journal of Herbal Medicine*. 2015, 5 (2): 57–70. https://doi.org/10.1016/j.hermed.2015.03.001.

Soto-Quintero A, Guarrotxena N, García O, and Quijada-Garrido I. Curcumin to Promote the Synthesis of Silver NPs and Their Self-Assembly with a Thermoresponsive Polymer in Core-Shell Nanohybrids. *Scientific Reports*. 2019, 9 (1): 1–14. https://doi.org/10.1038/s41598-019-54752-4.

Souza Ferreira S B, Braga G, Oliveira E L, Rosseto H C, Hioka N, Caetano W, and Bruschi M L. Colloidal Systems Composed of Poloxamer 407, Different Acrylic Acid Derivatives and Curcuminoids: Optimization of Preparation Method, Type of Bioadhesive Polymer and Storage Conditions. *Journal of Drug Delivery Science and Technology*. 2020, 57 (February): 101686. https://doi.org/10.1016/j.jddst.2020.101686.

Souza Ferreira S B, Da Silva J B, Junqueira M V, Borghi-Pangoni F B, Guttierres G R, and Bruschi M L. The Importance of the Relationship between Mechanical Analyses and Rheometry of Mucoadhesive Thermoresponsive Polymeric Materials for Biomedical Applications. *Journal of the Mechanical Behavior of Biomedical Materials*. 2017, 74. https://doi.org/10.1016/j.jmbbm.2017.05.040.

Souza Ferreira S B, Moço T D, Borghi-Pangoni F B, Junqueira M V, and Bruschi M L. Rheological, Mucoadhesive and Textural Properties of Thermoresponsive Polymer Blends for Biomedical Applications. *Journal of the Mechanical Behavior of Biomedical Materials*. 2015, 55: 164–178. https://doi.org/10.1016/j.jmbbm.2015.10.026.

Sudhakar Y, Kuotsu K, and Bandyopadhyay A K. Buccal Bioadhesive Drug Delivery—A Promising Option for Orally Less Efficient Drugs. *Journal of Controlled Release*. 2006, 114: 15–40. https://doi.org/10.1016/j.jconrel.2006.04.012.

Sueth-Santiago V, Mendes Silva G P, Decoté-Ricardo D, and Lima M E F. Curcumina, o Pó Dourado Do Açafrão-Da-Terra: Introspecções Sobre Química e Atividades Biológicas. *Química Nova*. 2015, 38 (4): 538–552. https://doi.org/10.1590/S0100-40422010000800018.

Sulaiman G M, Sammarrae K W A, Ad'hiah A H, Zucchetti M, Frapolli R, Bello E, Erba E, D'Incalci M, and Bagnati R. Chemical Characterization of Iraqi Propolis Samples and Assessing Their Antioxidant Potentials. *Food and Chemical Toxicology*. 2011, 49 (9): 2415–2421. https://doi.org/10.1016/j.fct.2011.06.060.

Sun C, Zeng X, Zheng S, Wang Y, Li Z, Zhang H, Nie L, Zhang Y, Zhao Y, and Yang X. Bio-Adhesive Catechol-Modified Chitosan Wound Healing Hydrogel Dressings through Glow Discharge Plasma Technique. *Chemical Engineering Journal*. 2021, 427 (May 2021): 130843. https://doi.org/10.1016/j.cej.2021.130843.

Sun Y, Du L, Liu Y, Li X, Li M, Jin Y, and Qian X. Transdermal Delivery of the In Situ Hydrogels of Curcumin and Its Inclusion Complexes of Hydroxypropyl-β-Cyclodextrin for Melanoma Treatment. *International Journal of Pharmaceutics*. 2014, 469 (1): 31–39. https://doi.org/10.1016/j.ijpharm.2014.04.039.

Tamburic S, and Craig D Q M. 1995. An investigation into the rheological, dielectrical and mucoadhesive properties of poly (acrylic acid) gel systems. *Journal of Controlled Release*. 1995, 37: 59–68.

Ternullo S, Werning L V S, Holsæter A M, and Škalko-Basnet N. Curcumin-in-Deformable Liposomes-in-Chitosan-Hydrogel as a Novel Wound Dressing. *Pharmaceutics*. 2020, 12 (1). https://doi.org/10.3390/pharmaceutics12010008.

Toledo L A S, Bavato M I, Rosseto H C, Cortesi R, and Bruschi M L. Pharmaceutical Films Made from the Waste Material from the Preparation of Propolis Extracts: Development and Characterization. *Brazilian Journal of Pharmaceutical Sciences*. 2015, 51 (4): 847–859. https://doi.org/10.1590/S1984-82502015000400011.

Toledo L A S, Rosseto H C, and Bruschi M L. Iron Oxide Magnetic Nanoparticles as Antimicrobials for Therapeutics. *Pharmaceutical Development and Technology*. 2018. https://doi.org/10.1080/10837450.2017.1337793.

Ugwoke M I, Agu R U, Vanbilloen H, Baetens J, Augustijns P, Verbeke N, Mortelmans L, Verbruggen A, Kinget R, and Bormans G. Scintigraphic Evaluation in Rabbits of Nasal Drug Delivery Systems Based on Carbopol 971p(??) and Carboxymethylcellulose. *Journal of Controlled Release*. 2000, 68 (2): 207–214. https://doi.org/10.1016/S0168-3659(00)00258-3.

Ulman A. 2013. An Introduction to Ultrathin Organic Films: From Langmuir—Blodgett to Self—Assembly. Cambridge: Academic press.

Vecchi C F, and Bruschi M L. Films Composed of Polyvinyl Alcohol, Polyvinylpyrrolidone and Poloxamer 407: In-Vitro Mucoadhesive Analysis. *Pharmaceutical Development and Technology*. 2019, 26 (2): 138–149.

Vecchi C F, Cesar G B, Souza P R, Caetano W, and Bruschi M L. Mucoadhesive Polymeric Films Comprising Polyvinyl Alcohol, Polyvinylpyrrolidone, and Poloxamer 407 for Pharmaceutical Applications. *Pharmaceutical Development and Technology*. 2021, 26 (2): 138–149. https://doi.org/10.1080/10837450.2020.1849283.

Vecchi C F, Said dos Santos R, Bassi da Silva J, Rosseto H C, Sakita K M, Svidzinski T I E, Bonfim-Mendonça P S, and Bruschi M L. Development and In Vitro Evaluation of Buccal Mucoadhesive Films for Photodynamic Inactivation of Candida Albicans. *Photodiagnosis and Photodynamic Therapy*. 2020, 32 (August): 101957. https://doi.org/10.1016/j.pdpdt.2020.101957.

Villanova J C O, Oréfice R L, and Cunha A S. Aplicações Farmacêuticas de Polímeros. *Polimeros*. 2010, 20 (1): 51–64. https://doi.org/10.1590/S0104-14282010005000009.

Villa Nova M, Ratti B A, Herculano L S, Bittencourt P R S, Novello C R, Bazotte R B, Lautenschlager S O S, and Bruschi M L. Design of Composite Microparticle Systems Based on Pectin and Waste Material of Propolis for Modified L-Alanyl-l-Glutamine Release and with Immunostimulant Activity. *Pharmaceutical Development and Technology*. 2019, 24 (1): 12–23. https://doi.org/10.1080/10837450.2017.1410556.

Wainwright M, and Preston P. Review Photodynamic Antimicrobial Chemotherapy (PACT). *Journal of Antimicrobial Chemotherapy*. 1998, 42: 13–28.

Wang H, Wang C, Zou Y, Hu J, Li Y, and Cheng Y. Natural Polyphenols in Drug Delivery Systems: Current Status and Future Challenges. *Giant*. 2020, 3: 100022. https://doi.org/10.1016/j.giant.2020.100022.

Wang X, Liu G, Ma J, Guo S, Gao L, Jia Y, Li X, and Zhang Q. In Situ Gel-Forming System: An Attractive Alternative for Nasal Drug Delivery. *Critical ReviewsTM in Therapeutic Drug Carrier Systems*. 2013, 30 (5): 411–434. https://doi.org/10.1615/CritRevTherDrugCarrierSyst.2013007362.

Wypych T C. 2009. Desenvolvimento e Avaliação de Sistemas Bucoadesivos Contendo Diclofenaco Sódico. Master Degree Dissertation. Federal University of Paraná. Brazil. 83 p. Available at: <https://acervodigital.ufpr.br/xmlui/bitstream/handle/1884/18605/R%20-%20D%20-%20THAIZ%20CRISTINA%20WYPYCH.pdf?sequence=1&isAllowed=y>.

Xu T, Ma Y, Huang J, Lai H, Yuan D, Tang X, and Yang L. Self-Organized Thermo-Responsive Poly (Lactic-Co-Glycolic Acid)-Graft-Pullulan Nanoparticles for Synergistic Thermo-Chemotherapy of Tumor. *Carbohydrate Polymers*. 2020, 237 (February): 116104. https://doi.org/10.1016/j.carbpol.2020.116104.

Yadav S K, Mishra M K, Tiwari A, and Shukla A. Emulgel: A New Approach for Enhanced Topical Drug Delivery. *International Journal of Current Pharmaceutical Research*. 2016, 9 (1): 15. https://doi.org/10.22159/ijcpr.2017v9i1.16628.

Zakerikhoob M, Abbasi S, Yousefi G, Mokhtari M, and Noorbakhsh M N. Curcumin-Incorporated Crosslinked Sodium Alginate-g-Poly (N-Isopropyl Acrylamide) Thermo-Responsive Hydrogel as an In-Situ Forming Injectable Dressing for Wound Healing: In Vitro Characterization and In Vivo Evaluation. *Carbohydrate Polymers*. 2021, 271 (June): 118434. https://doi.org/10.1016/j.carbpol.2021.118434.

Zhao L, Du J, Duan Y, Zang Y, Zhang H, Yang C, Cao F, and Zhai G. Curcumin Loaded Mixed Micelles Composed of Pluronic P123 and F68: Preparation, Optimization and In Vitro Characterization. *Colloids and Surfaces B: Biointerfaces*. 2012, 97: 101–108. https://doi.org/10.1016/j.colsurfb.2012.04.017.

Zhu Z, Zhai Y, Zhang N, Leng D, and Ding P. The Development of Polycarbophil as a Bioadhesive Material in Pharmacy. *Asian Journal of Pharmaceutical Sciences*. 2013, 8 (4): 218–227. https://doi.org/10.1016/j.ajps.2013.09.003.

9 Albumin Nanoparticles as a Versatile Carrier in Phytonanotechnology

Expanded Scope for Phytochemical Delivery

*Deepika Kumari, Akansha Jha, Purnima Mahato, Shuvadeep Bhowmik, Jatin Kumar Sahu, Madhu Kumari, Bapi Gorain, Pankaj Kumar Singh, Prashant Shukla, and *Monika Dwivedi*

9.1 INTRODUCTION

Phytochemicals are plant-derived compounds mostly present as secondary plant metabolites with various chemical compositions. Although these metabolites are unnecessary for plant growth and development, research has shown they are advantageous for maintaining human health and preventing disease. Phytochemicals positively impact human health by directly scavenging free radicals, chelating metals, inhibiting the generation of microtubules and microfilaments, and also constraining proteases and other processes. Thus, phytochemicals are non-nutritive, bioactive plant-based compounds that can defend against diseases (Ahmad et al. 2021). Due to their numerous uses, these bioactive constituents originating from plants have lately gained significant attention (Thakur et al. 2020).

Nanotechnology has been used in traditional drug delivery to create techniques specifically suited to enhance the therapeutic efficacy of natural bioactive substances. Nanocarriers provide an alternative to conventional formulation methods allowing site-specific targeted drug delivery, increased drug bioavailability, and reduced hazardous side effects. The phytochemicals are delivered to the target site with improved bioefficacy owing to the nanoparticles utilized as carriers. Nanoparticles are typically small nanostructures composed of materials created at the atomic or molecular level. As a result, they may travel through the human body with greater flexibility (Ahmad et al. 2021).

Furthermore, because of their basic structure, which is well-defined, it is possible to make many alterations on the surface, such as covalent drug attachment and targeting ligands. Many types of proteins are used for formulating protein nanoparticles, but albumin (non-glycosylatedprotein) is the most attractive, distinctive versatile carrier and shows a promising result in nanomedicine (Ahmad et al. 2021; Thakur et al. 2020). The uniqueness of albumin lies in its properties like excellent compatibility, biodegradability, safety profile, and well-defined basic structure, allowing the electrostatic adsorption of negatively charged like oligonucleotides or positively charged like ganciclovir molecules. Moreover, the techniques of producing albumin nanoparticles do not require tedious and troublesome processes and are manufactured under mild conditions, like coacervation, desolvation, and emulsification. They are also smaller (50–300 nm) than microparticles and have

DOI: 10.1201/9781003300588-9

more regulated release properties than liposomes, leading to improved compliance and acceptance by the patient. Industrially, albumin obtained from bovine serum is called bovine serum albumin, or BSA, albumin from eggs is called ovalbumin, and albumin from human serum is known as human serum albumin, or HSA. Albumin can also be sourced from soybeans, milk, and cereals. It is challenging for phytochemicals to pass through the endothelium lining and blood-brain barrier (BBB) in blood vessels, the gastrointestinal tract, and mucosa due to their enormous size and polar nature. Albumin nanoparticles encapsulate phytochemicals and release them to target tissues with improved stability and bioavailability. Albumin nanocarriers improve the therapeutic value of formulation and prevent therapy-related toxicity. Besides, the biocompatibility, excellent biological stability, and the possibility of oral, nasal, and other mode delivery make albumin nanoparticles a more captivating drug delivery strategy (Ahmad et al. 2021; Thakur et al. 2020).

9.2 PHYTOCONSTITUENTS FOR HEALTHCARE

The invention of the extraction method was seen as a breakthrough with the use of herbal remedies at the beginning of the 19th century (Sharma et al. 2011). Bioactive ingredient and nutraceutical research, development, and commercialization are of interest globally. Consumption of bioactive plant components has increased in the last 20 years due to understanding the link between nutrition, health, and disease (Ribaya-Mercado and Blumberg 2004).

Phytochemicals are present in most foodstuffs, including fruit and vegetables, nuts, leaves, flowers, and whole grains. These phytochemicals offer excellent therapeutic capability for treating various diseases, individually or in combination. They offer protection against numerous diseases such as cancers; hypertension; inflammation; cardiac-related disorders; viral, microbial, and parasitic infections; psychotic disorders; and osteoporosis. Antioxidant, antibacterial, anti-inflammatory, antifungal, anti-allergic, chemopreventive, hepatoprotective, hypolipidemic, neuroprotective, hypotensive, immuno-modulating, and carminative agents are a few functions of phytochemical compounds in maintaining human health (Ribaya-Mercado and Blumberg 2004; Prof and Gupta 2009). They induce apoptosis; act as a diuretic, CNS stimulant, and analgesic; and protect against UV radiation (Prof and Gupta 2009). The biggest problem in developing new phytochemical-based formulations is the lack of laws defining the identification, extraction, and modelling of individual components in various herbs. Herbal-drug-loaded nano-drug delivery systems, such as nanoparticles, liposomes, and SLNs, have been investigated (Gilani et al. 2020).

9.3 APPROACHES FOR PHYTONANOTECHNOLOGY

Nature has given us many drugs, including potent anti-inflammatory drugs, such as vindesine, vincristine, and vinblastine; *Vinca* alkaloids, such as vinorelbine; drugs such as paclitaxel, docetaxel, and podophyllotoxin; and derivatives such as etoposide and teniposide and its derivatives, such as topotecan, irinotecan, and anthracyclines. In addition, globally, half of approved anticancer drugs are either natural products or their derivatives obtained by utilizing the plant phytoconstituents and their derivatives. Due to their immense structural variety and therapeutic potential, these phytoconstituents remain valuable sources of potential drug candidates. Investigating and using nature, particularly for developing a wide range of medicinal substances, has expanded the search for new and better therapeutics delivery modes for various illnesses (Bowman and Leong 2006).

Plants also offer the benefits of being widely dispersed, easily accessible, and safer to handle. It is widely known that a single medicinal herb contains a variety of chemically complex but environmentally safe constituents. Identifying and isolating these compounds is laborious, but they are excellent sources for better clinical uses. Therefore, bioactive fractionation simplifies separating active fractions and identifying their active ingredients. Green synthesis allows these bioactive fractions to be developed in polymeric/lipid-based nanoparticles, Au NPs, and Ag NPs. Thus, applying nanotechnology to plant extracts has become a simple, functional, and efficient way to create plant-based nanoparticles

(Khan and Gurav 2018). Applications of nanotechnology in plant systems, or phytonanotechnology, for target-specific drug delivery, have gained considerable interest. As a result, phytonanotechnology has enormous promise for transforming drug delivery (Bowman and Leong 2006).

9.4 ALBUMIN AS A VERSATILE CARRIER IN PHYTOCOMPOUND DELIVERY

Albumin is a versatile distinctive protein that shows a promising result in nanomedicine because of its various advantages. Due to its properties, like high compatibility, non-immunogenicity, biodegradability, and high safety profile, it is now regarded as the most commendable carrier for drug delivery. According to research, it was proposed that albumin nanoparticles can penetrate the blood-brain barrier, which expands the area where therapeutic agents can be delivered.

Protein nanoparticles, especially of albumin, look to be a valuable material for drug administration since it avoids the adverse serum interactions typically after intravenous injection of complexes. High stability and binding affinity sites in albumin make it ideal for loading potent therapeutic drugs (Ahmad et al. 2021; Thakur et al. 2020; Yadav et al. 2020).

9.4.1 Proteins for Drug Delivery

Proteins are highly complex macromolecules and are the final products of a decoding process that begins with cellular DNA information. The common characteristic among all proteins is their composition as polymers, consisting of extended chains of α amino acids. Classification of proteins is done in several ways, but sorting based on chemical nature and solubility is most acceptable, as discussed in Figure 9.1. Proteins are classified as simple proteins composed of only amino acid

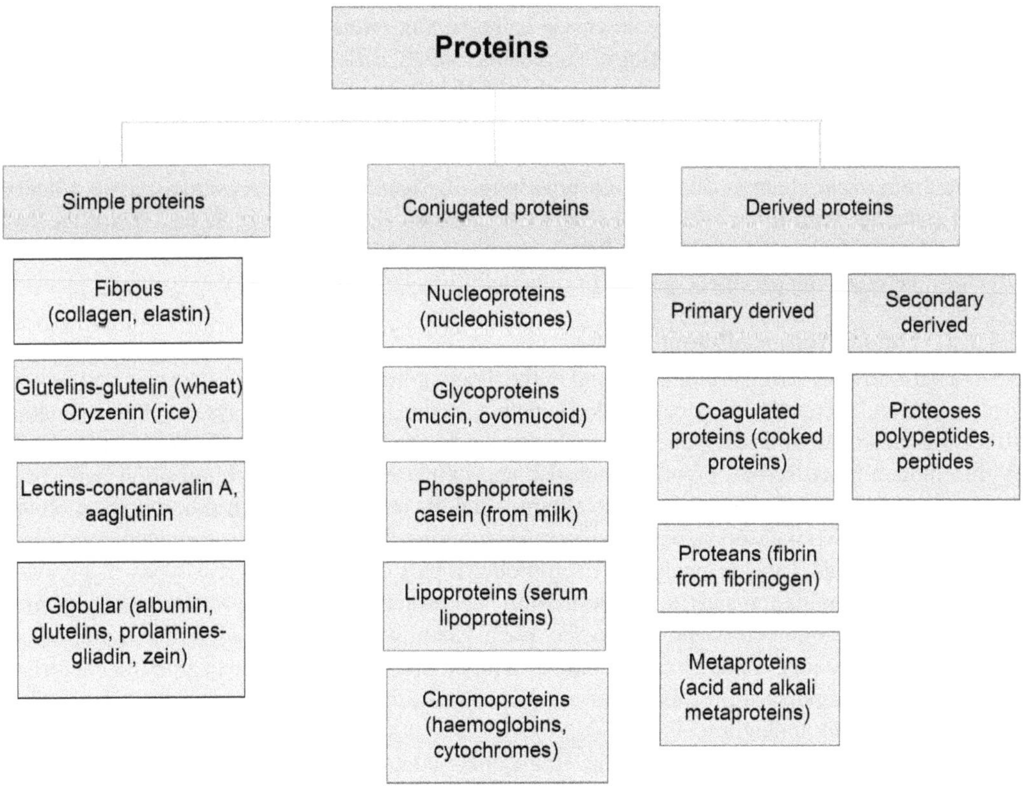

FIGURE 9.1 Classification of proteins based on chemical nature and solubility.

residues, conjugated proteins consisting of amino acids and a non-protein moiety, and derived proteins composed of degraded products of simple and derived proteins (Khan and Gurav 2018).

Albumin has been used significantly in the fabrication of multifunctional nanoparticles. Albumin hydrophobic pockets play a crucial role in efficient drug administration by making it easier for albumin to bond to hydrophobic or amphiphilic small molecules (Yadav et al. 2020).

9.4.2 Types of Albumins

Albumin is commercially produced using diverse sources, such as ovalbumin extracted from egg white, bovine serum (BSA), and human serum (HSA). While albumin can also be found in milk, soy, and legumes, HSA and BSA are the most frequently employed sources (Yadav et al. 2020).

9.4.2.1 Ovalbumin (OVA)

The most abundant protein in egg whites is ovalbumin (OVA), 55% of the total protein composition. It is a spherical acidic serpin that belongs to the serpin family. With a molecular weight of 4,247, the glycoprotein is a type of protein. kDa is made up of a single 386-amino-acidpolypeptide chain acid. It has a comparable isoelectric point to HSA and BSA, which is 4.8. Because of its immunogenicity, OVA is typically picked as one of the most popular models (Lv et al. 2015).

9.4.2.2 Bovine Serum Albumin (BSA)

Bovine serum is employed for the synthesis of BSA. Fraction V was also applied to the purification method developed by Cohn, which was based on various plasma protein solubility profiles. Both negatively and positively charged substances bind to positively and negatively charged moieties on BSA. It has a molecular weight of 69.323 kDa and an isoelectric point 4.7. It is a multipurpose carrier since it has a large loading capacity, is soluble in water, and can easily bind with both hydrophilic as well as hydrophobic drugs. The only downside is the human immunogenic reaction it causes. HSA and BSA have an 80% sequence analogy, a molecular weight difference of less than1%, and identical isoelectric values. BSA has a molecular mass of 66400 Da and is made up of 583 amino acids bonded in a single chain and cross-linked with seventeen cysteine residues, eight disulfide linkages, and one free thiol group (Jahanban-Esfahlan et al. 2019).

The amino acid chain is divided into nine loops by disulfide linkages, organized in a heart-shaped structure with three structurally different domains (I, II, and III). A and B are the two subdomains that make up each domain. BSA is a non-glycosylated globular protein produced by the liver without requiring prosthetic groups or other additives (Phan et al. 2015).

9.4.2.3 Human Serum Albumin (HSA)

HSA, a naturally occurring protein, is found in the blood of humans. It is a non-glycosylated protein synthesized and secreted by liver cells called hepatocytes into the circulatory system. Approximately 10 to 15 grams of albumin are generated and released by the liver every day (Caraceni et al. 2013). Within human blood serum, proteins comprise approximately 7% of the composition, with two-thirds of that protein fraction belonging to albumin and the remaining third residing in the globulin fraction. The approximate alpha, beta, and gamma globulins concentrations in normal human serum are 1.5%, 1.9%, and 1.1%, respectively. Each globulin part consists of diverse types of proteins. Beta globulin contains transferrin, ceruloplasmin, lipoproteins, and mucoproteins, while gamma proteins have macroglobulins and antibodies. Each globulin fraction is made up of a mixture of many different proteins. Alpha globulin comprises alpha lipoprotein, while beta globulin comprises transferrin and ceruloplasmin. Lipoproteins, mucoproteins, and gamma proteins are composed of macroglobulins and antibodies.

On the other hand, HSA comprises one single chain containing 585 amino acids with a highly flexible secondary structure comprising 67% helix formation. Additionally, it possesses seventeen disulfide bridges and six twists that serve as cross-linkers for the three homologous domains (Yadav et al. 2020; Caraceni et al. 2013).

X-ray crystallography shows that HSA has a three-dimensional heart-shaped structure (Quinlan et al. 2005). Each domain has subdomains (A and B) separated by four and six helices. Two crucial chemical binding sites exist in subdomain IIA: Sudlow site I and Sudlow site II (Sudlow et al. 1975). There are 35 cysteine residues in the albumin structure and 34 help form disulfide bonds. The extraordinary stability of albumin is a result of these qualities. Drug release benefits from the free cysteine residue (cys34) at position 34 because it enables covalent drug binding (Quinlan et al. 2005).

9.4.3 Properties of Albumin Nanoparticles

A longer lifetime of albumin is caused by the interaction between albumin and the neonatal Fc receptor (FcRn). Due to this interaction, albumin can avoid the pH-dependent mechanism FcRn receptors use to damage cells and recycle albumin in the blood stream. Albumin is desirable for lengthening the drug's half-lives and allowing targeted intracellular drug delivery because of the previously mentioned characteristics (Yadav et al. 2020). The functional amino and carboxylic groups in albumin allow for the functionalization of albumin nanoparticles with specific ligands or active therapeutics. Additionally, the stability of albumin nanoparticles allows for the regulated delivery of various drugs. For instance, siRNA and RNA interference agents based on plasmids have been effectively delivered using albumin nanoparticles (Jahanban-Esfahlan et al. 2016).

HSA is a protein that can bind ligands, making it a suitable candidate for loading various drugs intended for delivery through the circulatory system. Albumin nanoparticles generally allow for more precise drug release control than liposomal formulations. The main functions of this protein involve regulating blood osmotic pressure and serving as the primary carrier for hydrophobic compounds, including fatty acids and hormones. These essential roles make it a prime candidate for drug delivery purposes. Bovine serum albumin nanoparticles (BSA-NPs) are among the numerous nanocarriers studied for their drug-delivery capability in biomedical applications. BSA-NPs have significant promise as nanocarriers for delivering chondrogenic substances in cartilage tissue regeneration due to their high drug loading and continued release capacity (Larsen et al. 2016; Jahanban-Esfahlan et al. 2016).

9.4.4 Method of Preparation of Albumin Nanoparticle

Different physicochemical and chemical techniques are employed to fabricate albumin nanoparticles. An essential aspect of achieving this is the reproducibility of these methods, ensuring that each production process yields nanoparticles with consistent and predictable characteristics. Striving for reproducibility is vital to ensure that the traits of the generated nanoparticles can be anticipated and replicated reliably (Lv et al. 2015).

9.4.4.1 Physiochemical Methods

9.4.4.1.1 Nano Spray Drying

This technique is used for producing nanoparticles from solutions. The equipment consists of a chamber of heated nitrogen and carbon dioxide and an aqueous albumin solution that is sprayed. The size of the particles in this nanoparticle manufacturing method can be controlled by adjusting parameters such as the nozzle size and the spraying rate of the particles. To stabilize polymer particles, surfactants are often added to protein nanoparticles. Additionally, the process of solvent evaporation aids in maintaining the temperature of nanoparticle droplets and can be utilized for heat-sensitive samples (Lv et al. 2015).

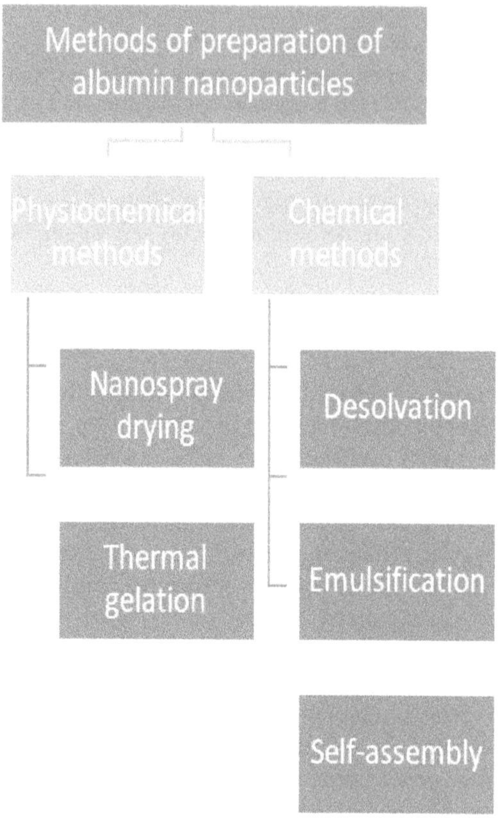

FIGURE 9.2 Schematic representation of various methods used for the preparation of albumin nanoparticles.

9.4.4.1.2 Thermal Gelation

In this technique, heat causes protein structural variations. Protein-protein interactions, such as disulfide-sulfhydryl, electrostatic, and hydrophobic interactions, are used to demonstrate hydrogen bond formation. It is also known as the sol-gel process with slight modification. This technique includes chemical processes like hydrolysis, gelation, and thermal therapy.

9.4.4.2 Chemical Methods

9.4.4.2.1 Desolvation

A desolvating agent, such as acetone ethanol or salt, is employed for this method. It is continuously added to an aqueous solution, drip-by-drip, until the desired turbidity stage is achieved (Storp et al. 2012). The tertiary structure of albumin is changed due to the desolvating agent, inducing phase separation (also known as coacervation) and protein association. As a result, the homogeneous solution separates into two distinct phases: one composed of the solvent and the other consisting of associated proteins forming the nanoparticle. Sometimes cross-linkers, such as glutaraldehyde, are added to stabilize the nanoparticles. Several crucial factors govern the nanoparticle formation process in this method. These characteristics include pH, protein content, cross-linker concentration, solvent phase, ionic strength, mixing speed, and the volume ratio of the aqueous phase to the solvent. An important consideration is also the quantity of desolvation agent used in the aqueous protein solution (Demirkurt et al. 2019).

To obtain desirable nanoparticles through this process, it is essential to consider the following point: when utilizing ethanol, the ratio of organic solvent to water should exceed 2. This ratio would be higher than 1 for acetone. In addition, the poly dispersity index of nanoparticles occurs with the

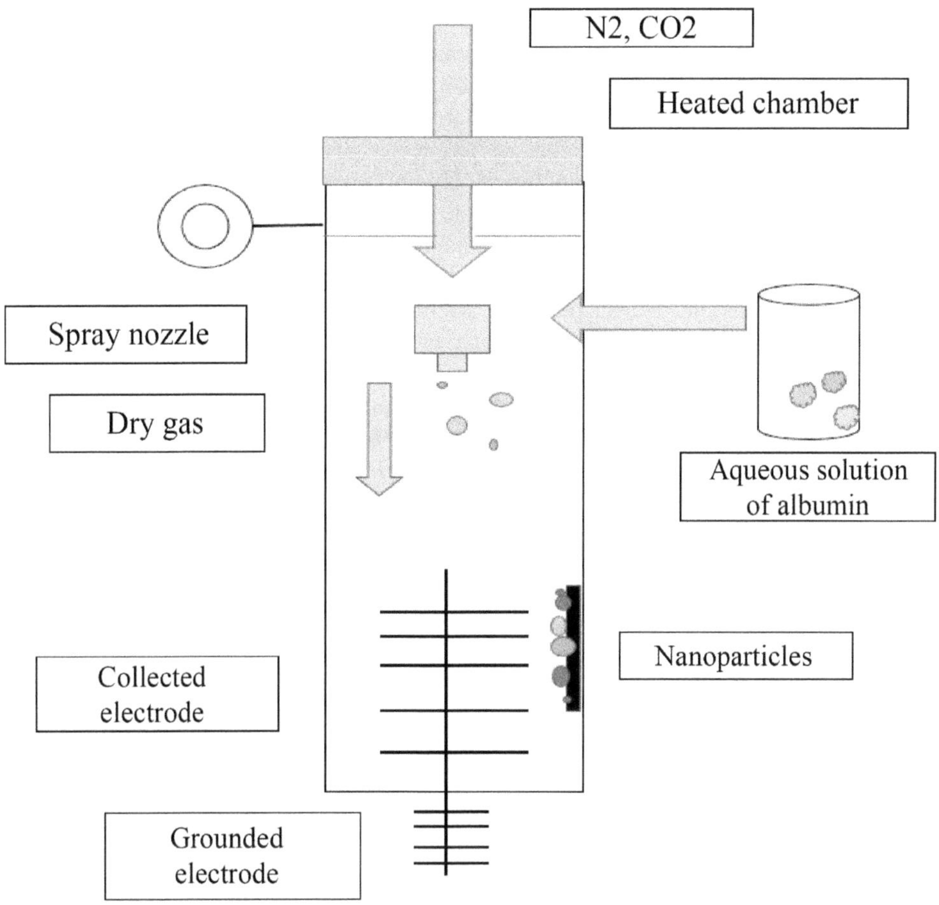

FIGURE 9.3 Diagrammatic representations of the nano spray drying method.

increase in the desolvating agent absorbed by the liquid protein. Because serum albumin's isoelectric point is close to circumstances around this value favors protein aggregation and nanoparticle production. pH values between 7 and 9 produce smaller nanoparticles, but the process yield is significantly reduced (Spada et al. 2021).

9.4.4.2.2 Emulsification

In this technique, the nonaqueous solution, the oil phase (cottonseed oil or organic solvent), is added to the aqueous, the albumin phase. To stabilize the high emulsion concentration of surfactant is added, the energetic stirring method is employed. Crosslinker-like glutaraldehyde is also added to harden the nanoparticle. Like desolvation, emulsification rate, mixing conditions, oil volume, water level, type and concentration of surfactants, and other factors all impact the formation of nanoparticles. A crude emulsion is formed by introducing a lipophilic drug to the albumin solution, which is subsequently subjected to high pressure through a jet to generate nanoparticles within the size range of 100–200 nm (Lluch et al. 2014).

9.4.4.2.3 Self-Assembly

To enhance the hydrophobicity of albumin, albumin nanoparticles are created by either disrupting disulfide bridges using mercapto ethanol or reducing the number of primary amine groups on the surface of the protein by adding a lipid-based substance.

9.4.4.3 NABT (Nanoparticle Albumin-Bound Technology)

This technology consists of a combination of emulsion and evaporation methods. American Bioscience Inc. invented this method, and it is now used to make albumin nanoparticles loaded with paclitaxel, which is sold as Abraxane® (nab-paclitaxel) (Fu et al. 2009). Nab-paclitaxel was the first FDA-approved and very successful chemotherapeutic treatment to use then a no particle albumin-bound route of drug delivery. This oil phase was combined with an aqueous phase containing albumin and 1% chloroform, and the resulting combination was homogenized to generate crude emulsion. High-pressure homogenization produces the final emulsion, followed by removing the solvent before transferring the mixture to a rotary evaporator. This process yields a transparent nanosuspension, wherein the drug-loaded albumin nanoparticles typically possess diameters ranging from tens to hundreds of nanometers (usually below 200 nm). Also, an aseptic filter membrane uses a pore density 0.22 μm to filter out the impurities, bacteria, and residual solvent. To make solid powders, nanoparticles are lyophilized (without any cryoprotectant). The NAB paclitaxel (ABI-007) has several advantages over the existing anticancer Taxol product (Cremophor's solvent). It does not require premedication and has an infusion time of 30 minutes, linear pharmacokinetic characteristics, and good patient acceptability. In this technique, the particle size depends upon factors like the concentration of albumin, the concentration of the drug (paclitaxel concentration), and the amount of ethanol added. It is generally preferred that a low fraction of ethanol be added in the organic phase to generate particles of desirable size. The oil phase consists of ethanol, drug (Paclitaxel), and chloroform, whereas the aqueous phase consists of human serum albumin presaturated with 1% chloroform. The ethanol is added to generate particles of smaller size. The preferred range for homogenization is 3,000–40,000 psi. The typical diameter of paclitaxel nanoparticle obtained is 140–160 nm (Fu et al. 2009).

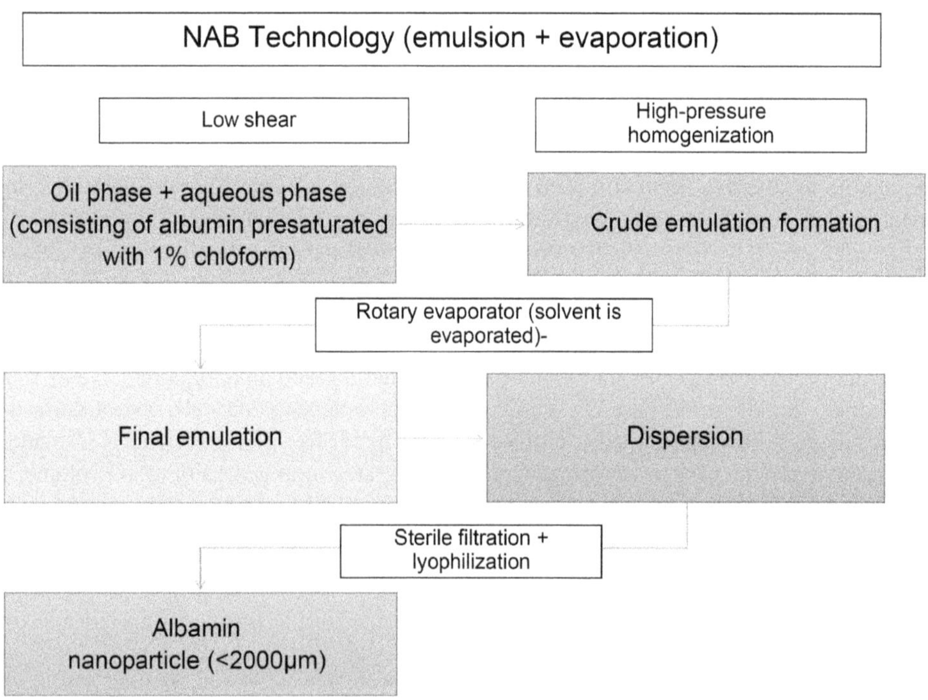

FIGURE 9.4 Schematic representation of steps involved in the NAB Technology process of ABI-007.

9.5 APPLICATION OF ALBUMIN CARRIER IN PHYTOCHEMICAL DELIVERY

9.5.1 Application of Albumin Nanoparticles in Cancer Treatment

In 2020, cancer was responsible for nearly 10 million deaths worldwide, emerging as the leading cause of mortality. The most prevalent types of cancer during that year were breast cancer (2.26 million cases), lung cancer (2.21 million cases), colon and rectal cancer (1.93 million cases), prostate cancer (1.41 million cases), non-melanoma skin cancer (1.20 million cases), and stomach cancer (1.20 million cases). Lung cancer accounted for 1.80 million cancer-related deaths in 2020, followed by colon and rectum (916,000), liver (830,000), stomach (769,000), and breast (685,000) deaths. Moreover, approximately 400,000 children are diagnosed with cancer each year. The prevalence of specific malignancies varies across countries, with cervical cancer being the most common type in 23 nations (https://www.who.int/news-room/fact-sheets/detail/cancer WHO factsheet [2020], last accessed on 25/07/2022).

Vegetables, fruits, and other edible plant parts contain natural substances called phytochemicals that fight cancer. Additional research on the pharmacological effects of natural anti-inflammatory compounds has revealed that these compounds frequently impact a variety of cancer cells, signaling pathways, and other processes. They are immunomodulatory, anti-inflammatory, and antioxidant in nature (Xie et al. 2016).

Over the past few decades, many phytochemicals have been extensively researched globally for their potential in cancer treatment. Despite the promise that phytochemicals have shown as cancer chemotherapeutic agents, several problems must be solved before their use becomes widespread. Some challenges include low solubility, poor cell penetration, hepatic disposition, low therapeutic index, and quick absorption by normal tissues. Phytochemicals often have an overall apparent volume of distribution, which causes significant drug accumulation in healthy tissue. Phytochemicals may also exhibit undesirable pharmacokinetic characteristics, such as a quick clearance rate and a brief elimination half-life (Rizwanullah et al. 2018). Anthracyclines, *Vinca* alkaloids, camptothecin, and its derivatives (Bowman and Leong 2006). Taxanes and their analogues, podophyllotoxin and its analogues, and many more substances are naturally occurring anticancer drugs.

The development of nanoencapsulation phytochemicals delivered by intravenous, transdermal, and ocular routes has been made possible by access to nanotechnology. The technique is widely used for encapsulating curcumin and its analogy (dibenzoyl methane) in nanoemulsion (Jahanban-Esfahlan et al. 2019). In 2005, the US Food and Drug Administration (FDA) approved ABI 007 (Abraxane®), marking the introduction of the first nanoparticle formulation based on human serum albumin. Abraxane, a 130-nm albumin-bound (nabTM) formulation of paclitaxel, emerged as a treatment for metastatic breast cancer. Extensive animal and human trials have revealed that Abraxane exhibits more excellent antitumor activity than conventional paclitaxel (Zhang et al. 2017).

9.5.1.1 Phytochemical Delivery in Different Types of Cancer

9.5.1.1.1 Albumin-Based Delivery System Using Gambogic Acid for Lung Cancer Treatment

Gambogic acid, a naturally derived compound released by the *Garcinia hanburyi* tree in the form of dry resin, possesses many biological properties. These properties encompass anticancer, anti-inflammatory, antioxidant, and antibacterial effects. Numerous studies have indicated that gambogic acid (GA) hampers invasion, metastasis, and angiogenesis while promoting apoptosis, autophagy, and cell cycle arrest. These molecular pathways are all thought to contribute to GA's clear anticancer benefits (Liu et al. 2020).

Clinical studies have demonstrated the efficacy of gambogic acid as a potent anticancer drug for treating lung cancer. Nonetheless, the medical effectiveness of gambogic acid is hindered by its limited water solubility and low chemical stability. By developing a more efficient delivery method based on human serum albumin, gambogic acid's solubility, chemical stability, and antitumor efficacy have all been enhanced. This improvement inhibits the development of various cancers, such as breast, pancreatic, prostate, and lung cancers and osteosarcoma. The modification of several signaling pathways, such as c-Jun N-terminal kinase-1 (JNK-1), protein kinase B (AKT)/mammalian target of rapamycin (mTOR), AKT/forkhead box protein O1 (FOXO1)/BIM, nuclear factor kappa-B (NF-B), and others, may be responsible for the inhibitory action. An integrated approach to tumor treatment is accomplished by modulating both intrinsic and extrinsic apoptotic pathways through the co-delivery of TRAIL and gambogic acid using nanoparticles based on folate-modified carboxymethyl-chitosan/polyethyleneimine/bovine serum albumin (Liu et al. 2020).

The nanoparticles (NPs) employed in the study consisted of either molecular or amorphous forms of gambogic acid (GA), and any new chemical bond formation was not observed between GA and HSA. The in vitro testing showed that the NPs exhibited excellent and more consistent drug release than the solution of GA and Arg. Stability investigations indicated that even after freeze-drying, the particle size and filtering effectiveness of GA-HSA NPs remained unchanged for up to 12 months and were stored in an aqueous solution for up to 8 weeks. Moreover, the chemical stability of GA within the HSA NPs was notably superior to that of the GA-Arg solution when diluted with pH 7.4 PBS or regular saline (Zhang et al. 2017).

9.5.1.1.2 Albumin-Stabilized Nanoparticle Formulation Paclitaxel (Abraxane®)

Natural taxane paclitaxel serves as the fundamental ingredient of Abraxane and a nanoparticle composite stabilized with human serum albumin and devoid of Cremophor toxins. Due to the toxic Cremophor's absence and shorter infusion time, this nanoformulation enables the delivery of paclitaxel at greater dosages (4–5 times), removing the need for premedication. This drug, combined with other treatments, is employed in managing several cancer types, including pancreatic, metastatic, lung, and bladder cancer. The novel nanoformulation interacts with other proteins overexpressed in cancer cells and the albumin receptor of endothelial cells to transport compounds into the cytoplasm. Compared to traditional paclitaxel alone, nanosized cancer treatments increased intratumoral accumulation and improved bioavailability, increasing the therapeutic cancer index in various cancer types. A pH-coacervation technique was employed to produce two types of nanoparticles: drug-free human serum albumin (HSA) nanoparticles and noscapine-loaded nanoparticles. The latter nanoparticles were cross-linked using glutaraldehyde. The coacervation process may be utilized to encapsulate noscapine in HSA nanoparticles effectively. The pH of the preparation appeared to impact the size and particle production, but it did not impact the ability to load drugs.

Additionally, in vitro drug delivery investigations suggested regulated delayed release patterns. Finally, utilizing noscapine-loaded human serum albumin (HSA) nanoparticles holds promise for delivering optimal doses of noscapine to target sites, achieving the highest therapeutic efficacy while minimizing potential adverse effects. Noscapine nanoparticle surfaces are being modified in other trials to target tumor cells specifically and better understand their effectiveness and harm to cancer cells (Dhupal and Chowdhury 2020; Sebak et al. 2010).

9.5.1.1.3 Forskolin-Loaded Human Serum Albumin Nanoparticles in Cancer Therapy

Tertiary alphahydroxy ketone, a plant constituent known as forskolin, was isolated from the *Coleus forskohlii* plant. In the investigation of cell physiology, it is frequently employed to increase cyclic AMP levels. These phytoconstituents have been shown to affect various critical human processes, including preventing the survival and growth of colon cancer cells. Recently, forskolin-polysorbate-80 nanoemulsions were fabricated and measured as methods for delivering forskolin in skin permeation in vitro.

Besides its diverse applications, forskolin exhibited its most potent anticancer function when combined with conjugate human serum albumin nanoparticles. This conjugation enhanced forskolin's effectiveness, bioavailability, and biodistribution, augmenting its anticancer properties. Forskolin has been well-known for its positive benefits, but the interaction of forskolin with the primary carrier protein HSA is unknown. Significant research has been done on HSA binding with forskolin because it affects the drug's pharmacokinetic, pharmacodynamic, and in vivo system functionality. Using biophysical and computational approaches, the molecular interactions of HSA with forskolin under physiological conditions (pH 7.4) were observed. Forskolin-loaded HSA NPs were also created using a modified desolvation technique to improve drug encapsulation, characterize the nanoparticles, and ensure their stability. HepG2 cell lines were used to investigate the cytotoxicity and drug release in vitro. These results give information on the unique effect of forskolin-loaded HSA NPs and the insights of HSA binding with forskolin, providing new insights into the effectiveness of drug loading, clearance rates, and tumor toxicity. Up to 100 µm concentration, there was around 27.7% higher cytotoxicity with NPs than with forskolin-loaded HSA NPs, and the insights of HSA binding with forskolin demonstrated that they were more effective in inhibiting cancer cells. The targeting of anticancer drugs has been improved, and their adverse effects have been reduced via the use of HSA-based delivery systems. Forskolin-loaded HSA NPs can be utilized to treat hepatocellular cancer (Nagati et al. 2020).

9.5.2 Application of Curcumin-Loaded Albumin Nanoparticles for the Treatment of Parkinson's Disease

Parkinson's disease, a neurodegenerative disorder, is characterized by impaired dopaminergic neurons' firing in the substantia nigra, a specific region in the basal ganglia. This leads to a reduction in dopamine release, contributing to the symptoms of the disease. Curcumin is the natural polyphenol obtained from the rhizomes of the *Curcuma longa*. In this, the nasal delivery of the drug is done to cross the blood-brain barrier, a non-invasive method to provide therapeutic chemicals to the brain (Shailja et al. 2020).

Albumin nanoparticles loaded with curcumin were successfully synthesized and designed using the pH coacervation method. The process involved using ethanol as a desolvating agent, glutaraldehyde as a cross-linker, and albumin was used as a polymer. The pH of the system was carefully maintained within the range of 7–8. Anhydrous glucose in 1% concentration was used as a cryoprotectant and was added to the nanoparticle suspension. The particle size and polydispersity index of the nanoparticles were analyzed with the help of dynamic light scattering with a zeta sizer instrument, yielding measurements of 147.5 nm and 0.330, respectively, indicating favorable characteristics. Both in vitro and in vivo methods were used for drug release studies. The in vitro study was done in phosphate buffer at pH 6.4 using a dialysis membrane, and the cumulative drug release percentage was determined to be 97.614%. For the ex vivo technique, the nasal membrane of the goat was used to keep a 5 ml suspension of drug-loaded nanoparticles. Following a 24-hour duration, the samples were subjected to scanning using a UV spectrophotometer at 289 nm. The ex vivo drug release within 24 hours from the albumin nanoparticles demonstrated a rate of 93.623%. The surface morphology and shape were determined by using scanning electron microscopy (SEM), which showed particles with spherical and smooth surfaces (Shailja et al. 2020).

The observation revealed that hydrophilic matrices undergo polymer swelling, leading to drug release through diffusion. This phenomenon aligns well with the Korsmeyer-Peppas equation, which describes drug release as a result of swelling of the polymer matrix and remains constant throughout the release of the drug within the body. The curcumin-loaded albumin nanoparticles can quickly go into the nasal mucosa. Also, these nanoparticles were found to be less toxic than other dosage forms and showed sustained drug release. The herbal-loaded nanoparticles were formulated to increase the therapeutic efficacy and decrease the dosage frequency (Shailja et al. 2020).

9.5.3 Application of Albumin Nanoparticles in Tissue Regeneration

Secondary metabolites produced by plants serve as a storehouse of phytochemicals for regenerative medicine applications because of their broad range of biological actions, which include antioxidant, antibacterial, anti-inflammatory, and tissue regeneration capacities. In addition, compared to traditional therapies that aim to lessen the long-lasting adverse effects of synthetic drugs, bioactive phytochemicals can be far more readily available, frequently more economical in large-scale manufacturing, and generally more tolerable. Unfortunately, their limited therapeutic impact results from their low absorption and lack of long-term stability. These restrictions can be addressed by nanotechnology-based delivery methods that increase the local efficacy of bioactive chemicals while decreasing the risk of adverse effects on healthy bone (Valentino et al. 2021).

Since the last decade, there has been an increased understanding of various phytochemicals, demonstrating a vast potential to substitute all the expensive recombinant growth factors and find good indigenous options that can act as replacements due to their limited half-lives and non-malignant properties. This branch of science combines traditional medicine with cutting-edge technologies, including clinically tested phytochemicals with cutting-edge technologies such as tissue engineering and nanotechnology, which are increasingly gaining significance in healthcare.

Additionally, advanced fabrication methods like electro spinning offer many opportunities for synthesizing next-generation biomaterials and tissue-engineered products with phytochemical integration ranging from nano to submicron levels to promote efficient site-specific delivery of therapeutic agents (Joseph et al. 2018). Because of graft rejection, pathogen transmission, and invasive surgical techniques, the usual treatment techniques have shown to be ineffective. Biomaterials, including different delivery vehicles, bioactive chemicals, stem cells, or demineralized bone matrix, may help generate bones by promoting and guiding the development of microenvironments that favor new bone formation because many bone transplants have adverse effects and constraints. An engineered bone replacement should have osteogenic (osteoinductive and osteoconductive) capabilities and be biocompatible with the host tissue when used as a template for new bone growth. Significant progress has been seen over the past few decades due to the utilization of nanotechnology in manufacturing materials that can naturally mimic bone characteristics (Valentino et al. 2021).

Much research is being done on biomaterials-based three-dimensional scaffolds for bone regeneration. A potential scaffold should have osteoconductive, osteoinductive, and osteogenic properties to enhance the process of bone regeneration. Due to the sustained release of recombinant human bone morphogenic protein-2, severe adverse reactions, including swelling of the mouth and neck, radiculitis, swelling of the organs around the release site, tissue swelling, and an increased risk of malignancy, are observed. Sterility in males was also observed. These growth factors are part of a family of natural growth factors involved in various signaling pathways; therefore, delivering them at the site of therapeutic location might cause severe adverse side effects on the surrounding tissues and organs. Resveratrol is a polyphenolic phytoestrogenic chemical obtained from plants. Resveratrol-loaded albumin nanoparticles (RNPs) boost the scaffold's therapeutic applicability in bone regeneration due to getting trapped inside the scaffold. Bone marrow generates the mesenchymal stem cells that enhance the proliferation and differentiation of osteoblasts in the presence of polyphenolic phytoestrogen resveratrol. The differentiation of mesenchymal stem cells from the bone marrow into the osteoblast line by enhancing alkaline phosphate activity is promoted by resveratrol.

Additionally, resveratrol stimulates osteoblast development and proliferation, improving bone production likelihood. The metabolism of resveratrol promptly takes place in the body resulting in the formation of sulfated and monoglucuronide derivatives that are excreted. Still, it may also aid in bone regeneration. At the bone loss site, oral delivery of resveratrol may not be as effective as the controlled release. A porous three-dimensional polycaprolactone (PCL) scaffold was strengthened using RNP to combat the adverse effects of resveratrol's quick body clearance. This permitted it to deliver resveratrol under regulated conditions. PCL-RNP was created, and its capacity to build bones and its in vitro drug release kinetics were assessed. The morphology of the scaffold made up of PCL-RNP was studied using scanning electron microscopy, and the scaffolds' qualitative properties were studied using Fourier transform infrared (FTIR) and X-ray diffraction. It was also physically characterized with a zeta sizer and atomic force microscopy. The mineralization ability of the scaffold was evaluated under in vitro conditions using human-bone-marrow-derived mesenchymal stem cells (hBMSCs).

According to an alkaline phosphatase experiment, PCL-RNP showed increased activity compared to a PCL scaffold (the positive control). Furthermore, von Kossa staining for calcium deposition revealed that PCL-RNP may have contributed to enhanced mineralization compared to PCL (negative and positive controls). The observations suggest prolonged and continuous resveratrol treatment may increase PCL-RNP's in vitro therapeutic potential. Throughout this investigation, scaffolds gradually released resveratrol over time. Due to its trapping in nanoparticles, resveratrol also underwent amorphization, accelerating its breakdown. Sixty-four percent of the drug was released from PCL-RNP over 12 days. A significant increase in proliferation in PCL-RNP indicated mineralization in research that targeted small bioactive molecules derived from plant sources in bone regeneration. The adverse effects associated with the therapeutic use of some natural growth factors can be decreased due to the prolonged release of these substances. Continuous and regulated distribution of resveratrol from PCL-RNP in bone repair is critically needed. Due to its beneficial effects on bone formation, it encourages both the conduction and induction of bone development, which is highly desired (Cinan et al. 2021).

9.5.4 Applications of Albumin Polymeric Nanoparticles

Combining proteins with polymers can increase albumin's ability to deliver drugs. This allows albumin carriers to encapsulate more drugs while displaying distinctive traits linked to synthetic polymers, such as stimuli-responsive features. Two unique kinds of nanoparticles produced from polymer-albumin conjugates exist. Albumin is used in the first class of nanoparticles as a bioactive surface coating, whereas albumin is used in the second class for its biocompatible qualities. Bioactive albumin-coated nanoparticles are produced by the self-assembly of albumin-polymer conjugates or by post-coating already-existing nanoparticles with albumin. Due to the coating of polymers, these nanoparticles can lead to albumin degradation and can cause an effect on drug release. A highly functional chain can be created by denaturing albumin, which can be modified using a polymer. Albumin nanoparticles can be designed flexibly to deliver different types of drugs, including those based on metals, drugs with hydrophobic nature, and even proteins with therapeutic potency and siRNA.

Mainly albumin nanoparticles act as transporters for several compounds in the blood plasma. Amino acids act as nourishing agents to the peripheral tissue when albumin is broken down. These characteristics make albumin a perfect material for nanoparticle formation for various drug delivery applications using HSA and BSA. These are widely investigated albumin types. Albumin, the most prevalent protein in the circulating blood, is a protein transporter of various endogenous and foreign substances. Albumin's characteristics make it an excellent material for the development of nanoparticles. The albumin by itself is also capable of producing these nanoparticles. The possibility of creating nanoparticles that can carry a variety of drugs, including hydrophobic anticancer agents, proteins, and nucleic acid therapeutics, is significantly increased by incorporating albumin with synthetic polymers. These two elements are combined to provide the main strategies for conjugating albumin and polymers and creating nanoparticles. In addition to enhanced drug-loading capacity, the resultant albumin-polymer combination nanoparticles also include traits including biodegradability, biocompatibility, and selectivity toward cancer cells. Numerous in vitro and in vivo investigations have shown how effective albumin-polymer nanoparticles are at delivering drugs under control and increasing the cytotoxicity of medicinal medicines. To completely appreciate the pharmacokinetic behavior of albumin-polymer drug carriers in vivo, more pharmacological research is still required (Jiang and Stenzel 2016).

9.5.5 Application of Albumin Carrier in Phytochemical Delivery for the Treatment of Liver Fibrosis

Liver fibrosis is caused by long-term damage and an extracellular matrix build-up, and no authorized medical treatment is currently available. Herbal medicine has been used to treat liver fibrosis more frequently than chemical substances. In experimental trials, several plant-derived hepatoprotective compounds, such as curcumin, silymarin, resveratrol, salvianolic acid A, and others, have demonstrated encouraging action in fighting CLDs. However, their clinical translation was ineffective due to low aqueous solubility, poor intestinal absorption, and low bioavailability. Therefore, their nanoformulations appear to be a viable technique to amplify the therapeutic activity and facilitate the sustained release of these plant chemicals. This will increase their bioavailability and give a controlled release at the site of absorption (Ezhilarasan 2021).

When compared to individual compounds, plant-derived substances' hepatoprotective effects are improved by nano-preparation. Following are a few examples from different sources regarding applying phytochemical nanocarriers for treating liver fibrosis.

The main active ingredient in DiFiore's seeds is quercetin, a polyphenolic flavonoid widely found in herbs. This study shows that polymeric quercetin nanocarriers coated with albumin can self-assemble and control inflammation without losing albumin specificity. Some studies have shown that quercetin can reduce TNF-α and IL-1 in a diabetes model and downregulate the NLRP3 inflammasome/IL-1 and IL-18 pathway in alcoholic liver injury (Saha et al. 2020). Oridonin-loaded GBSA NPs have been created for liver targeting in work by Li et al. (Karimi et al. 2016).

Additionally, berberine was delivered using BSA nanoparticles in research to inhibit the proliferation of activated hepatic stellate cells during liver fibrosis. As opposed to the free medication, it was demonstrated that such a nanoparticulate formulation produced therapeutic effects at lower doses and might be utilized as an antidote for fibrosis-related liver damage (Poilil Surendran et al. 2017).

A polyphenolic flavonoid called silymarin is produced by purifying the seeds and fruits of the *Silybum marianum* plant. Its anti-inflammatory and antioxidant properties are employed in treating chronic liver disorders. Silymarin is one of the powerful hepatoprotective drugs that has been extensively studied in CLD patients and experimental animals, despite its poor aqueous solubility and limited absorption. Combining NPs with drugs made from plants frequently results in increased bioavailability. To create the NPs biologically, reducing agents derived from plants, such as silymarin, quercetin, and berberine, were utilized (Cengiz et al. 2015).

The plant parts, such as the roots and stems of Berberidaceae, are used to extract a yellow-colored alkaloid called berberine (BBR). BBR is a natural herbal remedy used for many years to treat infectious infections. Antifungal, antibacterial, anti-inflammatory, and anticancer properties have been described for BBR. Liver fibrosis typically develops after a persistent or recurrent attack that is intense enough to trigger a wound-mending response and is characterized by an excessive build-up of connective tissue in the extracellular matrix (ECM). Studies show that BBR inhibits the proliferation of liver fibrosis-related active hepatic stellate cells (HSCs) in rats given carbon tetrachloride (CCl_4) treatment. Berberine, glucose, and BSA, a nontoxic, biocompatible, and biodegradable protein, were combined to develop a nanoparticulate formulation (Lam et al. 2015).

The herbal medicine derived from *Sophora alopecuroides* L. has an alkaloid called oxymatrine (OM), which has been shown to have antifibrotic properties. Oxymatrine successfully lowered collagen synthesis and deposition in CCl_4-induced hepatic fibrosis in rats (Giannitrapani et al. 2014).

9.6 CONCLUSION AND FUTURE ASPECTS OF ALBUMIN NANOPARTICLES IN PHYTOCHEMICAL DELIVERY

This chapter discusses recent developments in phytochemical delivery using albumin nanoparticles. Albumins act as versatile phytochemical delivery nanocarriers and have various applications such as in cancer therapy, Parkinsonism, bone tissue regeneration, and the treatment of liver fibrosis. It also finds its application as a polymeric nanoparticle to effectively deliver phytocompounds. Because of its unique qualities, including non-immunogenicity, biocompatibility, biodegradability, and nontoxicity, albumin has gained the attention of various researchers in recent years because of its enormous potential as a drug delivery system. The possibility of immunogenic responses is, most critically, eliminated because it is not a foreign substance and is the most prevalent plasma protein in the human body. Various phytoconstituent are loaded into the albumin, leading to improved pharmacokinetic properties. The human serum albumin holds the GRAS FDA excipient status, which promotes the approval of new therapeutic and diagnostic agents for ethical committees and regulatory agencies.

Additionally, adding HSA or cationized albumin with glycoprotein gp60 receptor integration improves the ability to cross the blood-brain barrier, resulting in improved delivery of drugs across this barrier.

The biocompatibility of albumin nanoparticles within the biological system, their capacity to target particular tumor locations, their low toxicity, the regulated release of chemotherapeutic drugs, and their low clearance rates are all noteworthy characteristics. Still, efforts are underway using albumin-based nanoparticles to explore novel and current therapeutics. Altogether, direct delivery of phytoconstituents to the target using albumin nanoparticles would enhance patients' overall quality of life, and further investigations are going in the development direction. Researchers may soon be able to create novel delivery methods with improved properties that subdue adverse effects caused by the therapeutic administration and augment therapeutic effects using albumin nanoparticles. Albumin nanoparticles demonstrated a new avenue for protein-based drug delivery as emerging carriers for therapeutic delivery.

REFERENCES

Ahmad, R., S. Srivastava, S. Ghosh, and S. K. Khare. 2021. 'Phytochemical Delivery through Nanocarriers: A Review', *Colloids Surf B: Biointerfaces*, 197: 111389.

Bowman, K., and K. W. Leong. 2006. 'Chitosan Nanoparticles for Oral Drug and Gene Delivery', *Int J Nanomed*, 1: 117–128.

Caraceni, P., M. Tufoni, and M. E. Bonavita. 2013. 'Clinical Use of Albumin', *Blood Transfus*, 11 Suppl 4: s18–25.

Cengiz, M., H. M. Kutlu, D. D. Burukoglu, and A. Ayhancı. 2015. 'A Comparative Study on the Therapeutic Effects of Silymarin and Silymarin-Loaded Solid Lipid Nanoparticles on D-GaIN/TNF-α-Induced Liver Damage in Balb/c Mice', *Food Chem Toxicol*, 77: 93–100.

Cinan, E., S. Cesur, M. Erginer Haskoylu, O. Gunduz, and E. Toksoy Oner. 2021. 'Resveratrol-Loaded Levan Nanoparticles Produced by Electrohydrodynamic Atomization Technique', *Nanomaterials(Basel)*, 11.

Demirkurt, B., G. Cakan-Akdogan, and Y. Akdogan. 2019. 'Preparation of Albumin Nanoparticles in Water-in-Ionic Liquid Microemulsions', *J Mol Liq*, 295: 111713.

Dhupal, M., and D. Chowdhury. 2020. 'Phytochemical-Based Nanomedicine for Advanced Cancer Theranostics: Perspectives on Clinical Trials to Clinical Use', *Int J Nanomed*, 15: 9125–9157.

Ezhilarasan, D. 2021. 'Advantages and Challenges in Nanomedicines for Chronic Liver Diseases: A Hepatologist's Perspectives', *Eur J Pharmacol*, 893: 173832.

Fu, Q., J. Sun, W. Zhang, X. Sui, Z. Yan, and Z. He. 2009. 'Nanoparticle Albumin-Bound (NAB) Technology is a Promising Method for Anti-Cancer Drug Delivery', *Recent Pat Anticancer Drug Discov*, 4: 262–272.

Giannitrapani, L., M. Soresi, M. L. Bondì, G. Montalto, and M. Cervello. 2014. 'Nanotechnology Applications for the Therapy of Liver Fibrosis', *World J Gastroenterol*, 20: 7242–7251.

Gilani, J. S., S. Beg, C. Kala, S. M. Nomani, K. D. Mahapatra, S. S. Imam, and M. Taleuzzaman. 2020. 'Chemically Nano-Engineered Theranostics for Phytoconstituents as Healthcare Application', *Curr Biochem Eng(Discontinued)*, 6: 53–61.

Jahanban-Esfahlan, A., S. Dastmalchi, and S. Davaran. 2016. 'A Simple Improved Desolvation Method for the Rapid Preparation of Albumin Nanoparticles', *Int J Biol Macromol*, 91: 703–709.

Jahanban-Esfahlan, A., A. Ostadrahimi, R. Jahanban-Esfahlan, L. Roufegarinejad, M. Tabibiazar, and R. Amarowicz. 2019. 'Recent Developments in the Detection of Bovine Serum Albumin', *Int J Biol Macromol*, 138: 602–617.

Jiang, Y., and M. Stenzel. 2016. 'Drug Delivery Vehicles Based on Albumin-Polymer Conjugates', *Macromol Biosci*, 16: 791–802.

Joseph, J., R. Sundar, A. John, and A. Abraham. 2018. 'Phytochemical Incorporated Drug Delivery Scaffolds for Tissue Regeneration', *Regen Eng Transl Med*, 4.

Karimi, M., S. Bahrami, S. B. Ravari, P. S. Zangabad, H. Mirshekari, M. Bozorgomid, S. Shahreza, M. Sori, and M. R. Hamblin. 2016. 'Albumin Nanostructures as Advanced Drug Delivery Systems', *Expert Opin Drug Deliv*, 13: 1609–1623.

Khan, T., and P. Gurav. 2018. 'Phyto Nanotechnology: Enhancing Delivery of Plant Based Anti-Cancer Drugs', *Front Pharmacol*, 8.

Lam, P. L., S. H. L. Kok, R. Gambari, T. W. Kok, H. Y. Leung, K. L. Choi, C. S. Wong, D. K. P. Hau, W. Y. Wong, K. H. Lam, Z. X. Bian, K. K. H. Lee, and C. H. Chui. 2015. 'Evaluation of Berberine/Bovine Serum Albumin Nanoparticles for Liver Fibrosis Therapy', *Green Chem*, 17: 1640–166.

Larsen, M. T., M. Kuhlmann, M. L. Hvam, and K. A. Howard. 2016. 'Albumin-based Drug Delivery: Harnessing Nature to Cure Disease', *Mol Cell Ther*, 4: 3.

Liu, Y., Y. Chen, L. Lin, and H. Li. 2020. 'Gambogic Acid as a Candidate for Cancer Therapy: A Review', *Int J Nanomedicine*, 15: 10385–10399.

Lluch, A., I. Alvarez, M. Muñoz, M. Á. Seguí, I. Tusquets, and L. García-Estévez. 2014. 'Treatment Innovations for Metastatic Breast Cancer: Nanoparticle Albumin-Bound (NAB) Technology Targeted to Tumors', *Crit Rev Oncol Hematol*, 89: 62–72.

Lv, L., Y. Chi, C. Chen, and W. Xu. 2015. 'Structural and Functional Properties of Ovalbumin Glycated by Dry-Heating in the Presence of Maltodextrin', *Int J Food Prop*, 18: 1326–1333.

Nagati, V., S. Nakkka, D. P. Yeggoni, and R. Subramanyam. 2020. 'Forskolin-Loaded Human Serum Albumin Nanoparticles and Its Biological Importance', *J Biomol Struct Dyn*, 38: 1539–1550.

Phan, H. T. M., S. Bartelt-Hunt, K. B. Rodenhausen, M. Schubert, and J. C. Bartz. 2015. 'Investigation of Bovine Serum Albumin (BSA) Attachment onto Self-Assembled Monolayers (SAMs) Using Combinatorial Quartz Crystal Microbalance with Dissipation (QCM-D) and Spectroscopic Ellipsometry (SE)', *PLoS One*, 10: e0141282.

Poilil Surendran, S., R. George Thomas, M. J. Moon, and Y. Y. Jeong. 2017. 'Nanoparticles for the Treatment of Liver Fibrosis', *Int J Nanomed*, 12: 6997–7006.

Prof, D., and K. R. Gupta. 2009. 'The Antioxidant Phytochemicals of Nutraceutical Importance', *The Open Nutraceuticals J*, 2: 20–35.

Quinlan, G. J., G. S. Martin, and T. W. Evans. 2005. 'Albumin: Biochemical Properties and Therapeutic Potential', *Hepatology*, 41: 1211–1219.

Ribaya-Mercado, J. D., and J. B. Blumberg. 2004. 'Lutein and Zeaxanthin and Their Potential Roles in Disease Prevention', *J Am Coll Nutr*, 23: 567s–87s.

Rizwanullah, M., S. Amin, S. R. Mir, K. U. Fakhri, and M. M. A. Rizvi. 2018. 'Phytochemical Based Nanomedicines against Cancer: Current Status and Future Prospects', *J Drug Target*, 26: 731–752.

Saha, S., J. Kundu, R. Verma, and P. Chowdhury. 2020. 'Albumin Coated Polymer Nanoparticles Loaded with Plant Extract Derived Quercetin for Modulation of Inflammation', *Materialia*, 9: 100605.

Sebak, S., M. Mirzaei, M. Malhotra, A. Kulamarva, and S. Prakash. 2010. 'Human Serum Albumin Nanoparticles as an Efficient Noscapine Drug Delivery System for Potential Use in Breast Cancer: Preparation and In Vitro Analysis', *Int J Nanomedicine*, 5: 525–532.

Shailja, C., J. Mayuri, and I. Mojahidul. 2020. 'Design, Development and Characterization of Curcumin Loaded Albumin Nanoparticles for the Treatment of Parkinson's Disease', *Int J Pharm Sci Drug Res*, 12.

Sharma, A. T., S. S. Mitkare, and R. S. Moon. 2011. 'Multicomponent Herbal Therapy: A Review', *Int J Pharm Sci Rev Res*, 6: 185–187.

Spada, A., J. Emami, J. A. Tuszynski, and A. Lavasanifar. 2021. 'The Uniqueness of Albumin as a Carrier in Nanodrug Delivery', *Mol Pharm*, 18: 1862–1894.

Storp, B., A. Engel, A. Boeker, M. Ploeger, and K. Langer. 2012. 'Albumin Nanoparticles with Predictable Size by Desolvation Procedure', *J Microencapsul*, 29: 138–146.

Sudlow, G., D. J. Birkett, and D. N. Wade. 1975. 'The Characterization of Two Specific Drug Binding Sites on Human Serum Albumin', *Mol Pharmacol*, 11: 824–832.

Thakur, M., K. Singh, and R. Khedkar. 2020. Amity Institute of Food Technology, Amity University, Noida, Uttar Pradesh, India. *Functional and Preservative Properties of Phytochemicals*, 341.

Valentino, A., F. D. Cristo, M. Bosetti, A. Amaghnouje, D. Bousta, R. Conte, and A. Calarco. 2021. 'Bioactivity and Delivery Strategies of Phytochemical Compounds in Bone Tissue Regeneration', *Appli Sci*, 11: 5122.

Xie, J., Z. Yang, C. Zhou, J. Zhu, R. J. Lee, and L. Teng. 2016. 'Nanotechnology for the Delivery of Phytochemicals in Cancer Therapy', *Biotechnol Adv*, 34: 343–353.

Yadav, N., S. Parveen, and M. Banerjee. 2020. 'Potential of Nano-Phytochemicals in Cervical Cancer Therapy', *Clin Chim Acta*, 505: 60–72.

Zhang, Y., Z. Yang, X. Tan, X. Tang, and Z. Yang. 2017. 'Development of a More Efficient Albumin-Based Delivery System for Gambogic Acid with Low Toxicity for Lung Cancer Therapy', *AAPS PharmSciTech*, 18: 1987–1997.

10 Quality-by-Design Perspectives of Phytosomes for Pharmaceutical Benefits

Sumant, Subh Naman, and Ashish Baldi

10.1 INTRODUCTION

The word "phyto" denotes "plant," whereas "some" indicates "cell-like." Phytosomes are used in dosage formulation technology to develop drugs and herbal goods with higher bioactivity and, as a result, yields improved results than ordinary herbal extracts (Dubey et al., 2007; Gupta et al., 2007). Phytosome formulations are a revolutionary paradigm with considerably better clinical benefits, improved bioavailability, and better distribution to the organ tissues, without any compromises with the nutrient safety. Some water-soluble phytomolecules (mainly flavonoids and other polyphenols) can be altered into lipid-soluble complexes through complexation of herbal medications with phospholipids. These complexes are called phytosomes. They have greater permeability compared to basic herbal extracts, which results in a higher potential to pass the cellular membranes and thus eventually reach the systemic circulation. They, therefore, have better pharmacological and pharmacokinetic properties that are beneficial in managing acute disorders and also beneficial in pharmaceutical and cosmetic formulations (Kumar et al., 2022; Moscarella et al., 1993). Phytosomes offer some advantages, such as improved absorption of bioactive ingredients and reduced amount requirement. Phytosomes require less dose to achieve the desired effects because the bioavailability of the herbal constituent is improved. Phospholipids in phytosomes provide additional nutritional advantages (Karimi et al., 2015; Kumar et al., 2019). Phytosomes have the potential to penetrate through skin easily and thus increases their efficiency (Amin & Bhat, 2012). Phosphatidylcholine (PC) shows hepatoprotective activity and thus exhibits synergistic action when hepatoprotective elements are used in the phytosomal complexes (Bhokare et al., 2016). They have a high lipid profile and increased skin penetration, making them suitable for widespread application in cosmetics (Gupta et al., 2007). Despite the many benefit of phytosomes, some downsides have been discovered; for example, phospholipids (lecithin) can promote growth in MCF-7 breast cancer cell line (Gándola et al., 2014; Kumar et al., 2021). The phytoconstituents can leach, lowering the target drug concentration and leading to instability of phytosomes, which are major disadvantages (Chivte et al., 2017).

10.2 COMPONENTS OF PHYTOSOMES

Phyto-phospholipid complexes, i.e., phytosomes, are formed by combining phospholipids in a stoichiometric ratio with active components derived from plants (Mane et al., 2020).

10.2.1 Phospholipids

The phospholipids are zwitter ionic and amphiphilic molecules, which are essential component of plasma membrane (Sharma & Baldi, 2018; Singh et al., 2017). The hydrophilic portion (head) is composed of a negative-charged phosphate group and the hydrophobic portion (tail) composed of

DOI: 10.1201/9781003300588-10

long-chain fatty acids leading to its amphiphilic nature. These tail and head of phospholipids are bridged by an alcohol or glycerol group, thus enabling formation of lipid bilayer in natural systems. The phosphatide can be classified in two types based on alcohol group, i.e., glycerophospholipids and sphingophospholipids (Litzinger & Huang, 1992). Simplified structure of phytosomes has been displayed as Figure 10.1. Various advantages of phytosomal formulation have been shown in Figure 10.2.

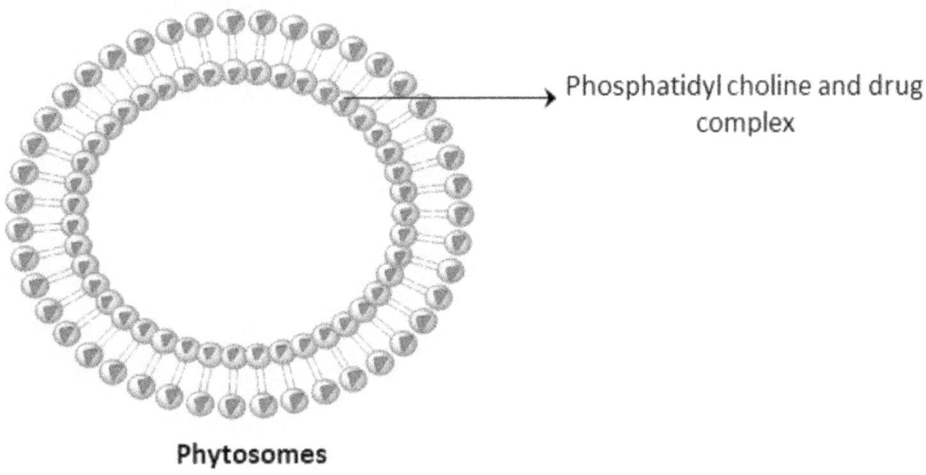

FIGURE 10.1 Structure of phytosomes.

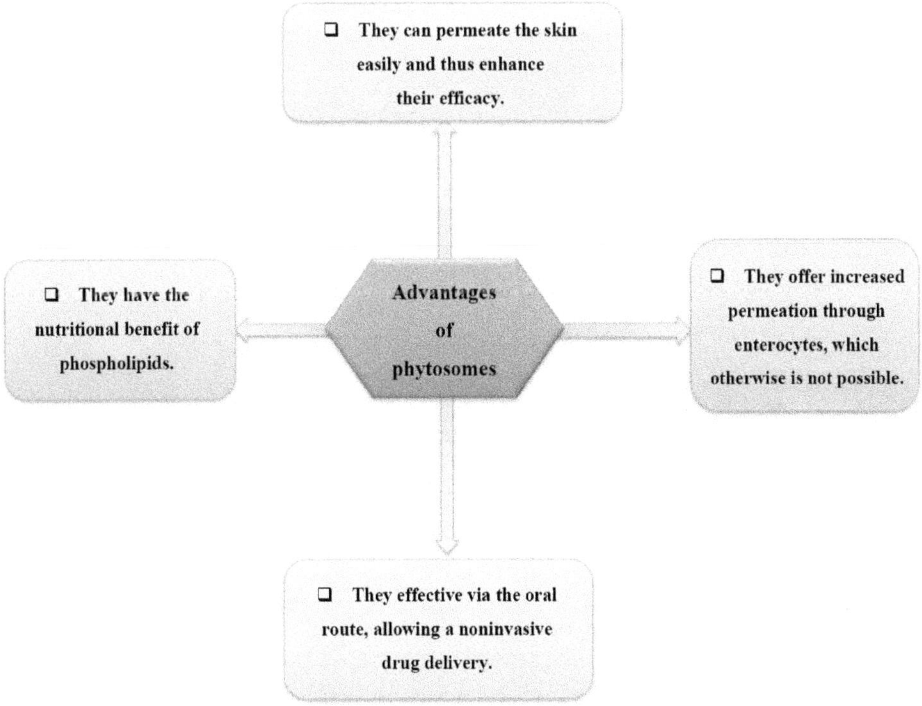

FIGURE 10.2 Advantages of phytosomes.

10.2.2 Phyto-Active Constituents

Phytosomes can be used to entrap both hydrophilic and lipophilic active ingredients to improve bioavailability of phytoconstituents (Semalty et al., 2009).

Liposomes and phytosomes are not similar; structurally, they are very dissimilar, as shown in Figure 10.3. There are some basic differences between these phospholipid-based delivery systems, which are described in Table 10.1.

TABLE 10.1
Differences Between Liposomes and Phytosomes (Verma & Prasad, 2011)

S. No	Phytosomes	Liposomes
1.	Hydrogen bonding takes place.	During liposome formation, no hydrogen bonds are formed.
2.	Complexation usually arises in a ratio of 1:1 or 2:1.	Complexation ratio is very high in case of liposomes (e.g., 1:500 or 1:1,000).
3.	Phytosomes are formed in the solution, having low dielectric constant.	Liposomes are synthesized in the presence of a solution having greater dielectric constant.
4.	Phytosomes are ideal to enhance bioactivity via non-invasive approach (e.g., oral route).	Liposomes are ineffective when used without invasion.
5.	Phytosomes offer more physical stability due to the formation of weak chemical bonds.	Liposomes are less stable as compared to phytosomes.
6.	Phytosomes are used as a dietary supplement	Liposomes are not preferred in case of dietary supplements.
7.	Phytosomes are better absorbed as compare to liposomes. Hence, they exhibit more bioavailability of active constituents.	Liposomes exhibit less bioavailability as compare to phytosomes.
8.	Phytosomes are better to liposomes in cosmetics.	Liposomes are less superior for cosmeceuticals.

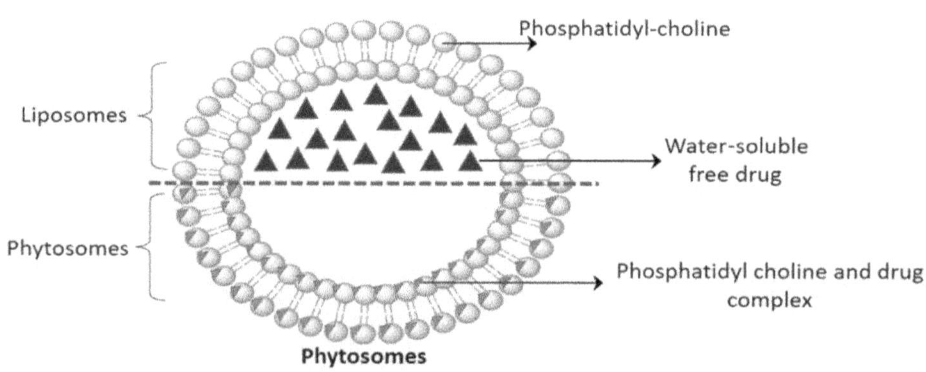

FIGURE 10.3 Diagrammatic representation of phytosomes and liposomes.

10.3 PREPARATION OF PHYTOSOMES

The general technique of production of phytosomes requires the following stages:

1. Phospholipids and substrates are combined in the presence of an aprotic solvent (e.g., dioxane and acetone) in an adequate ratio (ideally 1:1).
2. The precipitation method is used to isolate the complex. There are many ways to cause precipitation, including spray drying, lyophilization, and aliphatic hydrocarbons.
3. Phytosomes are dried.
4. Hydration creates a suspension of produced phytosomes (Bhattacharya, 2009).
5. The optimum phospholipid to PC ratio is 1:1 (Jose & Bombardelli, 1987). The common steps of the preparations of phytosomes are shown in Figure 10.4 (Jain & Jain, 2005).

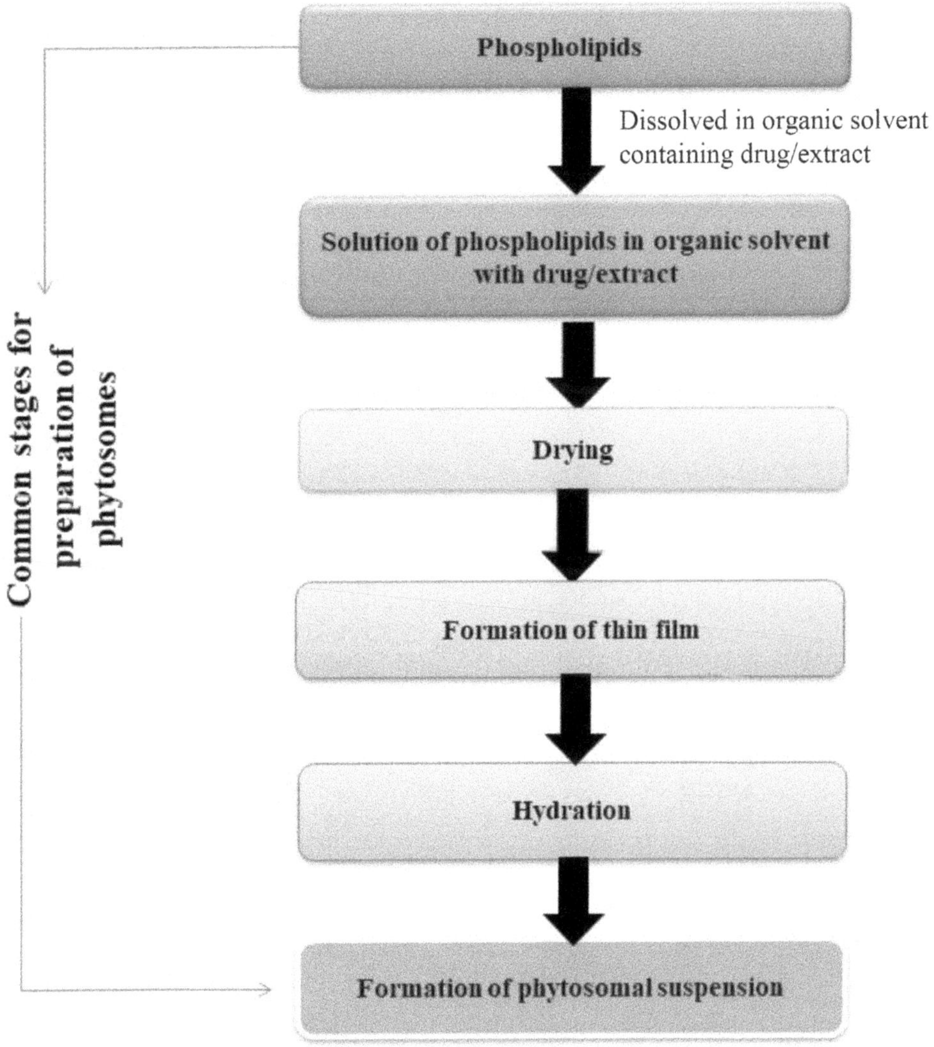

FIGURE 10.4 Common stages for preparation of phytosomes.

10.4 METHOD OF PREPARATION OF PHYTOSOMES

To prepare phytosomes (1:1) of plant constituents from selected plants, the solvent evaporation technique, antisolvent precipitation method, and rotary evaporation technique are most frequently used.

10.4.1 Antisolvent Precipitation Method

In this method, methanol is used to dissolve the amounts of plant extract and soy lecithin, whereas dichloromethane is used to dissolve the amounts of cholesterol. To get a concentration, the mixture is refluxed for 2 hours at a temperature of not more than 600°C. The precipitate is then formed by stirring in an antisolvent, hexane; it is then collected, filtered, and kept overnight in vacuum desiccators. The complex powder is chilled in an amber-colored glass container (Ittadwar & Puranik, 2017). Li and colleagues used antisolvent precipitation method to manufacture phospholipid complexes of puerarin and suggested that supercritical fluids is preferable to traditional procedures for the creation of medicinal phospholipid complexes (Li et al., 2008).

10.4.2 Rotary Evaporation Technique

A rotary round bottom flask, RBF, is used for dissolving the required quantity of plant extract and soy lecithin in THF, further stirred for 4 hours at a temperature of not more than 400°C, constant stirring through a magnetic stirrer thin layer of the material has been produced in which, n-hexane is added. Then, the obtained precipitate has been kept in an amber colored glass container and refrigerated to preserve it (Udapurkar et al., 2016). Detailed flow of rotatory evaporation technique has been shown in Figure 10.5.

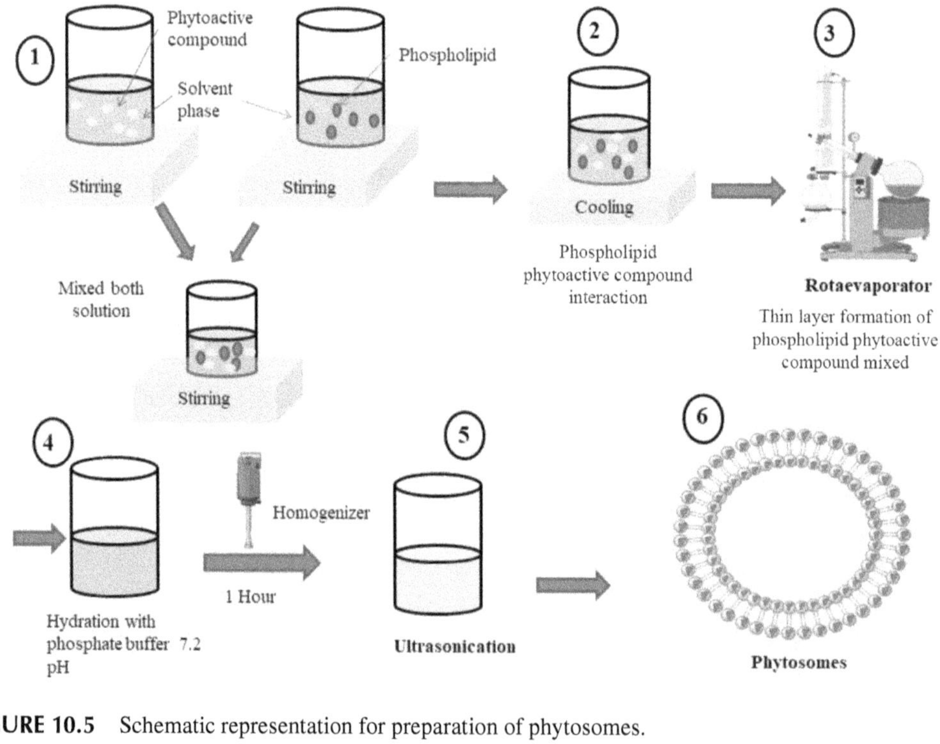

FIGURE 10.5 Schematic representation for preparation of phytosomes.

10.4.3 Solvent Evaporation Technique

In this method, acetone has been refluxed with the necessary amounts of plant extract and soy lecithin for 2 hours at a temperature of 50 to 600°C. The obtained mixture has been further concentrated to yield the precipitate, which is then collected and filtered. The precipitate of the phytosome complex is dried and put in an amber-colored glass container and refrigerated. Liu and coworkers prepared oleanolic acid–phospholipid complexes by solvent evaporation method (Liu, 2012). Yu and colleagues used solvent evaporation method to prepared a berberine-phospholipid complexes (P-BER) (Yu et al., 2016). Yanyu and coworkers developed a silybin-phospholipid complex using vacuum drying method and ethanol as solvent to enhance the oral bioavailability of silybin (Yanyu et al., 2006). Jiang and coworkers made phytosomes by using dissimilar phospholipids ratio of *herba epimedii* flavonoids and compared their dissolution profile with the tablets made with regular extract tablets. It is observed that cumulative dissolution rate of the produced phytosomes increases significantly (Jiang et al., 2001). Few of the researches related to phytosomes has been compiled in Table 10.2.

10.5 MECHANISM OF COMPLEX FORMATION IN PHYTOSOMES

Phospholipids are amphipathic in nature and have both non-polar and polar regions. The non-polar chain of phospholipid wraps around to lead a complex formation, which gave the lipophilic activity.

TABLE 10.2
Phytosomes and Their Biological Activity

S. No.	Phytosomal Formulation	Phytoconstituents Complexed with PC	Biological Activity	Reference
1.	Milk thistle phytosomes	Silybin from silymarin	For support of skin and liver	(Naik et al., 2006)
2.	Silymarin phytosomes	Silybin from silymarin	Silymarin phytosome demonstrated more feto-protectant efficacy against ethanol-induced behavioral impairments	(Busby et al., 2002)
3.	Silybin phytosomes	Silybin from *Silybum marianum*	Hepatoprotective	(Moscarella et al., 1993)
4.	Hesperetin phytosomes	Hydrogenated phosphatidylcholine loaded with hesperetin	Enhanced antioxidant activity	(Venkatesh et al., 2008)
5.	Green select phytosomes	Epigallocatechin 3-O-gallate from *Camelia sinensis* (green tea)	Beneficial against cancer, antioxidant activity	(Belcaro et al., 2013)
6.	Ginkgo select phytosomes	Ginkgo flavone glycosides from *Ginkgo biloba*	Neuroprotective activity	(Naik & Panda, 2008)
7.	Curcumin phytosomes	Naringenin (flavonoid from grape fruit, *Vitis vinifera*, and (flavonoid from turmeric, *Curcuma longa*)	Antioxidant activity higher in comparison of pure curcumin	(Maiti et al., 2007)
8.	Gallic acid phospholipid complex	—	Excellent DPPH radical scavenger and significant antioxidant activity	(Bhattacharyya et al., 2013)
9.	*Bacopa* phospholipid	—	Enhanced anti-amnesic action, improved absorption of bacopasides	(Habbu et al., 2013)

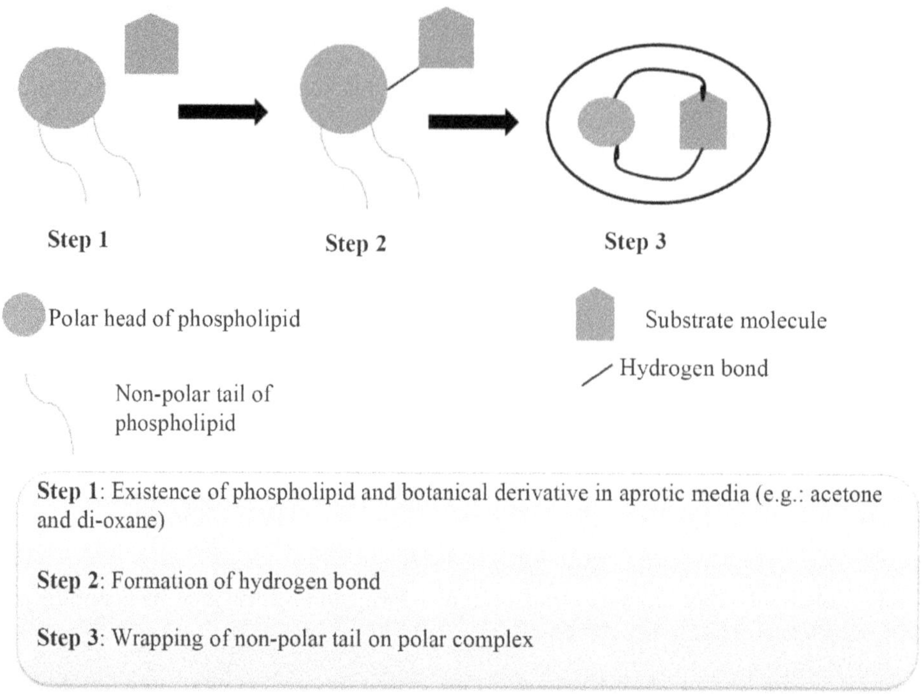

FIGURE 10.6 Step involved in formation of complex during phytosomes preparation.

Their polar end is made up of phosphate or amine groups that form weak hydrogen bonds with the polar groups on the substrate due to this they can easily permeate through the lipophilic enterocyte membrane, making it simpler for PC to pass across the Gastrointestinal-barrier (aqueous phase > enterocyte > blood) (Citernesi & Sciacchitano, 1995; Murray, 2004). The mechanism of action of phytosomes are also described in Figure 10.6.

10.6 CHARACTERIZATION OF PHYTOSOMES

There are several methods involved for the characterization of the phytosomes. Visualization of phytosomes can be done by transmission electron microscopy (TEM) and scanning electron microscopy (SEM) (El Maghraby et al., 2000) for determining the surface of phytosomes (Fry et al., 1978). Surface charge and size of the phytosome have been determined through zeta potential and particle size analyzer (Carini et al., 1994). Amount of active ingredient contained in the phytosome will be determined by drug content and entrapment efficiency (Cevc et al., 1995).

10.7 IMPLEMENTATION OF THE CONCEPT OF QbD IN PHYTOSOMAL PREPARATION

According to the ICH Q8 definition, "QbD [quality-by-design] is a systematic strategy to development that starts with established goals and promotes product, process knowledge, and process control, based on good science and QRM" (R2). The initial stage in implementing QbD for phytosomes is to establish a link between the quality target product profile (QTPP), critical material attribute (CMA), critical quality attribute (CQA), and critical process parameter (CPP). Figure 10.7

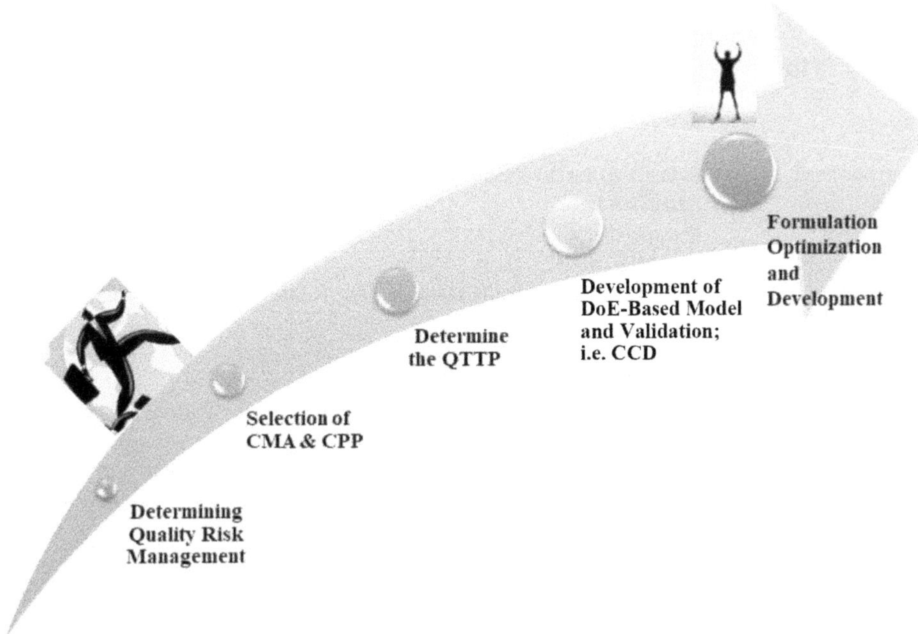

FIGURE 10.7 Steps involved in the development of phytosomal formulation based on QbD.

represents various steps involve in QbD based development of phytosomes as novel formulation (Yu et al., 2014). Phytosomes manufacturing is a highly sensitive process, and any mistakes made during the formulation process result in substandard products. The production of phytosomes is extremely delicate, so commercializing them still presents significant challenges. Execution of QbD in the formulation process minimizes the possibility of mistake and enhances the robustness of final phytosomal preparation. As the majority of regulatory authorities and the pharmaceutical industry are increasingly choosing their formulation decisions on the concept of QbD, this section discusses the various ways that QbD can be functional when creating novel formulations (Bajwa et al., 2022a, 2022b).

10.7.1 Step 1 of QbD: Identification of Quality Target Product Profile (QTPP) for Phytosomes

QTPP is one of the initial steps during the execution of QbD. Table 10.3 represents various QTTP factors responsible for phytosomal formulation.

10.7.2 Step 2 of QbD: Identification of Critical Quality Attributes (CQA) for Phytosomes

A very critical component of this stage is selecting the numerous process parameters and materials characteristics into CMA and CPP from the "plausible so many." After deciding the QTPPs for the phytosomes, it is critical to find the key quality aspects that have an impactful and severe consequence on the prepared phytosomal preparation. Table 10.4 shows various CQAs responsible for phytosomal formulation.

TABLE 10.3
Various QTPPs Involved During the Preparation of a Phytosome Formulation

QTPP Factors	Description	Remarks
Physical attributes of phytosomes	Particle size, surface morphology, and zeta potential	Differ as per targeted disease
Pharmacokinetics	Absorption, distribution, metabolism, and excretion (ADME)	Drug loaded phytosomal formulation required
Sterility	According to administration requirement, e.g., topical, oral, transdermal	Needed to increase patient compliance
Target patient population	Based on the requirements	Applicable to all patient groups
Appearance/dosage form	Phytosomes present in the aqueous suspension	Clear suspension
Action site	API based	Delivery of drug target site
Viscosity	Based on requirements	Influence the stability of phytosomes
Delivery route	Based on requirements	Based on drug selected for specific disease
Dissolution/solubility	In aqueous form	As needed, non-irritable
Indication/therapeutic effect	API based	Characteristics of API may require the use of phytosomes
Container-closure system	Store in suitable container	For final formulation stability in storage condition

TABLE 10.4
List of CQAs Screened Out After Analysis of QTPPs

CQAs	Description	Justification
Phytosomes classification	Conventional, immune, bio-responsive	Quality determination
Targeting	Route of administration	As per the therapeutic requirements
Vesicle size	100–1,200 nm	Permeability and targeted delivery, avoidance of first-pass metabolism (it needed)
Lamellas	As per requirement	Drug loading and permeability
Morphology	Shape and structure of phytosomes	Stability
Zeta potential	Higher value indicates greater stability	Stability
Polydispersity index	Less than 0.5 is acceptable	Stability
Surface area	Surface area to volume ratio	Impacts drug release

10.7.3 Step 3 of QbD: Application of Quality Risk Management (QRM)

QRM is process that helps to determine the severity and harm of a risk in a process or a system. The principle of QRM is defined in ICH Q9 guidelines. Risk assessment is an important part of QbD during the formulation of quality product that meets the desired output. There are various methods through which risk during a formulation of product can be accessed, such as hazard analysis and critical control point (HACCP) and failure mode effect analysis (FMEA). In this scenario, we applied the risk assessment thorough the HACCP method into the process of formulation of

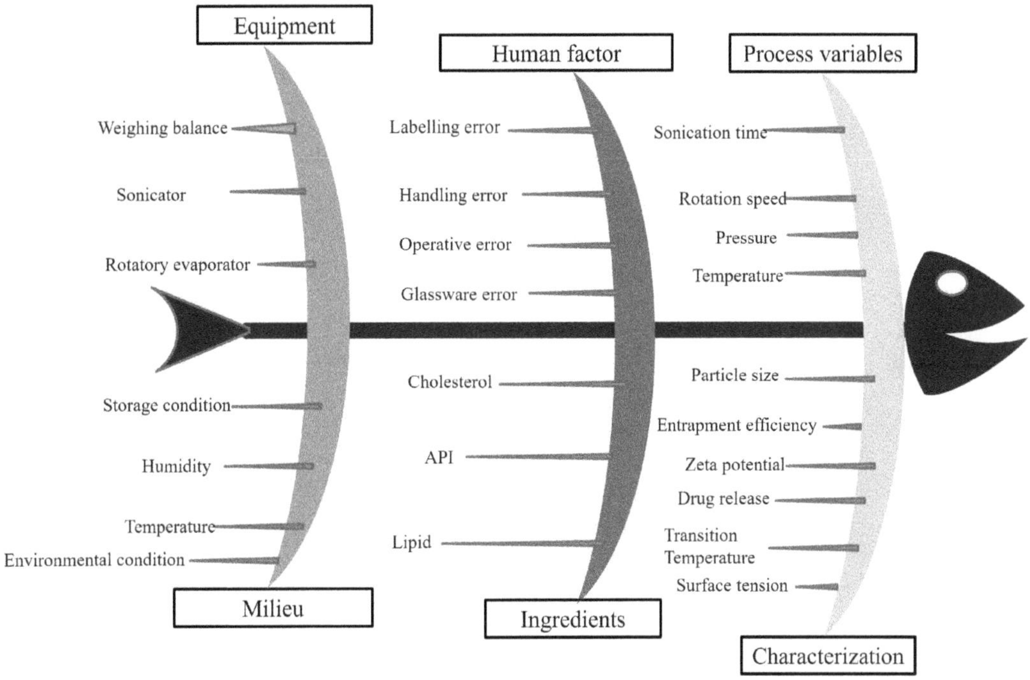

FIGURE 10.8 Ishikawa diagram for the risk assessment of phytosomal formulation.

phytosomes. In any risk evaluation method, first of all, the risk parameters that can affect the formulation of our product are listed in a common plot, commonly known as an Ishikawa fishbone diagram. The Ishikawa fishbone diagram for the phytosomal formulation is depicted in Figure 10.8.

After illustrating all the possible risk factor that can affect the formulation of the phytosome, all the risk factors will be aligned in three categories. For better understanding for the reader, all the three categories are marked in three different colors: green, red, and orange. Risk factors in green boxes are considered as low risk parameters and can be ignored during the formulation of the product at their lowest level. Risk factors in yellow have moderate risk and should be regularly checked during the formulation (Kumar & Baldi, 2015a, 2015b). Risk factors in red should be are high-risk and should be regularly optimized for obtaining the desired quality of the final product. Generally, risk factors in red should be carried for optimization through design of experiment (DoE) software for better optimization (Liu et al., 2013). The risk assessment matrix (RAM) is also built to track potential factors influencing the CQAs. Tables 10.5 and 10.6 show the potential risk factors in formulation of phytosomes in three categories, in order of significance.

After analyzing the factor/variable in the manner described in Table 10.5 and 10.6, the factors mentioned in red and orange boxes can be further investigated through different DoE-based optimization design.

10.7.4 Step 4: Development of DoE-Based Model and Validation

After screening or selection of essential factors (CMAs/CPPs) (as detailed) utilizing risk assessment and screening procedures, it is important that all of the identified critical elements must be optimized to get the required output. DoE aids in identifying the link between the response selected (also known as dependent variables) and the input factor (also known as independent variables). As a result, the DoE technique has proven to be useful against the earlier-used one-factor-at-a-time (OFAT) methodology (Naman et al., 2019). There are two types of design are mentioned in the

TABLE 10.5
Risk Assessment of the API Involved in the Phytosomal Preparation

QTPP→ CQA↓	Dosage Form	Therapeutic Effect	Dosage Form	Route of Administration	Targeting Pathway
Phospholipids	L	L	L	H	H
Cholesterol	L	L	L	L	L
Particle size	L	H	L	H	H
API	M	H	M	H	M
Zeta potential	M	H	M	M	M
Coating material	L	H	L	M	M
Aqueous phase	L	L	L	L	L
PDI	L	L	L	L	L

MA consisting of high risk to CQA; risk cannot be accepted because of efficacy/safety issues.
MA consisting of medium risk to CQA; risk acceptable but need investigation.
MA consisting of no/low risk to CQA; no investigation needed.

TABLE 10.6
Risk Assessment of Various Process Parameters Involved in the Phytosomal Preparation

CPP	CQAs →↓	Vesicle Size	PDI	Zeta Potential	Emulsification Efficiency	Drug Content	Drug Release	Stability
Process parameters	Rotation time	H	H	M	H	H	M	M
	Rotation speed	H	H	H	H	H	H	M
	Temperature	M	M	L	M	H	M	L
	Hydration time	H	M	H	H	M	M	L
	Sonication time	H	H	M	M	L	M	L

PP consisting of high risk to CQA; risk cannot be accepted because of efficacy/safety issues.
PP consisting of medium risk to CQA; risk acceptable but need investigation.
PP consisting of no/low risk to CQA; no investigation needed.

DoE software: screening and optimization design. Screening designs are generally applied for the screening of large of input factors for reducing the number of input factors. Optimization designs are generally applied for the optimization of the factors to required value. Various DoE designs, as mentioned in Figure 10.9, can used for optimization of phytosomes. Optimal response design, factorial, mixture, Box-Behnken, central composite, and full factorial are examples of second-order designs utilized for the screening of the optimization factor (Naman et al., 2022).

10.7.5 Step 5: Pilot Plant Scale-Up and Commercialization

This is the last and most significant phase because if a phytosomal product can't scale up on a larger scale, that formulation is useless if it not beneficial for commercial explorations (Kumar & Baldi,

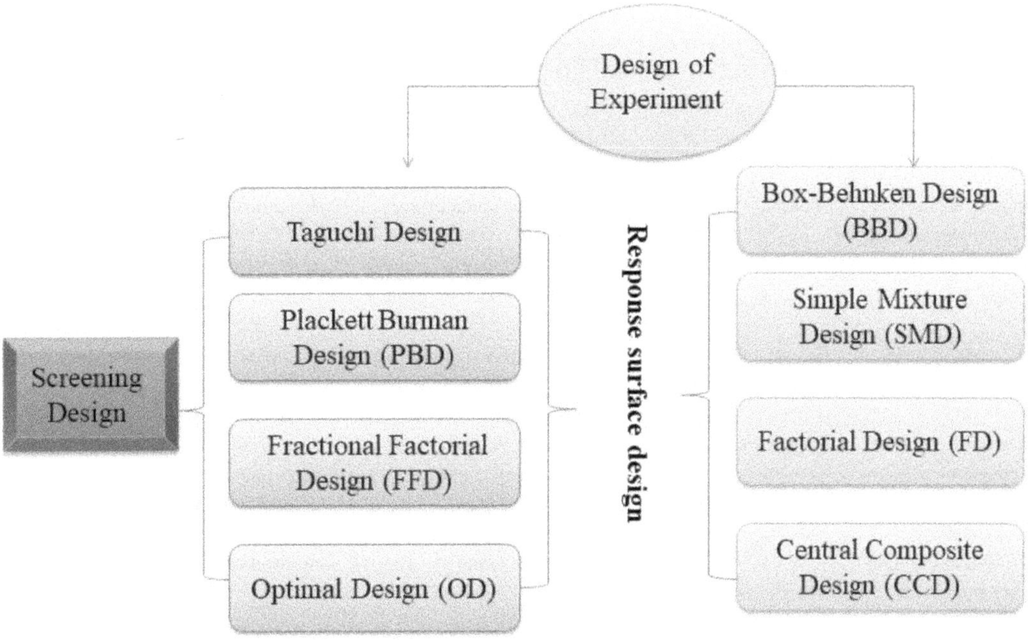

FIGURE 10.9 DoE-Based Various Response Surface and Screening Designs

2015a, 2015b). In such circumstances, QbD may play an important role in the quality manufacture of the therapeutic phytosomal products since it has a lower failure rate and is more patient friendly. Commercialization becomes simpler by applying QbD approach for development of formulation of phytosomes since QbD minimizes the likelihood of rejection and product recall from the market. Table 10.7 outlines previously reported studies that have been used in the idea of QbD during phytosome production (Bajwa et al., 2021).

10.7.6 Scale-Up Challenges in Phytosomes

Translating the production of goods at industrial scales is a difficult challenge for formulation scientists. A pharmaceutical company must ensure the quality of its product using a number of criteria before selling it: (1) safety, (2) scalability, (3) stability (product stability, long-term or chemical stability), (4) process optimization (reliability and reproducibility), and (5) GRAS status and regulatory clearance (Paliwal et al., 2014). Therefore, these issues need to be addressed properly to produce phytosomes at the industrial level.

10.8 THERAPEUTIC APPLICATIONS OF PHYTOSOMES

Potential medicinal applications of phytosomes have been investigated to study the several uses of phytosomes, particularly their capacity to increase the bioactivity of polar PC. Busby and coworkers studied that the utilization of a silymarin phytosome demonstrated higher feto-protectant effect against ethanol-induced behavioral impairments than simple silymarin (Busby et al., 2002). Silymarin phytosomes have been shown to have stronger anti-hepatotoxic action than silymarin alone (Tedesco et al., 2004). Maiti and group produced a quercetin-phospholipid combination using an easy, repeatable procedure and demonstrated that the formulation had higher bioactivity than the original molecule (Maiti et al., 2005). Yanyu and coworkers prepared silymarin phytosomes

TABLE 10.7
Review of DoE Involved in Phytosomal Preparation

Drug	Dosage Form	Design	Independent Variable	Dependent Variable	Reference
Phytosomes of *Terminalia arjuna*	Tablet	CCD	Reaction temperature, reaction time, drug ratio	Entrapment efficiency	(Saudagar et al., 2021)
Phytosomal gel of *Manjistha* extract	Gel	BBD	Speed of rotation (rpm), concentration of lecithin (%w/v)	Entrapment efficiency, vesicular size	(Taleuzzaman et al., 2021)
Aloe vera extract–loaded phytosomes	Gel	CCD	Speed of rotation (rpm), concentration of lecithin (%w/v)	Entrapment efficiency, vesicular size	(Jain et al., 2021)
Salicin phytosomal complex	Suspension	CCD	Salicin-phospholipon ratio, reaction temperature, and reaction time	Complexation rate and partition coefficient	(Ittadwar & Puranik, 2017)
p-Coumaric acid–phospholipid complex	Gel	BBD	Reaction temperature, reaction time (hs)	Complexation efficiency	(Biswas et al., 2021)
Soybean, (*Glycine max (L.) Merrill*)	Thermogel	Three-level factorial design	Drug ratio, reaction temperature, reaction time	Particle size (PS), percentage of drug released, entrapment efficiency (EE%)	(El-Menshawe et al., 2018)
Phytosomes containing *Cassia tora* extract (CTE)	Gel	CCD	Speed of rotation (rpm), concentration of lecithin (%w/v)	Vesicular size and entrapment efficiency	(Corrie et al., 2021)
Centella extract (SCE) complex of phospholipids	Naturosome	CCD	Phospholipid-drug ratio, reaction time, reaction temperature	Entrapment efficiency	(Saoji et al., 2016)

and its pharmacokinetics are observed in rats. The bioactivity of silybin in rats was significantly improved following oral administration of manufactured silybin phosphatide complex, leading to an enhancement in the lipophilic characteristic of silybin-phosphatide complex and an progression in the biological action of silybin (Yanyu et al., 2006). By combining hesperetin with hydrogenated phosphatidylcholine, novel phytosomes were prepared. At the same dosage level, a pharmacokinetic investigation found that the phytosomes had greater relative bioactivity than the parent molecule (Venkatesh et al., 2008; Kumar et al., 2015).

10.9 CONCLUSION AND FUTURE PERSPECTIVE

Phytochemicals may be administered through various routes for a variety of medicinal and aesthetic purposes. Nano-engineered drug delivery approaches are utilized to improve the permeation of bioactive phyto-compounds across biological membrane, hence enhancing their bioactivity, due to their limited absorption profile. Phytosomes as nanocarriers are among these cutting-edge nano-platforms. In comparison to the extract of phytochemicals alone, phytosomes have a greater ability to permeate epidermal layers due to their lipid content and nanovesicular structure. When it comes to vesicular shape, stability, and skin penetration, phytosomes are identical to liposomes. In phytosomes, the polar head of the phospholipid and the polar functionality of the bioactive components create an H-bond, which allows the phospholipid to interact with the phytochemicals. This is significant because it improves the skin penetration and stability of phytochemicals compared to liposomes. Implementing QbD for the production of phytosomal formulation with a complete knowledge of CMA and process parameters and their effect on final quality of the product, as well as insights into QRM strategies, as presented here, can lead to the large-scale production of phytosomal delivery systems. Using QbD principles for the development of phytosomal preparation may improve quality attributes and decrease the probability of rejection of phytosomal preparation during scale-up.

In recent years, various phytoconstituents have been explored for delivery through phytosomes in order to achieve therapeutically effective armamentarium for treatment of various diseases. By seeing various advantages associated with phytosomes as nano-cargoes, viz. simpler production methodologies, bio-acceptability, and easier scale-up approach, this delivery system will be further explored for a wide range of pharmaceutical benefits. To achieve synergistic effects, future studies may examine the use of phytosomes in combination with additional phytochemicals or the combination of drugs and phytochemicals in a single nanovesicular formulation with the help of QbD for industrial acceptance.

REFERENCES

Amin, T., & Bhat, S. V. (2012). A review on phytosome technology as a novel approach to improve the bioavailability of nutraceuticals. Int. J. Adv. Res. Technol., *1*(3), 43–57.

Bajwa, N., Mahal, S., Naryal, S., Singh, P. A., & Baldi, A. (2022a). Development of novel solid nanostructured lipid carriers for bioavailability enhancement using a quality by design approach. AAPS PharmSciTech, *23*(7), 1–14.

Bajwa, N., Naryal, S., Mahal, S., Singh, P. A., & Baldi, A. (2022b). Quality-by-design strategy for the development of arteether loaded solid self-micro emulsifying drug delivery systems. J. Drug Deliv. Sci. Technol., 103707.

Bajwa, N., Singh, P. A., Jyoti, K., Baldi, A., Chandra, R., & Madan, J. (2021). Scale-up, preclinical and clinical status of poly (lactide-co-glycolide) and its copolymers based drug delivery systems. Nanomater. Evol. Adv. Towar. Ther. Drug Deliv. (Part II), *13*, 246.

Belcaro, G., Ledda, A., Hu, S., Cesarone, M. R., Feragalli, B., & Dugall, M. (2013). Greenselect phytosome for borderline metabolic syndrome. Evidence-Based Complement. Altern. Med., *2013*.

Bhattacharya, S. (2009). Phytosomes: Emerging strategy in delivery of herbal drugs and nutraceuticals. Pharma Times, *41*(3), 9–12.

Bhokare, S. G., Dongaonkar, C. C., Lahane, S. V., Salunke, P. B., Sawale, V. S., & Thombare, M. S. (2016). Herbal novel drug delivery: A review. World J. Pharm. Pharm. Sci., *5*(8), 593–611.

Biswas, S., Mukherjee, P. K., Kar, A., Bannerjee, S., Jana, S. N., Haldar, P. K., & Sharma, N. (2021). Enhanced permeability and photoprotective potential of optimized p-Coumaric acid-phospholipid complex loaded gel against UVA mediated oxidative stress. J. Photochem. Photobiol. B Biol., *221*, 112246.

Busby, A., Grange, L. L., Edwards, J., & King, J. (2002). The use of a silymarin/phospholipid compound as a fetoprotectant from ethanol-induced behavioral deficits. J. Herb. Pharmacother., *2*(1), 39–47.

Carini, M., Aldini, G., Bombardelli, E., Morazzoni, P., & Morelli, R. (1994). Free radicals scavenging action and anti-enzyme activities of procyanidines from Vitis vinifera: A mechanism for their capillary protective action. Arzneimittelforschung., *44*(5), 592–601.

Cevc, G., Schätzlein, A., & Blume, G. (1995). Transdermal drug carriers: Basic properties, optimization and transfer efficiency in the case of epicutaneously applied peptides. J. Control. Release, *36*(1–2), 3–16.

Chivte, P. S., Pardhi, V. S., Joshi, V. A., & Rani, A. (2017). A review on therapeutic applications of phytosomes. J. Drug Deliv. Ther., *7*(5), 17–21.

Citernesi, U., & Sciacchitano, M. (1995). Phospholipid/active ingredient complexes. Cosmet. Toilet., *110*(11), 57–68.

Corrie, L., Gundaram, R., & Kukatil, L. (2021). Formulation and evaluation of cassia tora phytosomal gel using central composite design. Recent Innov. Chem. Eng. (Formerly Recent Patents Chem. Eng.), *14*(4), 347–357.

Dubey, D., Shrivastava, S., Kapoor, S., & Dubey, P. K. (2007). Phytosome: A novel dosage structure. Phar. Rev., 1–8.

El Maghraby, G. M. M., Williams, A. C., & Barry, B. W. (2000). Oestradiol skin delivery from ultradeformable liposomes: Refinement of surfactant concentration. Int. J. Pharm., *196*(1), 63–74.

El-Menshawe, S. F., Ali, A. A., Rabeh, M. A., & Khalil, N. M. (2018). Nanosized soy phytosome-based thermogel as topical anti-obesity formulation: An approach for acceptable level of evidence of an effective novel herbal weight loss product. Int. J. Nanomedicine, *13*, 307.

Fry, D. W., White, J. C., & Goldman, I. D. (1978). Rapid separation of low molecular weight solutes from liposomes without dilution. Anal. Biochem., *90*(2), 809–815.

Gándola, Y. B., Pérez, S. E., Irene, P. E., Sotelo, A. I., Miquet, J. G., Corradi, G. R., Carlucci, A. M., & Gonzalez, L. (2014). Mitogenic effects of phosphatidylcholine nanoparticles on MCF-7 breast cancer cells. Biomed. Res. Int., *2014*.

Gupta, A., Ashawat, M. S., Saraf, S., & Saraf, S. (2007). Phytosome: A novel approach towards functional cosmetics. J. Plant Sci., *2*(6), 644–649.

Habbu, P., Madagundi, S., Kulkarni, R., Jadav, S., Vanakudri, R., & Kulkarni, V. (2013). Preparation and evaluation of Bacopa–phospholipid complex for antiamnesic activity in rodents. Drug Invent. Today, *5*(1), 13–21.

Ittadwar, P. A., & Puranik, P. K. (2017). Novel umbelliferone phytosomes: Development and optimization using experimental design approach and evaluation of photo-protective and antioxidant activity. Int. J. Pharm. Pharm. Sci., *9*, 218–228.

Jain, N. K., & Jain, N. K. (2005). Liposomes as drug carriers, controlled and novel drug delivery. CBS Publ., *308*, 321–326.

Jain, P., Taleuzzaman, M., Kala, C., Kumar Gupta, D., Ali, A., & Aslam, M. (2021). Quality by design (Qbd) assisted development of phytosomal gel of aloe vera extract for topical delivery. J. Liposome Res., *31*(4), 381–388.

Jiang, Y. N., Yu, Z. P., Yang, Z. M., & Chen, J. M. (2001). Studies on preparation of herba epimedii total flavonoids phytosomes and their pharmaceutics. Zhongguo Zhong Yao Za Zhi= Zhongguo Zhongyao Zazhi= China J. Chinese Mater. Medica, *26*(2), 105–108.

Jose, M. M., & Bombardelli, E. (1987). Pharmaceutical compositions containing flavanolignans and phospholipida active principles. US Pat. EPO209037.

Karimi, N., Ghanbarzadeh, B., Hamishehkar, H., Keyvani, F., Pezeshki, A., & Gholian, M. M. (2015). Phytosome and liposome: The beneficial encapsulation systems in drug delivery and food application. Appl. Food Biotech., *2*(3), 17–27.

Kumar, B., Arora, M., & Baldi, A. (2015). Understanding industrial practices for pharmaceutical quality management-I, existing approaches and guidelines. Adv. Res. Pharm. Biol., *5*(1), 802–811.

Kumar, B., & Baldi, A. (2015a). Understanding industrial practices for pharmaceutical quality management-II, process validation. Adv. Res. Pharm. Biol., *5*, 812–824.

Kumar, S., & Baldi, A. (2015b). Formulation by design approach for fizzy granules using statistical optimization methodologies. Asian J. Pharm., *9*, S59–S67.

Kumar, S., Baldi, A., & Sharma, D. K. (2019). Phytosomes: A modernistic approach for novel herbal drug delivery—Enhancing bioavailability and revealing endless frontier of phytopharmaceuticals. J. Dev. Drugs, *9*, 1–8.

Kumar, S., Baldi, A., & Sharma, D. K. (2021). In vitro antioxidant assay guided ex vivo investigation of cytotoxic effect of phytosomes assimilating taxifolin rich fraction of *Cedrus deodara* bark extract on human breast cancer cell lines (mcf7). J. Drug Deliv. Sci. Technol., *63*, 102486.

Kumar, S., Baldi, A., & Sharma, D. K. (2022). Characterization and in vitro investigation of antiscabietic effect of phytosomes assimilating quercetin and naringenin rich fraction of *Pistacia integerrima* galls extract against *Sarcoptes scabiei*. J. Drug Deliv. Sci. Technol., *67*, 102851.

Li, Y., Yang, D.-J., Chen, S.-L., Chen, S.-B., & Chan, A. S.-C. (2008). Process parameters and morphology in puerarin, phospholipids and their complex microparticles generation by supercritical antisolvent precipitation. Int. J. Pharm., *359*(1–2), 35–45.

Litzinger, D. C., & Huang, L. (1992). Phosphatodylethanolamine liposomes: Drug delivery, gene transfer and immunodiagnostic applications. Biochim. Biophys. Acta (BBA)-Reviews Biomembr., *1113*(2), 201–227.

Liu, H.-C., Liu, L., & Liu, N. (2013). Risk evaluation approaches in failure mode and effects analysis: A literature review. Expert Syst. Appl., *40*(2), 828–838.

Liu, S. (2012). Preparation, characterization and in vitro anti-tumor activities of evodiamine phospholipids complex. Chinese Pharm. J., 517–523.

Maiti, K., Mukherjee, K., Gantait, A., Ahamed, H. N., & Mukherjee, P. K. (2005). Enhanced therapeutic benefit of quercetin–phospholipid complex in carbon tetrachloride–induced acute liver injury in rats: A comparative study. Iran. J. Pharm. Therapeutics, *4*(2), 84–90.

Maiti, K., Mukherjee, K., Gantait, A., Saha, B. P., & Mukherjee, P. K. (2007). Curcumin–phospholipid complex: Preparation, therapeutic evaluation and pharmacokinetic study in rats. Int. J. Pharm., *330*(1–2), 155–163.

Mane, K., Baokar, S., Bhujbal, A., Pharande, S., Patil, G., Patil, R., Jain, P., & Pandey, A. (2020). Phyto-phospholipid complexes (phytosomes): A novel approach to improve the bioavailability of active constituents. J. Adv. Sci. Res., *11*(03), 68–78.

Moscarella, S., Giusti, A., Marra, F., Marena, C., Lampertico, M., Relli, P., Gentilini, P., & Buzzelli, G. (1993). Therapeutic and antilipoperoxidant effects of silybin-phosphatidylcholine complex in chronic liver disease: Preliminary results. Curr. Ther. Res., *53*(1), 98–102.

Murray, M. T. (2004). *Phytosomes: Herbal Support–Increase the Absorption of Herbal Extracts*. Available Dr. Com/Articles/Silybin.html

Naik, S. R., & Panda, V. S. (2008). Hepatoprotective effect of Ginkgoselect Phytosome® in rifampicin induced liver injurym in rats: Evidence of antioxidant activity. Fitoterapia, *79*(6), 439–445.

Naik, S. R., Pilgaonkar, V. W., & Panda, V. S. (2006). Evaluation of antioxidant activity of Ginkgo biloba phytosomes in rat brain. Phyther. Res. An Int. J. Devoted to Pharmacol. Toxicol. Eval. Nat. Prod. Deriv., *20*(11), 1013–1016.

Naman, S., Madhavi, N., Singh, M., Singh, P. A., Bajwa, N., & Baldi, A. (2019). Formulation design and pharmaceutical considerations for paediatric patients: Current status and future dimensions. J. Drug Deliv. Ther., *9*(4), 601–606.

Naman, S., Naryal, S., Palliwal, R., Paliwal, S. R., & Baldi, A. (2022). Combating atherosclerosis with nanodrug delivery approaches: From bench side to commercialization. In *Drug Delivery Systems for Metabolic Disorders* (pp. 97–136). Elsevier.

Paliwal, R., Babu, R. J., & Palakurthi, S. (2014). Nanomedicine scale-up technologies: Feasibilities and challenges. AAPS PharmSciTech, *15*(6), 1527–1534.

Saoji, S. D., Raut, N. A., Dhore, P. W., Borkar, C. D., Popielarczyk, M., & Dave, V. S. (2016). Preparation and evaluation of phospholipid-based complex of Standardized Centella Extract (SCE) for the enhanced delivery of phytoconstituents. AAPS J., *18*(1), 102–114.

Saudagar, W. S., Sidram, G. P., Baburo, G. S., Agarwal, G., Agarwal, S., & Gadgeppa, B. O. (2021). Development and characterization of *Terminalia arjuna* phospholipid complex and its tablet formulation by Qbd approach. Int. J. Life Sci. Pharma Res., *11*(3), P14–P28.

Semalty, A., Semalty, M., Rawat, B. S., Singh, D., & Rawat, M. S. M. (2009). Pharmacosomes: The lipid-based new drug delivery system. Expert Opin. Drug Deliv., *6*(6), 599–612.

Sharma, A., & Baldi, A. (2018). Nanostructured lipid carriers: A review. J. Dev. Drugs, *7*(2), 1000191.

Singh, G., Kaur, L., Gupta, G. D., & Sharma, S. (2017). Enhancement of the solubility of poorly water soluble drugs through solid dispersion: A comprehensive review. Indian J. Pharm. Sci., *79*(5), 674–687.

Taleuzzaman, M., Sartaj, A., Kumar Gupta, D., Gilani, S. J., & Mirza, M. A. (2021). Phytosomal gel of Manjistha Extract (MJE) formulated and optimized with central composite design of Quality by Design (QbD). J. Dispers. Sci. Technol., 1–9.

Tedesco, D., Steidler, S., Galletti, S., Tameni, M., Sonzogni, O., & Ravarotto, L. (2004). Efficacy of silymarin-phospholipid complex in reducing the toxicity of aflatoxin B1 in broiler chicks. Poult. Sci., *83*(11), 1839–1843.

Udapurkar, P., Bhusnure, O., Kamble, S., & Biyani, K. (2016). Phyto-phospholipid complex vesicles for phytoconstituents and herbal extracts: A promising drug delivery system. Int. J. Herb. Med., *4*(5), 14–20.

Venkatesh, M., Mukherjee, K., Maiti, K., & Mukherjee, P. K. (2008). Enhancement of bioavailability of phytomolecules with value added formulation. Planta Med., *74*(9), PC11.

Verma, H., & Prasad, S. B. (2011). Phytosome: Phytolipid delivery system. Inventi Rapid NDDS, *1*(4), 1–4.

Yanyu, X., Yunmei, S., Zhipeng, C., & Qineng, P. (2006). The preparation of silybin–phospholipid complex and the study on its pharmacokinetics in rats. Int. J. Pharm., *307*(1), 77–82.

Yu, F., Li, Y., Chen, Q., He, Y., Wang, H., Yang, L., Guo, S., Meng, Z., Cui, J., & Xue, M. (2016). Monodisperse microparticles loaded with the self-assembled berberine-phospholipid complex-based phytosomes for improving oral bioavailability and enhancing hypoglycemic efficiency. Eur. J. Pharm. Biopharm., *103*, 136–148.

Yu, L. X., Amidon, G., Khan, M. A., Hoag, S. W., Polli, J., Raju, G. K., & Woodcock, J. (2014). Understanding pharmaceutical quality by design. AAPS J., *16*(4), 771–783.

Index

A

acrylic acid, 180–182, 190, 194
alginate, 180, 189
angiogenesis, 136–137
antibacterial, 179, 187
anticancer, 36, 179, 187
anti-inflammatory, 36, 179, 185, 187, 197, 198
anti-inflammatory effect
antioxidant, 61, 179, 184–187, 195, 197–200
antisolvent precipitation, 234
apigenin, 11, 81

B

baicalin, 13, 77, 115
bioavailability, 61, 230, 232, 233, 235
biological activity, 34–35, 40–41, 44
biopharmaceutical, 38
biopolymers, 63, 104

C

cancer, 136–137
cancer treatment, 62
Carbopol, 182, 189, 193–198
cardiovascular protective effect, 37
catechin, 8
curcuminoids, 134–135
Celastrol, 5
cellulose, 180, 183, 188–198
chitosan, 63, 180–197
cis-resveratrol, 34
clinical trials, 198
controlled release, 63
coumaric acid, 23, 179–201
critical process parameters (CPP), 236, 238
critical quality attributes (CQA), 236, 238
cubosomes, 81
curcuma longa, 134, 135, 150
curcumin
    biological effects, 135–138
cyanidin, 9

D

daidzein, 7, 119
dendrimers, 83
dosage formulation, 230, 234, 235
drug delivery, 231, 233, 234, 239
drug delivery system, 61

E

exosomes
    mammalian derived, 149–150
    plant-derived, 150
emodin, 15
encapsulation, 62
entrapment efficiency, 236, 240
epigallocatechin-3 Gallate, 17, 80, 120
ethosomes, 80
exosome delivery, 149–150
experimental colitis models, 142–144
extracellular vesicles, 87

F

fisetin, 19

G

gallic acid, 21, 61–71
gel core liposomes (GCLs), 148–149
genistein, 20, 85

H

hepatoprotective, 241
hesperidin, 4, 85
hybrid-based materials, 69
hydrogels, 65

I

inflammation, 135–136
inflammatory bowel disease (IBD), 138–141
inorganic nanoparticles, 47

J

JAK/STAT inhibitors, 25–26

L

lipid, 168–170, 173
lipid nanoparticles, 33, 40–42
lipid peroxidation, 136, 141, 142
liposomes, 40–41, 44–45, 66, 68, 76, 99–151, 160–161, 163–167, 232

M

magnesium glycyrrhizinate, 36
mangiferin, 24
metallic nanoparticles, 66, 82
microcarrier, 49–52
microcarrier for the delivery of resveratrol, 49
microemulsions, 162–163
microneedle, 50–51
microparticle, 50
microsponge, 51

N

nano polymersomes, 88
nanocapsules, 162–163
nanocarrier, 7, 40, 74–90, 164–171

nanocrystals, 33, 40, 42, 49, 78
nanoemulsion, 40, 46, 75, 162,
nanomedicine, 242
nanoparticles, 63, 66, 134–151, 160–174, 191–192, 195–196
nanostructured lipid carriers (NLCs), 86, 161, 167–174
naringenin, 10, 77, 119
neuroprotective effect, 38
niosomes, 40–41, 44–45, 78, 163, 166, 199

P

phospholipids, 230
pH-sensitive liposomes, 148
phyto-active constituents, 232
phytochemicals, 164, 170
phytoconstituent, 74, 99, 102, 117, 214, 230
phytosomes, 79, 230–243
pilot plant scale-up, 240
poloxamer, 179–198
poly (N-propyl acrylamide) page, 179
poly-lactic-co-glycolic acid nanoparticles, 85
polymer, 179–201
polymeric nanoparticles, 40–44, 89
polymeric systems, 63
polyphenols, 179, 184, 187–189, 192, 197
preparation methods, 233–235

Q

quality by design (QbD), 236–238, 241
quality risk management, 238

R

reactive oxygen species (ROS), 136, 148
resveratrol, 2, 33–52, 119, 166, 224
resveratrol bioavailability, 38–39
risk management, 239
rosmarinic Acid, 14
rotary evaporation, 234
rotary evaporation technique, 234

S

solid lipid nanoparticles, 83, 161
solvent evaporation, 235
solvent evaporation technique, 235
stevioside, 16
synthesis of resveratrol, 34

T

therapeutic applications, 241
toxicity, 66
trans-resveratrol, 34, 39, 42, 45, 49
turmeric, 118, 134, 184, 235

V

validation, 239
vesicle size, 238, 239

W

wound healing, 137, 145

Z

zeta potential, 236, 239, 240